Teuvo Suntio

**Dynamic Profile of
Switched-Mode Converter**

Related Titles

Batarseh, I.

Power Electronic Circuits

2006
ISBN: 978-0-471-12662-1

Mohan, N., Robbins, W. P., Undeland, T. M.

Power Electronics

Converters, Applications and Design, Media Enhanced. International Edition

2003
ISBN: 978-0-471-42908-1

3.6.2.3	Input-to-Output Transfer Functions	*111*
3.6.2.4	Input Admittance	*112*
3.6.2.5	Ideal Input Admittance	*114*
3.6.2.6	Short-Circuit Input Admittance	*116*

4 Average and Small-Signal Modeling of Peak-Current-Mode Control *121*

4.1	Introduction	*121*
4.2	PCM-Control Principle	*122*
4.3	Modeling in CCM	*124*
4.3.1	Duty-Ratio Constraints for Buck, Boost, and Buck–Boost Converters	*126*
4.3.1.1	Buck Converter	*126*
4.3.1.2	Boost Converter	*126*
4.3.1.3	Buck–Boost	*128*
4.3.1.4	General CCM Transfer Functions	*129*
4.3.2	Specific Transfer Functions for the Basic Converters	*131*
4.3.2.1	Buck Converter	*131*
4.3.2.2	Boost Converter	*133*
4.3.2.3	Buck–Boost Converter	*134*
4.3.3	Origin and Consequences of Mode Limit in CCM	*136*
4.4	Modeling in DCM	*139*
4.4.1	Duty-Ratio Constraints for Basic Converters	*140*
4.4.1.1	Buck Converter	*140*
4.4.1.2	Boost Converter	*142*
4.4.1.3	Buck–boost Converter	*142*
4.4.2	Small-Signal PCMC State Spaces for the Basic Converters	*143*
4.4.3	Origin and Consequences of Mode Limit in DCM	*144*
4.5	Dynamic Review	*146*
4.5.1	Buck Converter	*147*
4.5.1.1	Control-to-Output Transfer Function	*148*
4.5.1.2	Output Impedance	*151*
4.5.1.3	Input-to-Output Transfer Function	*152*
4.5.1.4	Input Admittance	*153*
4.5.1.5	Ideal Input Admittance	*155*
4.5.1.6	Short-Circuit Input Admittance	*155*
4.5.2	Boost Converter	*157*
4.5.2.1	Control-to-Output Transfer Function	*159*
4.5.2.2	Output Impedance	*161*
4.5.2.3	Input-to-Output Transfer Function	*163*
4.5.2.4	Input Admittance	*164*
4.5.2.5	Ideal Input Admittance	*165*
4.5.2.6	Short-Circuit Input Admittance	*166*

5	**Average and Small-Signal Modeling of Average-Current-Mode Control** *169*
5.1	Introduction *169*
5.2	ACM-Control Principle *169*
5.3	Modeling with Full-Ripple-Current Feedback *171*
5.4	Dynamic Review *175*
5.4.1	Control-to-Output Transfer Function *177*
5.4.2	Output Impedance *177*
5.4.3	Input-to-Output Transfer Function *179*
5.4.4	Input Admittance *183*
5.5	Effect of Current-Loop High-Frequency Pole *183*

6	**Average and Small-Signal Modeling of Self-Oscillation Control** *189*
6.1	Introduction *189*
6.2	Self-Oscillation Modeling *189*
6.2.1	Averaged Direct-on-Time Model *190*
6.2.2	Small-Signal Direct-on-Time Model *193*
6.2.3	Small-Signal PCM Models *194*
6.3	Dynamic Review *198*
6.3.1	Buck Converter *198*
6.3.1.1	Control-to-Output Transfer Function *199*
6.3.1.2	Output Impedance *199*
6.3.1.3	Input-to-Output Transfer Function *201*
6.3.1.4	Input Admittances *201*
6.3.2	Flyback Converter *202*
6.3.2.1	Control-to-Output Transfer Function *203*
6.3.2.2	Output Impedance *206*
6.3.2.3	Input-to-Output Transfer Function *206*
6.3.2.4	Input Admittance *207*
6.3.2.5	Ideal and Short-Circuit Admittances *208*

7	**Dynamic Modeling and Analysis of Current-Output Converters** *211*
7.1	Introduction *211*
7.2	Dynamic Models for Current-Output Converter *212*
7.2.1	Modified-State-Space-Averaging Technique *213*
7.2.2	General Dynamic Models *215*
7.3	Load and Supply Interactions *216*
7.4	Cascaded Voltage-Current Loops *218*
7.5	Dynamic Review *219*

8	**Interconnected Systems** *225*
8.1	Introduction *225*
8.2	Theoretical Interaction Formulation *226*
8.2.1	Load and Supply Interactions *227*

Teuvo Suntio

Dynamic Profile of Switched-Mode Converter

Modeling, Analysis and Control

WILEY-VCH Verlag GmbH & Co. KGaA

The Author

Prof. Teuvo Suntio
Tampere University of Technology
Dept. of Electrical Energy Engineering
Tampere, Finland
teuvo.suntio@tut.fi

Cover-Picture
Spiesz Design, Neu-Ulm

All books published by Wiley-VCH are carefully produced. Nevertheless, authors, editors, and publisher do not warrant the information contained in these books, including this book, to be free of errors. Readers are advised to keep in mind that statements, data, illustrations, procedural details or other items may inadvertently be inaccurate.

Library of Congress Card No.: applied for

British Library Cataloguing-in-Publication Data
A catalogue record for this book is available from the British Library.

Bibliographic information published by the Deutsche Nationalbibliothek
The Deutsche Nationalbibliothek lists this publication in the Deutsche Nationalbibliografie; detailed bibliographic data are available in the Internet at http://dnb.d-nb.de.

© 2009 WILEY-VCH Verlag GmbH & Co. KGaA, Weinheim

All rights reserved (including those of translation into other languages). No part of this book may be reproduced in any form – by photoprinting, microfilm, or any other means – nor transmitted or translated into a machine language without written permission from the publishers. Registered names, trademarks, etc. used in this book, even when not specifically marked as such, are not to be considered unprotected by law.

Typesetting Laserwords, Chennai, India

Printing Strauss GmbH, Mörlenbach

Binding Litges & Dopf Buchbinderei GmbH, Heppenheim

Printed in the Federal Republic of Germany
Printed on acid-free paper

ISBN: 978-3-527-40708-8

Contents

Preface *XI*

1 Introduction *1*
1.1 Introduction *1*
1.2 Dynamic Modeling of Switched-Mode Converters *4*
1.3 Dynamic Analysis of Interconnected Systems *6*
1.4 Canonical Equivalent Circuit *8*
1.5 Load-Response-Based Dynamic Analysis *9*
1.6 Content Review *12*

2 Basis for Dynamic Analysis and Control Dynamics *17*
2.1 Introduction *17*
2.2 Dynamic Representations at Open Loop *17*
2.2.1 State Space *19*
2.2.2 Two-Port Models *21*
2.2.3 Control-Block Diagrams *23*
2.3 Dynamic Representations at a Closed Loop *23*
2.3.1 Voltage-Output Converter *26*
2.3.2 Current-Output Converter *27*
2.4 Load and Source Effects *28*
2.4.1 Voltage-Output Converter *29*
2.4.2 Current-Output Converter *31*
2.5 An Example LC Circuit *33*
2.5.1 Voltage-Output Circuit *33*
2.5.2 Current-Output Circuit *35*
2.6 Review of Basic Mathematical Tools *37*
2.6.1 Linearization *37*
2.6.2 Transfer Functions *38*
2.6.2.1 Single Zero *39*
2.6.2.2 Single Pole *40*
2.6.2.3 Second-Order Transfer Function *40*
2.6.2.4 Example *43*
2.6.3 Stability and Performance *45*

| vi | Contents

2.6.3.1 Stability 46
2.6.3.2 Loop-Gain-Related Dynamic Indices 48
2.6.3.3 Right-Half-Plane Zero and Pole 50
2.6.4 Matrix Algebra 50
2.6.4.1 Addition of Matrices 53
2.6.4.2 Multiplication by Scalar 53
2.6.4.3 Matrix Multiplication 54
2.6.4.4 Matrix Determinant 54
2.6.4.5 Matrix Inversion 55
2.7 Operational and Control Modes 55

3 Average and Small-Signal Modeling of Direct-On-Time Controlled Converters *59*
3.1 Introduction 59
3.2 Direct-on-Time Control 60
3.3 Generalized Modeling Technique 62
3.3.1 Buck Converter 64
3.3.2 Boost Converter 66
3.3.3 Buck–Boost Converter 68
3.4 Fixed-Frequency Operation in CCM 70
3.4.1 Synchronous Buck Converter 71
3.4.2 Dynamic Descriptions of Buck, Boost, and Buck–Boost Converters 76
3.4.2.1 Diode-Switched Buck (Figure 3.6a) 76
3.4.2.2 Diode-Switched Boost (Figure 3.8a) 77
3.4.2.3 Synchronous Boost (Figure 3.8b) 79
3.4.2.4 Diode-Switched Buck–Boost (Figure 3.10a) 80
3.4.2.5 Synchronous Buck–Boost (Figure 3.10b) 81
3.4.3 Steady-State and Small-Signal Equivalent Circuits 82
3.5 Fixed-Frequency Operation in DCM 85
3.5.1 Buck Converter 87
3.5.2 Dynamic Models for Boost and Buck–Boost Converters 92
3.5.2.1 Boost Converter (Figure 3.8a) 92
3.5.2.2 Buck–Boost Converter (Figure 3.10a) 94
3.6 Dynamic Review 95
3.6.1 Buck Converter 96
3.6.1.1 Control-to-Output Transfer Function 96
3.6.1.2 Output Impedance 98
3.6.1.3 Input-to-Output Transfer Function 100
3.6.1.4 Input Admittance 103
3.6.1.5 Ideal Input Admittance 104
3.6.1.6 Short-Circuit Input Admittance 106
3.6.2 Boost Converter 106
3.6.2.1 Control-to-Output Transfer Function 108
3.6.2.2 Output Impedance 109

8.2.2	Internal and Input–Output Stabilities	*230*
8.2.3	Output Voltage Remote Sensing	*234*
8.2.4	Input EMI Filter	*236*
8.3	Review of Methods to Reduce the Interactions	*238*
8.3.1	Input-Voltage Feedforward	*238*
8.3.2	Output-Current Feedforward	*240*
8.4	Experimental Dynamic Review	*241*
8.4.1	Load and Supply Interactions	*243*
8.4.2	Remote Sensing	*251*
8.4.3	System Stability	*255*
9	**Control Design Issues**	*261*
9.1	Introduction	*261*
9.2	Feedback-Loop-Design Constraints	*265*
9.2.1	Phase and Gain Margins	*266*
9.2.2	RHP Zeros and Poles	*268*
9.2.3	Minimum and Maximum Loop Crossover Frequencies	*268*
9.2.4	Internal Gain of an Operational Amplifier	*270*
9.3	Controller Implementations	*271*
9.4	Optocoupler Isolation	*272*
9.5	Shunt-Regulator-Based Control Systems	*274*
9.5.1	Dynamic Model	*274*
9.5.2	Two-Loop Control System	*281*
9.6	Simple Control-Design Method	*284*
9.6.1	Control Design Example: VMC Buck Converter	*285*
9.6.2	Control Design Example: PCMC Buck Converter	*290*
9.6.3	Control Design Example: VMC Boost Converter	*295*
9.6.4	Control Design Example: PCMC Boost Converter	*298*
9.7	Conclusions	*302*
10	**The Fourth-Order Converter – Superbuck**	*307*
10.1	Introduction	*307*
10.2	Basic Dynamics	*309*
10.2.1	Averaged Models	*311*
10.2.1.1	Averaged State Space	*311*
10.2.1.2	Steady-State Operating Point	*311*
10.2.1.3	Boundary Conduction Mode	*312*
10.2.2	Small-Signal Models	*312*
10.2.2.1	Small-Signal State Space	*312*
10.2.2.2	Transfer Functions	*313*
10.2.3	RHP Poles	*315*
10.2.4	Design Considerations	*317*
10.3	Coupled-Inductor Superbuck	*318*
10.3.1	Small-Signal Models	*319*

10.3.2 RHP Poles *322*
10.3.3 Input-Current-Ripple Reduction *322*
10.3.4 Design Considerations *325*
10.4 PCM-Controlled Superbuck *325*
10.4.1 Small-Signal Models *326*
10.4.2 Design Considerations *330*
10.4.2.1 Inductor-Current-Feedback Compensation *330*
10.4.2.2 Avoiding RHP Poles *330*
10.5 Coupled-Inductor PCM-Controlled Superbuck *331*
10.5.1 Small-Signal Models *331*
10.5.2 Design Considerations *336*
10.6 Dynamic Review *337*
10.6.1 Superbuck I: 15–20 V/10 V/2.5 A *337*
10.6.2 Superbuck II: 6–9 V/3.4 V/12 A *345*
10.7 Summary *348*

Index *351*

Preface

Twenty-five years in industry engaged in the practical design of switched-mode converters and the systems based on them and especially the multitude of problems we faced convinced me that there has to be something we do not really understand. After joining the academy, I decided to find out what are the reasons for the problems we were confronted with almost daily. After ten years in academy, I understand many of the problems but there still seems to be a lot to solve. The biggest challenge is to convince the other designers and academics that the electrical circuits have an internal dynamic sole which dictates the way the circuits behave and contribute to the behavior of the systems composed of them. The plain sole has to be known for being able to understand and predict the behavior and especially to avoid the undesired consequences such as instability and deteriorated transient performance. The book is intended to introduce the dynamic features the different converter topologies may incorporate and the changes the different control methods and operation modes may create in them in addition to the introduction of the methods to model analytically the dynamic behavior and to design the controllers.

Many individuals have helped me to create the book and especially to understand the extent of the problems and to find the solutions for them: Professor Dr. D. R. Vij as the consultant editor and the staff of Wiley-VCH have provided me the opportunity to publish the book and patient guidance during the process. Dr. Kai Zenger has guided me into the secrets of control engineering but a lot of work is still left. My former doctorate students Dr. Idris Gadoura, Dr. Ander Tenno, Dr. Mikko Hankaniemi, Dr. Ali Altowati, and Dr. Matti Karppanen have contributed substantially to the solutions presented in the book and especially to their experimental validation. I am very grateful to my wife Sirpa for her love and patience during the process.

Teuvo Suntio

1
Introduction

1.1
Introduction

The switched-mode converters can be divided into two main classes such as voltage- (Figure 1.1) and current-sourced (Figure 1.2) converters [1], where either the output voltage (Figures 1.1a and 1.2b) or output current (Figures 1.1b and 1.2a) is kept constant [2]. As a consequence, there are four different main types of converters namely voltage-to-voltage, voltage-to-current, current-to-current, and current-to-voltage converters having different dynamic features. The most usual converter is the voltage-to-voltage converter (Figure 1.1a) because most of the energy sources are voltage sources and the loads current sinks [3]. Sometimes storage batteries are connected at the output of the voltage-sourced converter, which requires to limiting the maximum output current for preventing the converters from damage due to the extremely low internal impedance of a storage battery [4–8]. The operation at current-limiting mode changes the voltage-to-voltage converter to voltage-to-current converter (Figure 1.1b). Current-sourced converters can be used to interface solar arrays and magnetic energy storage systems due to the current-output nature of those energy sources [9, 10]. Such a basic converter is naturally the current-to-current converter (Figure 1.2a). If the maximum-output voltage limiting is used, the current-to-current converter changes to a current-to-voltage converter (Figure 1.2b).

Every switched-mode converter has a unique dynamic profile or internal dynamics, which would determine the obtainable transient dynamics and robustness of stability as well as the converter's sensitivity to the external source and load impedances [11–13]. The dynamic profile can be changed by means of certain internal feedback or feedforward arrangements but not much in practice by means of the feedback-loop control design. The internal dynamics can be characterized by means of a certain set of open-loop transfer functions constituting the circuit theoretical two-port parameters known as G (Figure 1.1a), Y (Figure 1.1b), H (Figure 1.2a), or Z (Figure 1.2b) depending on the input source and the type of the converter output [11–15]. The different sets do characterize only one main type of a converter and are not interchangeable

Dynamic Profile of Switched-Mode Converter. Teuvo Suntio
© 2009 WILEY-VCH Verlag GmbH & Co. KGaA, Weinheim
ISBN: 978-3-527-40708-8

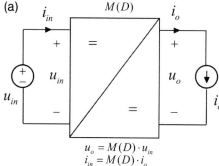

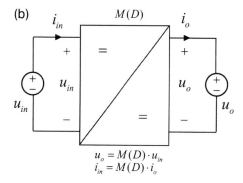

Figure 1.1 Voltage-sourced converter (a) at voltage-output mode and (b) at current-output mode.

but the parameters within the main converter class (i.e., G and Y, H and Z) can be computed from each other. In addition with the open-loop transfer functions, certain admittance or impedance parameters have to be defined for obtaining the full picture of the internal dynamic profile [11].

The term internal means that the transfer functions constituting the sets are to be such that all the effects of the source and load impedances are removed from them. The analytical models can be easily derived to be such, when knowing the correct load yielding the internal models (Figures 1.1. and 1.2). The dynamic parameter sets for the voltage-to-voltage and current-to-current converters can also be usually measured by means of frequency response analyzers but certain internal control modes may change the open-loop converter such that it cannot operate at the defined load or the required ideal load is not available. In such cases, a resistive load has to be used and the internal models have to be solved computationally [11, 16, 17]. It is, however, extremely important to obtain those internal models because they only characterize the converter not the source- or load-affected models.

A large number of power electronics text books are available such as [18–26], which tend to give a comprehensive picture of all the issues related to the design of switched-mode converters both in AC and DC applications. Therefore, it is understandable that the dynamic issues are typically not treated adequately. The exceptions are [27] and [28], which mainly concentrates on the dynamic

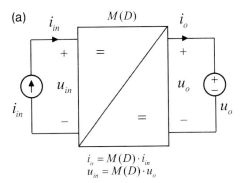
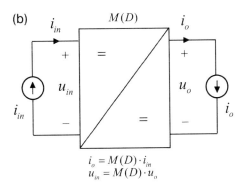

Figure 1.2 Current-sourced converter (a) at current-output mode and (b) at voltage-output mode.

issues. The main deficiency of the dynamic analyses in the aforementioned text books is the inclusion of the load usually as a resistor in the presented dynamic models, which may effectively hide the true dynamics and thereby made the output of the system-level interaction analyses useless. A describing example of the misunderstanding such a treatment can cause is the prevailing understanding that the damping of the resonant behavior in a converter would decrease, when the resistive load is decreased [29]. The phenomenon is naturally true from the external point of view but the internal dynamics does not, however, change if the operating point is maintained. Therefore, it may be a big surprise when the converter behaves nicely in the laboratory but dynamic problems arise when connected into a real application. Such an experience might be very common among the industrial switched-mode-converter designers leading easily to frustration and blaming the customer of abusing the converter.

The main goal of the book is to provide the reader with the tools by means of which the challenging dynamics of the systems comprising of switched-mode converters can be made more understandable and the design of them more deterministic. It is natural that the key element is the building block of such a system – the switched-mode converter. The most fundamental issue behind the ideas provided in the book is the observation that each electrical device or circuit has its unique internal dynamic profile similar to the psychological profile of a human being [11]: the profile determines how the device or circuit would behave as a part of the system under different external interactions and how it would affect the other subsystems within the overall system. The internal profile cannot be basically changed by applying external feedback control but only by providing internal feedback or feedforward from the input, output and/or state variables constituting the dynamic constellation of the device. An illustrative example is the application of inductor current to produce the duty ratio in a peak-current-mode-controlled (PCMC) converter [30], which changes profoundly the converter dynamics compared to the corresponding direct-duty-ratio or voltage-mode-controlled (VMC) converter, where the duty ratio is produced using a constant ramp voltage: The resonant nature of the VMC converter disappears, the input-noise attenuation may be substantially increased, the internal open-loop output impedance is increased but the nonminimum nature if existing in the VMC converter would not be removed. A multitude of similar examples can be given, which actually proves the existence of such a profile.

During the time of writing the book, the analog control is still dominating but digital control with all the opportunities involved in it is evidently coming and may dominate the future converter applications. The fact is, however, that the power stage does not change and, therefore, the basic dynamic profile related to the power stage does not change. The digital control with the physical resolution and time limitations may cause more dynamic problems or equally also improvements, which can be revealed and analyzed using the methods and information based on the corresponding continuous-time processes treated in this book.

The issues related to the dynamic profiles are briefly discussed and clarified in the subsequent subsections in order to make the reader familiar with the issues treated in the subsequent chapters. Even if we discussed on the current-sourced converters in the beginning of the chapter, we will limit our discussions on the voltage-sourced converters within the rest of the book.

1.2
Dynamic Modeling of Switched-Mode Converters

The dynamic analysis of the voltage-output switched-mode converters dates back to the early 1970s [31], when the foundation for the state-space-averaging (SSA) method [32] was laid down. It was observed that the dynamics associated with the direct-duty-ratio or VMC converter in continuous conduction mode (CCM) could be quite accurately captured up to half the switching frequency by averaging the converter variables within a switching cycle and computing the small-signal models from the corresponding averaged state space by means of linearization. The dynamic behavior of a converter was represented by means of the canonical equivalent circuit shown in Figure 1.3 for the

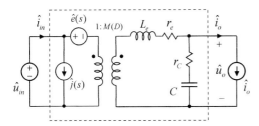

Figure 1.3 Small-signal canonical equivalent circuit for a two-memory-element converter.

two-memory-element converters, where the different circuit elements are defined according to a specific converter. It may be obvious that the equivalent circuit in Figure 1.3 provides real physical insight into the dynamic processes inside the converter and has, therefore, promoted the acceptance of the theoretical method providing the model. Similar equivalent circuit to Figure 1.3 can also be naturally constructed for the higher order converters.

The first attempt to model the dynamics associated with a VMC converter operating in discontinuous mode (DCM) is presented in [33] but it failed to capture the true full-order dynamics due to the lack of proper understanding of the dynamical processes inside a converter. The accurate small-signal models for the DCM operation were developed in the late 1990s [34]. A unified method based on the SSA method was finally developed in the early 2000s providing consistent modeling tools for fixed and variable-frequency operation both in DCM, CCM, and even in the combination of them [35]. The pulsewidth modulation (PWM) process would not produce linear responses but only at rather low frequencies (i.e., ~1/10 of switching frequency) for sinus excitations [36–38]. Therefore, the responses measured through the PWM input (i.e., control-to-input and control-to-output) may have more phase lag than the models derived using the SSA method would predict. Further studies on the topic are needed in order to find the correct dynamic behavior of the converter also at the frequencies approaching half the switching frequency. This is important because the desired loop crossover frequencies tend to approach ever higher frequencies beyond those typically used in the past.

The small-signal models of the VMC operation are important because the other control modes would usually only change the dynamics associated with the duty-ratio generation and, therefore, the corresponding dynamic models can be derived from the VMC state-space representation by substituting the perturbed duty ratio with the developed relation between the new control variable and the duty ratio known as duty-ratio constraints [22].

In reality, the controlled variable is usually the length of the on-time of the main switch [35]. Under fixed-frequency operation, the dynamical information incorporated into the on-time is equal to that of the duty ratio because of constant cycle time. Under variable-frequency operation, the duty ratio is nonlinear and, therefore, the on-time has to be used as the control variable under the VMC mode of operation. A comprehensive survey of the modeling issues can be found from [39].

1.3
Dynamic Analysis of Interconnected Systems

The first system-level analysis was actually applied to a system comprising of an EMI (electromagnetic interference) filter and a regulated converter in the mid-1970s [40]. The analysis yielded the design rules for the EMI-filter design. The equivalent circuit of Figure 1.3 was effectively utilized. It was noticed that the EMI filter would influence the converter dynamics through its output impedance if certain impedance overlaps take place. The developed design rules were straightforward: avoid impedance overlap with a substantial margin. It was also stated that the stability of the converter can be deduced based on the ratio of the filter output impedance and the closed-loop input impedance of the converter by applying the *Nyquist* stability criterion. The impedance ratio is commonly known as minor-loop gain according to [40].

The analyses of interconnected regulated systems [41] have been based on the minor-loop-gain concept. It was concluded that the design rules given in [40] are too conservative and they may lead to unnecessary costs if applied as such. Typically, the design rules are given as a certain forbidden region in a complex plane out of which the minor-loop gain should stay to avoid instability and performance degradation. The shaded area in Figure 1.4 is the minimum area out of which the minor-loop gain should stay for stability to exist. This criterion is known as ESAC criterion [41]. The design rules of [40] would define the forbidden region as the area outside the circle having a radius of 1/GM, where GM stands for gain margin related to the minor-loop gain.

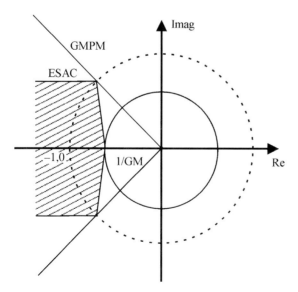

Figure 1.4 Forbidden regions.

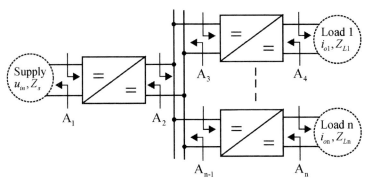

Figure 1.5 Different system interfaces.

It has turned out that avoiding the stated forbidden regions does not actually ensure that the transient performance of the converter would stay intact but more detailed considerations should be carried out [42, 43]: The practical power systems consist of several interfaces at which the minor-loop gain can be defined as depicted in Figure 1.5 (i.e., $A_1 - A_N$) containing also different information on the dynamics of the overall system. The interfaces that exist at the direct input or output of the power stage of the converter would contain the most useful information as actually has been demonstrated in [40]. The existence of stability can be concluded equally based on any of the defined minor-loop gains within the system [44]. The existence of stability even with a good margin (i.e., no impedance overlap) does not necessarily ensure that the transient performance of the associated converters is acceptable [42, 43].

Typically the converters are equipped with EMI filters or capacitors at the input side further complicating the performance analysis based on the measurable information at system level (Figure 1.6) due to hiding effects of those components [43]. The converter modules may be also provided with output-voltage remote sensing [45]. The application of the remote sensing may profoundly change the dynamics of the associated converter depending on what kind of external passive circuit elements are connected inside the converter (Figure 1.6) as demonstrated in [46].

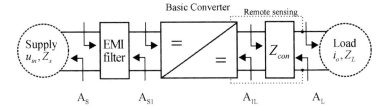

Figure 1.6 Internal system interfaces inside a converter.

As a summary we can state that the stability analysis of an interconnected-regulated system would be deterministic but to conclude whether the transient performance of the system is satisfactory or not is a more complicated issue and perfect information on it is difficult to obtain. Therefore, the methods by means of which the interactions can be reduced or totally eliminated are of great importance and worth to be considered as explained and demonstrated in [47–50].

1.4
Canonical Equivalent Circuit

The equivalent circuit introduced in Figure 1.3 as a canonical circuit is not a true canonical equivalent circuit, because it can only represent the dynamics of a two-memory-element VMC converter operating in CCM. A two-port model (Figure 1.7), where the input port is a Norton equivalent circuit and the output port a Thevenin equivalent circuit, would provide a real canonical representation of the dynamics associated with a voltage-input–voltage-output converter [11, 51]: The input and output-port parameters constitute of a set known as G-parameters, which can be proved to exist always and thus they can be defined for any voltage-input-voltage-output electrical system [14]. In practice, the set composes of the well-known transfer functions typically used to characterize switched-mode converters. It is essential that the transfer functions are defined in such a way that the source and load effects are removed in order to represent the real internal dynamics. Such transfer functions are commonly known as unterminated transfer functions [51].

In addition, the dynamic representation of the current-output converter can be derived from the two-port model of the voltage-output converter by transforming its Thevenin output port to an equivalent Norton representation shown in Figure 1.8 [52, 53]. The parameters of the current-output model (i.e., admittance or Y-parameters) can be derived as a function of the well-known voltage-output transfer functions (i.e., the G-parameters).

It is known that the source and load impedances, Figure 1.9, may affect the dynamics of a converter [11]. Mathematical formulation describing analytically the effects can be solved by using either circuit theory [15] or by applying the extra element theorem (EET) described in [16]. The EET method would provide useful formulation but may be difficult to apply.

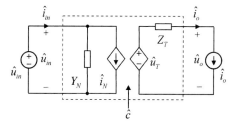

Figure 1.7 Canonical equivalent circuit of voltage-output converter.

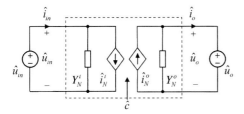

Figure 1.8 Canonical equivalent circuit for current-output converter.

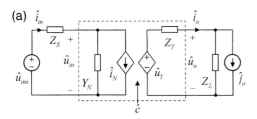

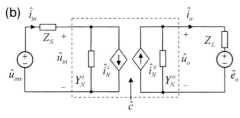

Figure 1.9 Nonideal source and load: (a) voltage-output converter and (b) current-output converter.

1.5
Load-Response-Based Dynamic Analysis

The frequency response of the voltage or current loop can be measured injecting an excitation signal into the corresponding loop, which has to be disconnected in such a way that only the DC signal can pass through it. This means that the internal circuitry of the converter has to be manipulated. Therefore, the methods not requiring the tear-down approach would be desirable such as the load-transient analysis [54–56]. The transient-based-analysis technique is commonly used in the control engineering textbooks such as [54]. It should, however, be noticed that the control engineering textbooks usually discuss on the transients resulting from applying excitations into the reference input, which is usually not available in the power electronic converters. The transients resulting from applying excitations into the disturbance inputs (i.e., input voltage or load current) also contain the effect of the corresponding open-loop transfer function in addition with the loop behavior. Those effects usually dominate and they would hide the information from the loop behavior. An illustrative example of such a phenomenon is shown in Figure 1.10 [16], where a certain converter under three different control principles is subjected to a constant-current-type load change. Some similarities and differences are clearly observable:

10 | *1 Introduction*

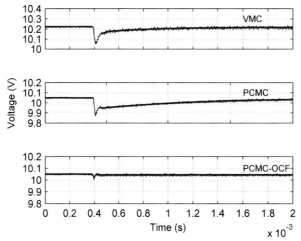

Figure 1.10 Output-voltage responses of a buck converter under VMC, PCM, and PCM with output-current-feedforward (OCF) control to an output-current change.

The setup time of the PCM converter seems to be very long compared to the VMC converter interpreted easily as a substantial difference in the voltage-loop crossover frequencies if following the information given for example in [56]. The output transient of the PCMC-OCF converter is extremely small and recovers quickly. This could be interpreted as a sign of very high control bandwidth. The output-voltage loop gains of the converters are, however, designed in a comparable manner as shown in Figure 1.11. Therefore, Figure 1.10 clearly demonstrates that the time-domain transients

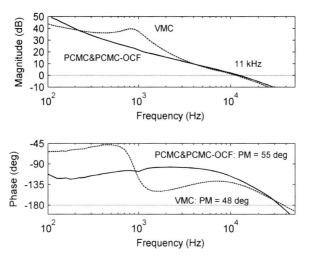

Figure 1.11 Measured voltage-loop gains (PM = phase margin).

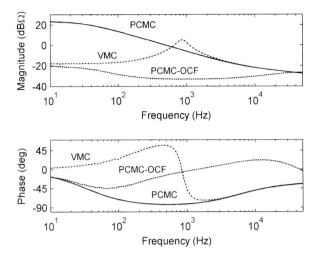

Figure 1.12 Measured open-loop output impedances.

do not provide the desired information on the voltage-loop properties because of the dominating effect of the disturbance input.

The measured open-loop output impedances of the converters are shown in Figure 1.12 and the corresponding closed-loop output impedances in Figure 1.13, where the voltage-loop-gain effect is observable. The closed-loop impedances provides the explanations for the observed load transients as explained in detail in [57–59], but basically the origin of the observed differences is the internal open-loop output impedances (Figure 1.12).

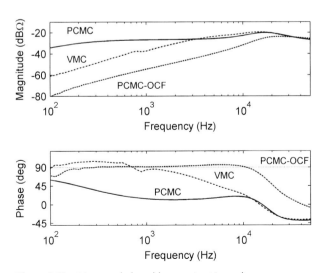

Figure 1.13 Measured closed-loop output impedances.

The internal open- and closed-loop output impedances are important sources of information, because they can be used to predict the dynamic behavior of the converter in different load environment and eventually to choose a best type of converter for the specific application [12].

1.6
Content Review

The content of the subsequent chapters is briefly reviewed in order to clarify the message each chapter contains:

The conceptual and theoretical basis of the book is provided in Chapter 2 in a simple and practical manner without using difficult mathematical treatments. The same theoretical formulas are repeated in the associate chapters if deemed to be necessary for understanding the message. The definition of different stability concepts, the influence of zeros and poles in the transfer functions, and the definition of the open-loop condition in a converter are especially important to be fully understood in order to understand the messages the book will provide.

The unified dynamic modeling of the direct-on-time control is provided in Chapter 3. The method is applied more in detail to the basic converters (i.e., buck, boost, and buck–boost) in the fixed-frequency mode of operation both in continuous (CCM) and discontinuous conduction modes (DCM). The variable-frequency operation is treated separately in Chapter 6. Extensive dynamic review is provided at the end of the chapter based both on experimental and theoretical evidence. The emphasis is in introducing the dynamical changes the operation in CCM and DCM would provide. The modeling of the direct-on-time control is important, because the dynamical models of the other control modes would be mainly derived based on it. The presented methods are also easily applicable to modeling of higher order converters.

The dynamic modeling of peak-current-mode control (PCMC) is provided in Chapter 4 and applied to the same converters as in Chapter 3. The origin of the peculiar phenomena observed in the operation of the PCM-controlled converters is fully explained. Extensive dynamic review is provided at the end of the chapter based both on experimental and theoretical evidence. The dynamic differences of the VM- and PCM-controlled buck and boost converters in CCM are compared. The PCM control is widely applied in controlling the switched-mode converters due to providing advantageous features but its modeling is the subject of intensive discussions. Unanimously accepted modeling method does not exist. The method presented in the book is based on the natural processes taking place in the converter without any kind of curve fitting or similar approaches. Therefore, the resulting dynamic models naturally represent the dynamics of the converter well and also provide the scientifically sound explanations for the phenomena observed in a PCM-controlled converter.

The dynamic modeling of average-current-mode control (ACMC) is provided in Chapter 5 and applied to a buck converter in CCM. The ACM control has naturally similarities to the PCM control, because both of the control methods use the inductor current for producing the duty ratio. Therefore, the presented modeling method applies the PCMC modeling presented in Chapter 4. The full ripple-feedback case is treated in detail. The effect of the high-frequency pole in the inductor-current-loop amplifier is discussed more in detail. The comparisons between VM, PCM, and ACM control in a buck converter are provided in order to highlight the dynamic changes and features the ACM control would provide.

The dynamic modeling of self-oscillation control (i.e., variable-frequency operation) is provided in Chapter 6 and applied to the basic converters. The dynamic review is provided for buck and buck–boost or flyback converters. The self-oscillation control is usually a form of PCM control requiring similar modeling steps as the fixed-frequency PCM control introduced in Chapter 4.

The dynamics associated with the current-output converters is treated in Chapter 7. The dynamic review is provided for a buck converter under VM and PCM controls based on the experimental and theoretical evidence. The observed peculiar dynamical behavior is fully explained. The chapter concentrates on the single-feedback-loop case but the method to solve the dynamic modeling of cascaded cases is also provided.

The dynamic issues related to the interconnected systems are treated in Chapter 8. The theoretical interaction formalism is briefly reviewed. The concepts of intermediate and input–output stabilities are consistently defined by applying system theory and shown to be related to the impedance ratio known as minor-loop gain. The analysis of the output-voltage remote-sensing and EMI-filter effects are briefly discussed and the theoretical formulation to treat them is provided. Practical evidence is provided to support the theoretical findings.

The control-related issues such as different controller implementations, factors affecting the transient, responses and limiting the maximum loop crossover frequency are discussed in Chapter 9. In addition, the dynamic constraints related to the simple control systems based on the adjustable shunt regulator TL431 are treated. Finally, a consistent method to shape the loop gain in order to achieve the desired loop dynamics is proposed and verified experimentally.

Dynamic modeling and analysis of a fourth-order converter known as two-inductor buck, current-sourced buck, and superbuck converter is provided in Chapter 10 as an example of the higher order converters and the possible dynamic anomalies involved in them. The superbuck converter has similar features as the conventional buck converter but its input and output currents are continuous. The input-current ripple can be further reduced by coupling the inductors. Consistent and easy-to-apply analysis methods for the coupled-inductor technique are given and applied to the VM- and PCM-controlled converter. Practical evidence on the dynamics of the PCM-controlled superbuck converter is provided.

References

1. S. Cuk, 'General topological properties of switching structures,' in *Proc. IEEE Power Electronics Specialists Conf.*, **1979**, pp. 109–130.
2. Y. Huang and C.K. Tse, 'Circuit theoretical classification of parallel connected DC–DC converters,' *IEEE Trans. Circuits Syst. – I: Regular Papers*, vol. 54, no. 5, **2007**, pp. 1099–1108.
3. R. Tymerski and V. Vorperian, 'Generation and classification of PWM DC-to-DC converters,' *IEEE Trans. Aerosp. Electron. Syst.*, vol. 24, no. 6, **1988**, pp. 743–754.
4. V.J. Thottuvelil, 'Modeling and analysis of power converter systems with batteries,' in *Proc. IEEE International Telecommunications Energy Conf.*, **1997**, pp. 517–522.
5. T. Suntio, A. Glad, and P. Waltari, 'Constant-current vs. constant-power protected rectifier as a DC UPS system's building block,' in *Proc IEEE International Telecommunications Energy Conf.*, **2006**, pp. 227–233.
6. T. Suntio, I. Gadoura, J. Lempinen, and K. Zenger, 'Practical design issues of multiloop controller for a telecom rectifier,' in *Proc. IEEE Telecommunications Energy Special Conf.*, **2000**, pp. 197–201.
7. J. Lempinen and T. Suntio, 'Small-signal modeling for design of robust variable-frequency flyback battery chargers for portable device applications,' in *Proc. IEEE Applied Power Electronics Conf.*, **2001**, pp. 548–554.
8. A. Tenno, R. Tenno, and T. Suntio, 'Method for battery impedance analysis,' *J. Electrochem. Soc.*, vol. 151, no. 6, **2004**, pp. A806–A824.
9. S. Lyi and R.A. Dougal, 'Dynamics and multiphysics model of solar array,' *IEEE Trans. Energy Conv.*, vol. 17, no. 2, **2002**, pp. 285–294.
10. D. Shmilovitz and S. Singer, 'A switched-mode converter suitable for superconductive magnetic energy storage (SMES) systems,' in *Proc. IEEE Applied Power Electronics Conf.*, **2002**, pp. 630–634.
11. M. Hankaniemi, Dynamical profile of switched-mode converter – fact of fiction, PhD Thesis, Tampere University of Technology, **2007**, Publication No. 687.
12. T. Roinila, M. Hankaniemi, T. Suntio, M. Sippola, and M. Vilkko, 'Dynamical profile of a switched-mode converter – Reality or imagination,' in *Proc. IEEE International Telecommunications Energy Conf.*, **2007**, pp. 420–427.
13. M. Hankaniemi, T. Suntio, and M. Sippola, 'Characterization of regulated converters to ensure stability and performance in distributed power supply system,' in *Proc. IEEE International Telecommunications Energy Conf.*, **2005**, pp. 533–538.
14. P.G. Maranesi, V. Tavazzi, and V. Varanoli, 'Two-port modeling of PWM voltage regulators at low frequencies,' *IEEE Trans. Indust. Electron.*, vol. 35, no. 3, **1988**, pp. 444–450.
15. C.K. Tse, *Linear Circuit Analysis*, Addison-Wesley Longman, Harlow, UK, **1998**.
16. R.D. Middlebrook, 'Null double injection and the extra element theorem,' *IEEE Trans. Edu.*, vol. 32, no. 3, pp. 167–180, **1989**.
17. T. Suntio, M. Hankaniemi, and M. Karppanen, 'Analysing the dynamics of regulated converters,' *IEE Proc. Electr. Power Appl.*, vol. 153, no. 6, **2006**, pp. 905–910.
18. R. Severn and G. Bloom, *Modern DC-to-DC Switch Mode Power Converter Circuits*, Van Nostrand Reinhold, New York, USA, **1985**.
19. N. Mohan, T.M. Undeland, and W.P. Robbins, *Power Electronics – Converters, Applications, and Design*, John Wiley & Sons, New York, USA, **1989**, 2nd Edition.
20. J.G. Kassakian, M.F. Schlecht, and G.C. Verghese, *Principles of Power Electronics*, Addison-Wesley, Reading, MA, USA, **1991**.
21. P.T. Krein, *Elements of Power Electronics*, Oxford University Press, New York, USA, **1998**.

22. R.W. Erickson and D. Maksimovic, *Fundamentals of Power Electronics*, Kluwer, Norwell, MA, USA, **2001**, 2nd Edition.
23. M.H. Rashid, *Power Electronics – Circuits, Devices and Applications*, Pearson Education, Upper Saddle River, NJ, USA, **2004**, 3rd Edition.
24. S. Maniktala, *Switching Power Supply Design & Optimization*, McGraw-Hill, New York, USA, **2004**.
25. S. Ang and A. Oliva, *Power-Switching Converters*, Taylor & Francis, Boca Raton, FL, USA, **2005**, 2nd Edition.
26. I. Batarseh, *Power Electronic Circuits*, John Wiley & Sons, New York, USA, **2004**.
27. A.S. Kislovski, R. Redl, and N.O. Sokal, *Dynamic Analysis of Switching-Mode DC/DC Converters*, Van Nostrand Reinhold, New York, USA, **1991**.
28. D.M. Mitchell, *DC–DC Switching Regulator Analysis*, DMMitchell Consultants, Cedar Rapids, IA, USA, **1992**.
29. J.Y. Zhu, 'Interpreting small signal behavior of the synchronous buck converter at light load,' *IEEE Power Electron. Lett.*, vol. 3, no. 4, **2005**, pp. 144–147.
30. C.W. Deisch, 'Simple switching control method changes power converter into a current source,' in *Proc. IEEE Power Electronics Special Conf.*, **1978**, pp. 300–306.
31. G.W. Wester and R.D. Middlebrook, 'Low-frequency characterization of switched-mode dc–dc converters,' *IEEE Trans. Aerosp. Electron. Syst.*, vol. AES-9, no. 3, **1973**, pp. 376–385.
32. R.D. Middlebrook and S. Cuk, 'A general unified approach to modeling switching-converter power stages,' *Int. J. Electron.*, vol. 42, no. 6, **1977**, pp. 521–550.
33. S. Cuk and R.D. Middlebrook, 'A general unified approach in modeling switching DC-to-DC converters in discontinuous conduction mode, ' in *Proc. IEEE Power Electronics Special Conf.*, **1977**, pp. 36–57.
34. J. Sun, D.M. Mitchell, M.F. Greuel, P.T. Krein, and R.M. Bass, 'Average modeling of PWM converters in discontinuous modes,' *IEEE Trans. Power Electron.*, vol. 16, no. 4, **2001**, pp. 482–492.
35. T. Suntio, 'Unified average and small-signal modeling of direct-on-time control,' *IEEE Trans. Indust. Electron.*, vol. 53, no. 1, **2006**, pp. 287–295.
36. D.M. Mitchell, 'Pulsewidth modulator phase shift,' *IEEE Trans. Aerosp. Electron. Syst.*, vol. AES-16, no. 3, **1980**, pp. 272–278.
37. J. Sun, 'Small-signal modeling of variable-frequency pulsewidth modulators,' *IEEE Trans. Aerosp. Electron. Syst.*, vol. 38, no. 3, **2002**, pp. 1104–1108.
38. Y. Qiu, M. Xu, K. Yao, J. Sun, and F.C. Lee, 'Multifrequency small-signal model for buck and multiphase buck converters,' *IEEE Trans. Power Electron.*, vol. 21, no. 5, **2006**, pp. 1185–1192.
39. D. Maksimovic, A.M. Stankovic, V.J. Tottuvelil, and G.C. Verghese, 'Modeling and simulation of power electronic converters,' *Proc. IEEE*, vol. 89, no. 6, **2001**, pp. 898–912.
40. R.D. Middlebrook, 'Input filter considerations in design and application of switching regulators,' in *Proc. IEEE Industry Application Society Annual Conf.*, **1976**, pp. 366–382.
41. S.D. Sudhoff, S.F. Glover, P.T. Lamm, D.H. Schmucker, and D.E. Delisle, 'Admittance space stability analysis and power electronic systems,' *IEEE Trans. Aerosp. Electron. Syst.*, vol. 36, no. 3, **2000**, pp. 965–973.
42. M. Hankaniemi, M. Karppanen, and T. Suntio, 'Load imposed instability and performance degradation in a regulated converter,' *IEE Proc. Electr. Power Appl.*, vol. 153, no. 6, **2006**, pp. 781–786.
43. M. Karppanen, M. Sippola, and T. Suntio, 'Source-imposed instability and performance degradation in a regulated converter,' in *Proc. IEEE Power Electronics Specialists Conf.*, **2007**, pp. 194–200.
44. K. Zenger, A. Altowati, and T. Suntio, 'Dynamic properties of interconnected

power systems – A system theoretic approach,' in *Proc. IEEE Industrial Electronics and Applications Conf.*, **2006**, pp. 835–840.
45. C. Gezgin, W.C. Bowman, and V.J. Thottuvelil, 'A stability analysis tool for DC–DC converters,' in *Proc. IEEE Applied Power Electronics Conf.*, **2003**, pp. 1014–1020.
46. M. Karppanen, T. Suntio, and M. Sippola, 'Impact of output-voltage remote sensing on converter dynamics,' *Int. Rev. Electr. Eng.*, vol. 2, no. 2, **2007**, pp. 196–202.
47. T. Suntio, K. Kostov, T. Tepsa, and J. Kyyrä, 'Using input invariance as a method to facilitate system design in DPS applications,' *J. Circuits, Syst., Comput.*, vol. 13, no. 4, **2004**, pp. 707–723.
48. M. Karppanen, M. Hankaniemi, and T. Suntio, 'Load and supply invariance in a regulated converter,' in *Proc. IEEE Power Electronics Specialists Conf*, **2006**, pp. 2663–2668.
49. M. Karppanen, M. Hankaniemi, T. Suntio, and M. Sippola, 'Dynamical characterization of peak-current-mode-controlled buck converter with output-current feedforward,' *IEEE Trans. Power Electron.*, vol. 22, no. 2, **2007**, pp. 444–451.
50. M. Karppanen, T. Suntio, and M. Sippola, 'Dynamical characterization of input-voltage-feedforward-controlled buck converter,' *IEEE Trans. Indust. Electron.*, vol. 54, no. 2, **2007**, pp. 1005–1013.
51. B.H. Cho, 'Modeling and analysis of spacecraft power systems,' PhD Thesis, Virginia Polytechnic Institute and State University, USA, **1985**, 181 pp.
52. M. Hankaniemi and T. Suntio, 'Small-signal models for constant-current regulated converters,' in *Proc. IEEE Industrial Electronics Society Annual Conf.*, **2006**, 2037–2042.
53. M. Hankaniemi, M. Sippola, and T. Suntio, 'Analysis of the load interactions in constant-current-controlled buck converter,' in *Proc. IEEE International Telecommunications Energy Conf.*, **2006**, pp. 343–348.
54. K. Ogata, *Modern Control Engineering*, Prentice-Hall, Upper Saddle River, NJ, USA, **1997**, 3rd Edition.
55. J.C. Basio and S.R. Matos, Design of PI and PID controllers with transient performance specifications,' *IEEE Trans. Educ.*, vol. 45, no. 4, **2002**, pp. 364–370.
56. W.H. Tuttle, 'Relating converter transient response characteristics to feedback loop control design,' in *Proc. Powercon 11*, **1984**, pp. 10.1–10.12.
57. B. Choi, 'Step load response of a current-mode-controlled DC-to-DC converter,' *IEEE Trans. Aerosp. Electron. Syst.*, vol. 33, no. 4, **1997**, pp. 1115–1121.
58. C. Gezgin, 'Predicting load transient response of output voltage in DC–DC converters,' in *Proc. IEEE Applied Power Electronics Conf.*, **2004**, pp. 1339–1343.
59. J. Betten and R. Kollman, 'Easy calculation yields load transient response,' *Power Electronics Technology*, February **2005**, pp. 40–48.

2
Basis for Dynamic Analysis and Control Dynamics

2.1
Introduction

The chapter introduces the basis behind the dynamic analysis and control dynamics, defines the concepts of open- and closed-loop systems, and presents the sets of transfer functions for voltage- and current-output converters used to characterize their dynamics (i.e., *G*- and *Y*-parameters) in matrix form as a two-port network and control-engineering block diagrams. The basic dynamic representation is given to correspond to the true internal dynamics of the associated converter. The methods to address the effect of a nonideal source and load are shortly introduced. The control-engineering block diagrams are used to derive the closed-loop representations for the output and input dynamics. The stability and performance indices such as control bandwidth, loop crossover frequency, phase and gain margins, sensitivity and complementary sensitivity functions as well as instability, unconditional and conditional stabilities are reviewed. The meaning and consequences of zeros and poles in the transfer functions are explained. Especially, the right-half-plane zeros and poles are discussed in the light of the control bandwidth constraints they provide. A short review of matrix algebra is also provided.

2.2
Dynamic Representations at Open Loop

The open-loop converter is the basis for the models utilized in the control design. The open- loop condition is defined to be such a condition where the external feedback from the controlled output variable (i.e., usually the output voltage or current) does not affect the dynamic behavior of the converter [1]. The open-loop representations of the voltage-output and current-output converters are shown in Figure 2.1. The voltage-output converter is such a converter (Figure 2.1a) where the control system regulates the output voltage (i.e., Figure 2.2, the external feedback is from the output voltage u_o). The current-output converter is such a converter (Figure 2.1b) where the control

Dynamic Profile of Switched-Mode Converter. Teuvo Suntio
© 2009 WILEY-VCH Verlag GmbH & Co. KGaA, Weinheim
ISBN: 978-3-527-40708-8

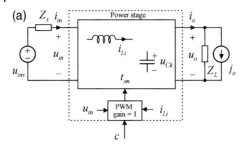
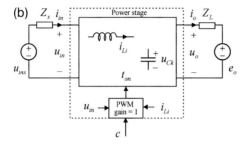

Figure 2.1 Converters at open loop: (a) voltage-output converter and (b) current-output converter.

system regulates the output current [3, 4]. Sometimes the converter may change the operation mode from the voltage output to the current output [4] to protect itself from damage due to excess load current generated for example by a storage battery [5] when charged. The outer active feedback loop determines the mode of operation. The converter may have internal feedback or feedforward loops connected in order to change its dynamic behavior, but despite these connections the converter is defined to operate at open loop, whenever the outer external feedback loop is open as depicted in Figure 2.1.

Typical representatives of such a condition are the peak-current-mode control (PCMC) [6] where the pulsewidth modulation (PWM) is accomplished using the up-slope of the inductor current, the average-current-mode (ACM) control [7], where the inductor current affects the modulation process through

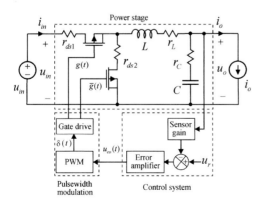

Figure 2.2 A second-order converter known as the buck or step-down converter.

an average filter, and the input-voltage-feedforward (IVFF) control [8], where the PWM ramp is modulated by means of the input voltage. All these control modes are known to change the open-loop dynamics of the associated converter compared to the corresponding direct-on-time or voltage-mode (VM) controlled converter [9–11], where the PWM is accomplished using a constant-slope PWM signal.

The dynamic description of the converter is usually given in such a way that the load impedance (i.e., Z_L in Figure 2.1, a resistor) is included in the derived transfer functions [12–14]. Such models are load-affected models, which do not characterize the true internal dynamics of the converter and consequently do not suffice for analyzing and predicting the converter behavior in its practical environment, that is, as a subsystem in an interconnected system. The models corresponding to the internal dynamics are called unterminated models [15–17]. They can be usually measured and/or derived by using an ideal voltage source (i.e., $Z_s = 0$ in Figure 2.1) at the input of the converter, and an ideal current sink (i.e., $Z_L = \infty$ in Figure 2.1a) or an ideal voltage source (i.e., $Z_L = 0$ in Figure 2.1b) as a load. The open-loop transfer functions of the current-output converter cannot be usually measured due to the short-circuit nature of the required ideal voltage source at output but only in case of the converters having current-output nature at open loop such as peak-current-mode or average-current-mode-controlled converters [6, 7]. The practical electronic loads providing the voltage-source capability have usually nonzero internal impedance (i.e., $Z_L > 0$, Figure 2.1b), which would produce load-affected transfer functions [2, 3].

2.2.1
State Space

The variables associated with the converter dynamics are called *state variables*, which are usually the currents (i_{Li}) and voltages (u_{Ck}) of the inductors and capacitors incorporated in the power stage (Figure 2.1), *input variables*, which are the input voltage (u_{in}), output current (i_o), and the control variable (c) in the case of the voltage-output converter (Figure 2.1a), and the input voltage (u_{in}), the output voltage (u_o), and the control variable (c) in the case of the current-output converter (Figure 2.1b) as well as the *output variables*, which are the input current (i_{in}) and the output voltage (u_o) in the case of the voltage-output converter (Figure 2.1a) and the input current (i_{in}) and the output current (i_o) in the case of the current-output converter (Figure 2.1b). The ports, to which the input variables are applied, are also known as disturbance inputs [1].

The switched-mode converters are variable-structure systems where the topological circuit structure varies depending on the state of the semiconductor switches in the power stage producing nonlinearity (Figure 2.2). The dynamics associated with them can be effectively captured, however, by averaging the converter behavior within one switching cycle and linearizing the averaged behavior at the desired operating point [9–11]. The modeling issues are discussed more in detail in the subsequent chapters.

The small-signal state-space representation is constructed by expressing the *derivatives* of the *state variables* and the *output variables* as a function of the *state and input variables* as illustrated in (2.1), where the linearized state space is developed corresponding to a second-order voltage-output converter (i.e., the power stage consists of one inductor and one capacitor),

$$\begin{bmatrix} \dfrac{d\hat{i}_L}{dt} \\ \dfrac{d\hat{u}_C}{dt} \end{bmatrix} = \begin{bmatrix} a_{11} & a_{12} \\ a_{21} & a_{22} \end{bmatrix} \begin{bmatrix} \hat{i}_L \\ \hat{u}_C \end{bmatrix} + \begin{bmatrix} b_{11} & b_{12} & b_{13} \\ b_{21} & b_{22} & b_{23} \end{bmatrix} \begin{bmatrix} \hat{u}_{in} \\ \hat{i}_o \\ \hat{c} \end{bmatrix}$$

$$\begin{bmatrix} \hat{i}_{in} \\ \hat{u}_o \end{bmatrix} = \begin{bmatrix} c_{11} & c_{12} \\ c_{21} & c_{22} \end{bmatrix} \begin{bmatrix} \hat{i}_L \\ \hat{u}_C \end{bmatrix} + \begin{bmatrix} d_{11} & d_{12} & d_{13} \\ d_{21} & d_{22} & d_{23} \end{bmatrix} \begin{bmatrix} \hat{u}_{in} \\ \hat{i}_o \\ \hat{c} \end{bmatrix}$$

(2.1)

and in (2.2), where the linearized state space represents a similar current-output converter, respectively. The hat over the variables denotes small-signal behavior, that is, a small perturbation is applied to the variables at the defined operating point.

$$\begin{bmatrix} \dfrac{d\hat{i}_L}{dt} \\ \dfrac{d\hat{u}_C}{dt} \end{bmatrix} = \begin{bmatrix} a_{11} & a_{12} \\ a_{21} & a_{22} \end{bmatrix} \begin{bmatrix} \hat{i}_L \\ \hat{u}_C \end{bmatrix} + \begin{bmatrix} b_{11} & b_{12} & b_{13} \\ b_{21} & b_{22} & b_{23} \end{bmatrix} \begin{bmatrix} \hat{u}_{in} \\ \hat{u}_o \\ \hat{c} \end{bmatrix}$$

$$\begin{bmatrix} \hat{i}_{in} \\ \hat{i}_o \end{bmatrix} = \begin{bmatrix} c_{11} & c_{12} \\ c_{21} & c_{22} \end{bmatrix} \begin{bmatrix} \hat{i}_L \\ \hat{u}_C \end{bmatrix} + \begin{bmatrix} d_{11} & d_{12} & d_{13} \\ d_{21} & d_{22} & d_{23} \end{bmatrix} \begin{bmatrix} \hat{u}_{in} \\ \hat{u}_o \\ \hat{c} \end{bmatrix}$$

(2.2)

The state spaces in (2.1) and (2.2) are typically given in matrix form [18] by

$$\dot{\mathbf{x}}(t) = \mathbf{A}\mathbf{x}(t) + \mathbf{B}\mathbf{u}(t)$$
$$\mathbf{y}(t) = \mathbf{C}\mathbf{x}(t) + \mathbf{D}\mathbf{u}(t)$$

(2.3)

where $\mathbf{x}(t)$ is the *state-variable vector*, the dot over $\mathbf{x}(t)$ denotes derivative, $\mathbf{u}(t)$ is *the input vector*, and $\mathbf{y}(t)$ is the *output vector*. The entries in the coefficient matrices **A**, **B**, **C**, and **D** are scalars for a time-invariant linear system [18]. The time-domain state space (2.3) can be solved in frequency domain by applying *Laplace* transform with zero initial conditions yielding the input-to-output description as

$$\mathbf{y}(s) = \left[\mathbf{C}[s\mathbf{I} - \mathbf{A}]^{-1} \mathbf{B} + \mathbf{D} \right] \mathbf{u}(s)$$

(2.4)

where **I** is an identity matrix and superscript '−1' denotes the matrix inverse. The matrix $\mathbf{G}(s) = \mathbf{C}[s\mathbf{I} - \mathbf{A}]^{-1}\mathbf{B} + \mathbf{D}$ is called the *transfer matrix* constituting the G- or Y-parameter representation of the converter input–output dynamics. The use of G-parameters is well justified, because their existence is always quarantined [19].

2.2 Dynamic Representations at Open Loop

To solve the dynamics associated with the peak and average-current-mode control may require to solve also the input-to-state description by applying *Laplace* transformation to the state equation in (2.3) yielding

$$\mathbf{x}(s) = [s\mathbf{I} - \mathbf{A}]^{-1}\mathbf{B}\mathbf{u}(s) \qquad (2.5)$$

The matrix $\mathbf{\Phi}(s) = [s\mathbf{I} - \mathbf{A}]^{-1}$ is called *state-transition matrix*, and its determinant $\Delta(s) = |s\mathbf{I} - \mathbf{A}|$ defines the denominator of the transfer functions in (2.4) and (2.5). The roots of $\Delta(s)$ are known as the poles of the transfer function defining the basic nature of the internal dynamics of the open-loop system, and also its order (i.e., the order of the system is usually the number of state variables).

The open-loop transfer matrices of voltage-output (2.6) and current-output (2.7) converters are usually of the form

$$\mathbf{G}(s) = \begin{bmatrix} Y_{\text{in}-o} & T_{\text{oi}-o} & G_{\text{ci}} \\ G_{\text{io}-o} & -Z_{o-o} & G_{\text{co}} \end{bmatrix} \qquad (2.6)$$

$$\mathbf{G}(s) = \begin{bmatrix} Y_{\text{in}-o} & T_{\text{oi}-o} & G_{\text{ci}} \\ G_{\text{io}-o} & -Y_{o-o} & G_{\text{co}} \end{bmatrix} \qquad (2.7)$$

where

$Y_{\text{in}-o}$ = input admittance (i.e., $\frac{\hat{i}_{\text{in}}}{\hat{u}_{\text{in}}}$)

$T_{\text{oi}-o}$ = reverse or output-to-input transfer function (i.e., $\frac{\hat{i}_{\text{in}}}{\hat{i}_o}$ or $\frac{\hat{i}_{\text{in}}}{\hat{u}_o}$)

G_{ci} = control-to-input transfer function (i.e., $\frac{\hat{i}_{\text{in}}}{\hat{c}}$)

$G_{\text{io}-o}$ = forward, input-to-output, line-to-output transfer function or audio-susceptibility (i.e., $\frac{\hat{u}_o}{\hat{u}_{\text{in}}}$ or $\frac{\hat{i}_o}{\hat{u}_o}$)

Z_{o-o} = output impedance (i.e., $\frac{\hat{u}_o}{\hat{i}_o}$)

Y_{o-o} = output admittance (i.e., $\frac{\hat{i}_o}{\hat{u}_o}$)

G_{co} = control-to-output transfer function (i.e., $\frac{\hat{u}_o}{\hat{c}}$ or $\frac{\hat{i}_o}{\hat{c}}$).

The open-loop transfer matrices in (2.6) and (2.7) describe fully the dynamics associated with the open-loop converter. Physically the *G*- and *Y*- parameters of the voltage- and current-output modes differ from each other but the output impedance and output admittance are related as $Z_{o-o} = Y_{o-o}^{-1}$. The differences are described more in detail in the next subsection.

2.2.2
Two-Port Models

The internal dynamics of the voltage-output converter can also be represented equally to (2.6) by means of a linear two-port network, Figure 2.3, where the input port is a *Norton* equivalent circuit corresponding to

$$\hat{i}_{\text{in}} = Y_{\text{in}-o}\hat{u}_{\text{in}} + T_{\text{oi}-o}\hat{i}_o + G_{\text{ci}}\hat{c} \qquad (2.8)$$

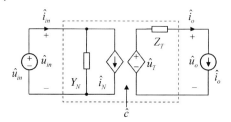

Figure 2.3 Two-port model of a voltage-output converter.

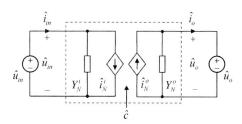

Figure 2.4 Two-port model for a current-output converter.

and the output port a *Thevenin* equivalent circuit corresponding to

$$\hat{u}_o = G_{io-o}\hat{u}_{in} - Z_{o-o}\hat{i}_o + G_{co}\hat{c} \tag{2.9}$$

Similarly, the dynamics of the current-output converter can be represented by using a two-port network, Figure 2.4, where the input port is a Norton equivalent circuit corresponding to

$$\hat{i}_{in} = Y^i_{in-o}\hat{u}_{in} + T^i_{oi-o}\hat{u}_o + G^i_{ci}\hat{c} \tag{2.10}$$

and the output port a similar Norton equivalent circuit corresponding to

$$\hat{i}_o = G^i_{io-o}\hat{u}_{in} - Y_{o-o}\hat{u}_o + G^i_{co}\hat{c} \tag{2.11}$$

The relation between the voltage (2.6) and current-output (2.7) parameters can be found by applying the Thevenin-to-Norton transformation technique defined in [19] to the output port of the voltage-output-converter model in Figure 2.3 yielding [2]

$$\mathbf{G}^i(s) = \begin{bmatrix} Y^i_{in-o} & T^i_{oi-o} & G^i_{ci} \\ G^i_{io-o} & -Y_{o-o} & G^i_{co} \end{bmatrix}
= \begin{bmatrix} Y_{in-o} + \dfrac{G_{io-o}T_{oi-o}}{Z_{o-o}} & -\dfrac{T_{oi-o}}{Z_{o-o}} & G_{ci} + \dfrac{G_{co}T_{oi-o}}{Z_{o-o}} \\ \dfrac{G_{io-o}}{Z_{o-o}} & -\dfrac{1}{Z_{o-o}} & \dfrac{G_{co}}{Z_{o-o}} \end{bmatrix} \tag{2.12}$$

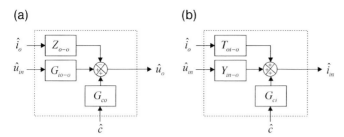

Figure 2.5 Control-block diagrams at open loop for a voltage-output converter: (a) output dynamics and (b) input dynamics.

where the superscript 'i' denotes the transfer function of the current-output converter. Equation (2.12) shows that the internal dynamics of the voltage-output converter would change radically, when it enters into the current-output mode of operation [3].

2.2.3
Control-Block Diagrams

The input and output dynamics of the converter can also be represented by using the control-engineering block diagrams [1] shown for a voltage-output converter in Figure 2.5 and for a current-output converter in Figure 2.6, respectively. The key for constructing the representations is Eqs. (2.8)–(2.11). The block diagrams are useful in deriving the closed-loop dynamic representations.

2.3
Dynamic Representations at a Closed Loop

The closed-loop converter is naturally a converter, where the external feedback loop or loops are connected as depicted in Figure 2.7. In practice, the converter

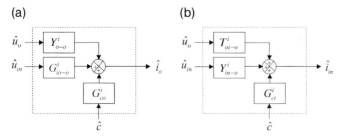

Figure 2.6 Control-block diagrams for a current-output converter: (a) output dynamics and (b) input dynamics.

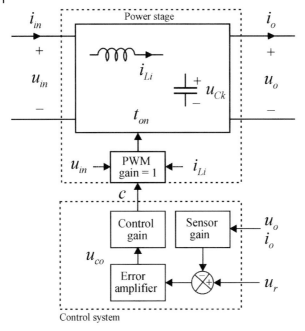

Figure 2.7 Closed-loop converter.

may have a multiloop control system, where the voltage loop is the main or inner loop, and the current loop is the outer loop activated in the case of excess load current by reducing the voltage-loop reference in order to make the output power or current constant as described in detail in [4]. The overload protection is typically needed in the applications where the converter has to recharge a storage battery [5]. The multiloop control system is naturally implemented in a current-output converter in such a way that the output voltage and current are interchanged compared to the voltage-output converter.

The loop gain of a converter is denoted by $L_i(s)$, where the subscript 'i' defines whether the loop is related to the output voltage (i.e., 'v') or the output current (i.e., 'c'), respectively. The internal loop gain consists of the dynamic elements along the closed loop from the feedback variable back to the feedback variable (Figure 2.7) such as the sensor gain, error amplifier, the PWM process, and the power stage depending on the application: In the isolated converters, the PWM process is usually located in the primary side of the converter, and the control system in the secondary side requiring also to isolate the control loop [20–22] for safety reasons. A typical control-loop isolation medium is an optocoupler, which would affect the dynamical behavior and has to be carefully considered [22]. Sometimes the simple control

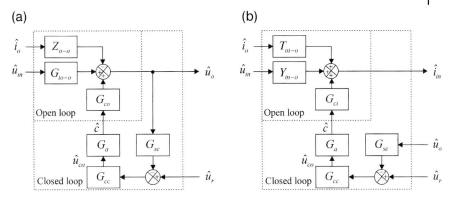

Figure 2.8 The closed-loop control block diagrams of the voltage-output converter for (a) the output dynamics and (b) the input dynamics.

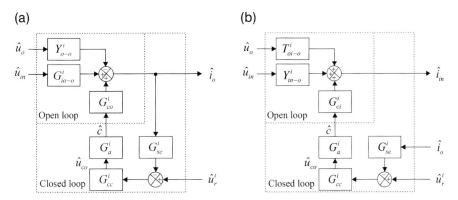

Figure 2.9 The closed-loop control-block diagrams of the current-output converter for (a) output dynamics and (b) input dynamics.

systems based for example on the popular shunt regulator TL431 are actually extremely complicated systems requiring special attention as discussed in [23–26].

The closed-loop dynamic representation of the converter can be solved most conveniently applying the corresponding open-loop block diagrams defined in Figures 2.5 and 2.6 as shown in Figures 2.8 and 2.9. The loop gain can be given generally by

$$L_i(s) = G_{se} G_{cc} G_a G_{co} \qquad (2.13)$$

where the transfer functions correspond to the control-system blocks defined in Figure 2.7 starting from the sensor gain G_{se}, respectively.

2.3.1
Voltage-Output Converter

The closed-loop output voltage (i.e., $\hat{u}_o$, Figure 2.8a) can be computed to be

$$\hat{u}_o = \frac{G_{io-o}}{1 + G_{se}G_{cc}G_aG_{co}} \cdot \hat{u}_{in} - \frac{Z_{o-o}}{1 + G_{se}G_{cc}G_aG_{co}} \cdot \hat{i}_o$$
$$+ \frac{G_{cc}G_aG_{co}}{1 + G_{se}G_{cc}G_aG_{co}} \cdot \hat{u}_r \quad (2.14)$$

The application of the loop-gain definition (2.13) yields

$$\hat{u}_o = \frac{G_{io-o}}{1 + L_v(s)} \cdot \hat{u}_{in} - \frac{Z_{o-o}}{1 + L_v(s)} \cdot \hat{i}_o + \frac{1}{G_{se}} \cdot \frac{L_v(s)}{1 + L_v(s)} \cdot \hat{u}_r \quad (2.15)$$

The closed-loop input current (i.e., $\hat{i}_{in}$, Figure 2.8b) can be computed to be

$$\hat{i}_{in} = Y_{in-o} \cdot \hat{u}_{in} + T_{oi-o} \cdot \hat{i}_o - G_{se}G_{cc}G_aG_{ci} \cdot \hat{u}_o + G_{cc}G_aG_{ci} \cdot \hat{u}_r \quad (2.16)$$

The output voltage in (2.16) has to be substituted with (2.14) yielding with the application of the loop-gain definition (2.13)

$$\hat{i}_{in} = \left(Y_{in-o} - \frac{G_{io-o}G_{ci}}{G_{co}} \cdot \frac{L_v(s)}{1 + L_v(s)} \right) \cdot \hat{u}_{in}$$
$$+ \left(T_{oi-o} + \frac{Z_{o-o}G_{ci}}{G_{co}} \cdot \frac{L_v(s)}{1 + L_v(s)} \right) \cdot \hat{i}_o + \frac{G_{ci}}{G_{se}G_{co}} \cdot \frac{L_v(s)}{1 + L_v(s)} \cdot \hat{u}_r \quad (2.17)$$

If the voltage reference (u_r) is constant as is usually the case in single-loop converters, (2.15) and (2.17) reduce to

$$\hat{u}_o = \frac{G_{io-o}}{1 + L_v(s)} \cdot \hat{u}_{in} - \frac{Z_{o-o}}{1 + L_v(s)} \cdot \hat{i}_o$$
$$\hat{i}_{in} = \left(Y_{in-o} - \frac{G_{io-o}G_{ci}}{G_{co}} \cdot \frac{L_v(s)}{1 + L_v(s)} \right) \cdot \hat{u}_{in} \quad (2.18)$$
$$+ \left(T_{oi-o} + \frac{Z_{o-o}G_{ci}}{G_{co}} \cdot \frac{L_v(s)}{1 + L_v(s)} \right) \cdot \hat{i}_o$$

which defines the usual closed-loop transfer matrix of the voltage-output converter to be

$$\mathbf{G}(s) = \begin{bmatrix} Y_{in-c} & T_{oi-c} \\ G_{io-c} & -Z_{o-c} \end{bmatrix}$$
$$= \begin{bmatrix} Y_{in-o} - \dfrac{G_{io-o}G_{ci}}{G_{co}} \cdot \dfrac{L_v(s)}{1 + L_v(s)} & T_{oi-o} + \dfrac{Z_{o-o}G_{ci}}{G_{co}} \cdot \dfrac{L_v(s)}{1 + L_v(s)} \\ \dfrac{G_{io-o}}{1 + L_v(s)} & -\dfrac{Z_{o-o}}{1 + L_v(s)} \end{bmatrix}$$
$$(2.19)$$

2.3 Dynamic Representations at a Closed Loop

In the case of the multiloop operation, the full-order representations in (2.15) and (2.17) have to be applied. The two-port network defined in Figure 2.3 would equally represent also the closed-loop voltage-output converter, when the input port is defined using (2.17) and the output port using (2.15), corresponding to the following transfer matrix:

$$\mathbf{G}(s) = \begin{bmatrix} Y_{in-c} & T_{oi-c} & G_{ci-c} \\ G_{io-c} & -Z_{o-c} & G_{co-c} \end{bmatrix}$$

$$= \begin{bmatrix} Y_{in-o} - \dfrac{G_{io-o}G_{ci}}{G_{co}} \cdot \dfrac{L_v(s)}{1+L_v(s)} & T_{oi-o} + \dfrac{Z_{o-o}G_{ci}}{G_{co}} \cdot \dfrac{L_v(s)}{1+L_v(s)} & \dfrac{G_{ci}}{G_{se}G_{co}} \cdot \dfrac{L_v(s)}{1+L_v(s)} \\ \dfrac{G_{io-o}}{1+L_v(s)} & -\dfrac{Z_{o-o}}{1+L_v(s)} & \dfrac{1}{G_{se}} \cdot \dfrac{L_v(s)}{1+L_v(s)} \end{bmatrix}$$

(2.20)

where G_{co-c} and G_{ci-c} are the reference-to-output and reference-to-input transfer functions, respectively. In the case of the cascaded control system, the current-output model can be derived from (2.20) utilizing (2.12).

2.3.2
Current-Output Converter

The closed-loop output current (i.e., $\hat{i}_o$, Figure 2.9a) can be computed to be

$$\hat{i}_o = \dfrac{G^i_{io-o}}{1+G^i_{se}G^i_{cc}G^i_aG^i_{co}} \cdot \hat{u}_{in} - \dfrac{Y^i_{o-o}}{1+G^i_{se}G^i_{cc}G^i_aG^i_{co}} \cdot \hat{u}_o$$
$$+ \dfrac{G^i_{cc}G^i_aG^i_{co}}{1+G^i_{se}G^i_{cc}G^i_aG^i_{co}} \cdot \hat{u}^i_r \qquad (2.21)$$

The application of the loop-gain definition (2.13) yields

$$\hat{i}_o = \dfrac{G^i_{io-o}}{1+L_c(s)} \cdot \hat{u}_{in} - \dfrac{Y^i_{o-o}}{1+L_c(s)} \cdot \hat{u}_o + \dfrac{1}{G^i_{se}} \cdot \dfrac{L_c(s)}{1+L_c(s)} \cdot \hat{u}^i_r \qquad (2.22)$$

The closed-loop input current (i.e., $\hat{i}_{in}$, Figure 2.9b) can be computed to be

$$\hat{i}_{in} = Y^i_{in-o} \cdot \hat{u}_{in} + T^i_{oi-o} \cdot \hat{u}_o - G^i_{se}G^i_{cc}G^i_aG^i_{ci} \cdot \hat{i}_o + G^i_{cc}G^i_aG^i_{ci} \cdot \hat{u}^i_r \qquad (2.23)$$

The output current in (2.23) has to be substituted with (2.22) yielding with the application of the loop-gain definition (2.13)

$$\hat{i}_{in} = \left(Y^i_{in-o} - \dfrac{G^i_{io-o}G^i_{ci}}{G^i_{co}} \cdot \dfrac{L_c(s)}{1+L_c(s)} \right) \cdot \hat{u}_{in}$$
$$+ \left(T^i_{oi-o} + \dfrac{Y^i_{o-o}G^i_{ci}}{G^i_{co}} \cdot \dfrac{L_c(s)}{1+L_c(s)} \right) \cdot \hat{u}_o + \dfrac{G^i_{ci}}{G^i_{se}G^i_{co}} \cdot \dfrac{L_c(s)}{1+L_c(s)} \cdot \hat{u}^i_r$$

(2.24)

If the current reference (u_r^i) is constant as is usually the case in single-loop converters, (2.22) and (2.24) reduce to

$$\hat{i}_o = \frac{G_{io-o}^i}{1+L_c(s)} \cdot \hat{u}_{in} - \frac{Y_{o-o}^i}{1+L_c(s)} \cdot \hat{u}_o$$

$$\hat{i}_{in} = \left(Y_{in-o}^i - \frac{G_{io-o}^i G_{ci}^i}{G_{co}^i} \cdot \frac{L_c(s)}{1+L_c(s)} \right) \cdot \hat{u}_{in}$$

$$+ \left(T_{oi-o}^i + \frac{Y_{o-o}^i G_{ci}^i}{G_{co}^i} \cdot \frac{L_c(s)}{1+L_c(s)} \right) \cdot \hat{u}_o \qquad (2.25)$$

which defines the usual closed-loop transfer matrix of the current-output converter to be

$$\mathbf{G}^i(s) = \begin{bmatrix} Y_{in-c}^i & T_{oi-c}^i \\ G_{io-c}^i & -Y_{o-c}^i \end{bmatrix}$$

$$= \begin{bmatrix} Y_{in-o}^i - \frac{G_{io-o}^i G_{ci}^i}{G_{co}^i} \cdot \frac{L_c(s)}{1+L_c(s)} & T_{oi-o}^i + \frac{Y_{o-o}^i G_{ci}^i}{G_{co}^i} \cdot \frac{L_c(s)}{1+L_c(s)} \\ \frac{G_{io-o}^i}{1+L_c(s)} & -\frac{Y_{o-o}^i}{1+L_c(s)} \end{bmatrix}$$

(2.26)

In the case of multiloop operation, the full-order representations in (2.22) and (2.24) have to be applied. The two-port network defined in Figure 2.4 would equally represent also the closed-loop current-output converter when the input port is defined using (2.24) and the output port using (2.22) yielding the following input-to-output transfer matrix:

$$\mathbf{G}(s) = \begin{bmatrix} Y_{in-c}^i & T_{oi-c}^i & G_{ci-c}^i \\ G_{io-c}^i & -Y_{o-c}^i & G_{co-c}^i \end{bmatrix}$$

$$= \begin{bmatrix} Y_{in-o}^i - \frac{G_{io-o}^i G_{ci}^i}{G_{co}^i} \cdot \frac{L_c(s)}{1+L_c(s)} & T_{oi-o}^i + \frac{Y_{o-o}^i G_{ci}^i}{G_{co}^i} \cdot \frac{L_c(s)}{1+L_c(s)} & \frac{G_{ci}^i}{G_{se}^i G_{co}^i} \cdot \frac{L_c(s)}{1+L_c(s)} \\ \frac{G_{io-o}^i}{1+L_c(s)} & -\frac{Y_{o-o}^i}{1+L_c(s)} & \frac{1}{G_{se}^i} \cdot \frac{L_c(s)}{1+L_c(s)} \end{bmatrix}$$

(2.27)

2.4
Load and Source Effects

The previous sections treated the pure internal dynamics from which all the effects of nonideal load and source were removed. In practice, the open-loop transfer functions describing the pure internal dynamics can be usually

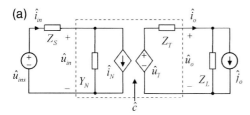
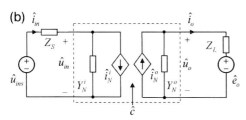

Figure 2.10 Two-port models with nonideal source (i.e., Z_s, $\hat{u}_{ins}$) and nonideal load (i.e., Z_L, $\hat{j}_o$): (a) voltage-output converter and (b) current-output converter.

measured by supplying the converter by means of an ideal voltage source, and loading the converter with an ideal current sink or voltage source [26]. The peak-current-mode controlled converter [6] is a current source at open loop, and therefore, the voltage-output transfer functions cannot be measured as described above but using a resistor as load [27]. The internal dynamics determines the external interactions, and therefore, knowing it is of prime importance. Actually computational methods have to be applied in such cases as instructed in [27].

It is well known that the nonideal source and load may affect the dynamics of the converter [28, 29]. Consequently, its dynamical performance in the practical applications may differ significantly from that measured in the laboratory. The two-port models introduced earlier can be used to solve analytically the load and source effects. Circuit theory provides the media to solve the required equations easily. Basically, the origins of the load and source effects are the impedances associated with the load and source as depicted in Figure 2.10. We provide here only the analytical equations describing the interactions, and the more detailed analysis would be provided later in Chapter 8.

2.4.1 Voltage-Output Converter

The two-port model of the voltage-output converter (Figure 2.10a) is characterized by means of $\hat{u}_o$ and $\hat{i}_{in}$ as follows:

$$\hat{u}_o = G_{io-o}\hat{u}_{in} - Z_{o-o}\hat{i}_o + G_{co}\hat{c}$$
$$\hat{i}_{in} = Y_{in-o}\hat{u}_{in} + T_{oi-o}\hat{i}_o + G_{ci}\hat{c}$$

(2.28)

The load effect on the converter dynamics can be solved by computing $\hat{i}_o$ from Figure 2.10a, which yields

$$\hat{i}_o = \frac{G_{io-o} \cdot \hat{u}_{in} + G_{co} \cdot \hat{c} - Z_L \cdot \hat{j}_o}{Z_{o-o} + Z_L} \quad (2.29)$$

Substituting it in (2.28) with (2.29) yields the desired load-affected dynamic model of the voltage-output converter as follows:

$$\begin{bmatrix} \hat{i}_{in} \\ \hat{u}_o \end{bmatrix} = \begin{bmatrix} Y_{in-o} + \dfrac{G_{io-o} T_{oi-o}}{Z_{o-o} + Z_L} & \dfrac{Z_L T_{oi-o}}{Z_{o-o} + Z_L} & G_{ci} + \dfrac{G_{co} T_{oi-o}}{Z_{o-o} + Z_L} \\ \dfrac{G_{io-o}}{1 + \dfrac{Z_{o-o}}{Z_L}} & -\dfrac{Z_{o-o}}{1 + \dfrac{Z_{o-o}}{Z_L}} & \dfrac{G_{co}}{1 + \dfrac{Z_{o-o}}{Z_L}} \end{bmatrix} \begin{bmatrix} \hat{u}_{in} \\ \hat{j}_o \\ \hat{c} \end{bmatrix} \quad (2.30)$$

The source effect on the converter dynamics can be solved by computing $\hat{u}_{in}$ from Figure 2.10a, which yields

$$\hat{u}_{in} = \frac{1}{1 + Z_s Y_{in-o}} \cdot \hat{u}_{ins} - \frac{Z_s}{1 + Z_s Y_{in-o}} \cdot (T_{oi-o} \cdot \hat{i}_o + G_{ci} \cdot \hat{c}) \quad (2.31)$$

Substituting it in (2.28) yields the source-affected dynamic model for the voltage-output converter as follows:

$$\begin{bmatrix} \hat{i}_{in} \\ \hat{u}_o \end{bmatrix} = \begin{bmatrix} \dfrac{Y_{in-o}}{1 + Z_s Y_{in-o}} & \dfrac{T_{oi-o}}{1 + Z_s Y_{in-o}} & \dfrac{G_{ci}}{1 + Z_s Y_{in-o}} \\ \dfrac{G_{io-o}}{1 + Z_s Y_{in-o}} & -\left(Z_{o-o} + \dfrac{Z_s G_{io-o} T_{oi-o}}{1 + Z_s Y_{in-o}}\right) & G_{co} - \dfrac{Z_s G_{io-o} G_{ci}}{1 + Z_s Y_{in-o}} \end{bmatrix}$$

$$\times \begin{bmatrix} \hat{u}_{ins} \\ \hat{i}_o \\ \hat{c} \end{bmatrix} \quad (2.32)$$

The form of (2.32) is not most convenient but can be transformed to correspond the formulation given in [30] yielding (2.33), where the special admittances $Y_{in-\infty}$ (ideal input admittance) and Y_{in-sc} (short-circuit input admittance) are defined in (2.34), respectively.

$$\begin{bmatrix} \hat{i}_{in} \\ \hat{u}_o \end{bmatrix} = \begin{bmatrix} \dfrac{Y_{in-o}}{1 + Z_s Y_{in-o}} & \dfrac{T_{oi-o}}{1 + Z_s Y_{in-o}} & \dfrac{G_{ci}}{1 + Z_s Y_{in-o}} \\ \dfrac{G_{io-o}}{1 + Z_s Y_{in-o}} & -\dfrac{1 + Z_s Y_{in-sc}}{1 + Z_s Y_{in-o}} \cdot Z_{o-o} & \dfrac{1 + Z_s Y_{in-\infty}}{1 + Z_s Y_{in-o}} \cdot G_{co} \end{bmatrix} \begin{bmatrix} \hat{u}_{ins} \\ \hat{i}_o \\ \hat{c} \end{bmatrix} \quad (2.33)$$

$$Y_{in-\infty} = Y_{in-o} - \frac{G_{io-o} G_{ci}}{G_{co}}$$

$$Y_{in-sc} = Y_{in-o} + \frac{G_{io-o} T_{oi-o}}{Z_{o-o}} \quad (2.34)$$

2.4.2
Current-Output Converter

The two-port model of the current-output converter (Figure 2.10b) is characterized by means of $\hat{i}_o$ and $\hat{i}_{in}$ as follows, where the superscript 'i' denotes the current-output transfer function:

$$\begin{aligned} \hat{i}_o &= G^i_{io-o} \cdot \hat{u}_{in} - Y^i_{o-o} \cdot \hat{u}_o + G^i_{co} \cdot \hat{c} \\ \hat{i}_{in} &= Y^i_{in-o} \cdot \hat{u}_{in} + T^i_{oi-o} \cdot \hat{u}_o + G^i_{ci} \cdot \hat{c} \end{aligned} \tag{2.35}$$

The load effect on the converter dynamics can be found by computing $\hat{u}_o$ from Figure 2.10b, which yields

$$\hat{u}_o = \frac{Z_L}{1 + Z_L Y^i_{o-o}} \cdot \left(G^i_{io-o} \cdot \hat{u}_{in} + G^i_{co} \cdot \hat{c} \right) + \frac{1}{1 + Z_L Y^i_{o-o}} \cdot \hat{e}_o \tag{2.36}$$

Substituting it in (2.35) yields the desired load-affected dynamic models of the current-output converter as follows:

$$\begin{bmatrix} \hat{i}_{in} \\ \hat{i}_o \end{bmatrix} = \begin{bmatrix} Y^i_{in-o} + \dfrac{Z_L G^i_{io-o} T^i_{oi-o}}{1 + Y^i_{o-o} Z_L} & \dfrac{T^i_{oi-o}}{1 + Y^i_{o-o} Z_L} & G^i_{ci} + \dfrac{Z_L G^i_{co} T^i_{oi-o}}{1 + Y^i_{o-o} Z_L} \\ \dfrac{G^i_{io-o}}{1 + \dfrac{Z_L}{(Y^i_{o-o})^{-1}}} & -\dfrac{1}{Z_L + (Y^i_{o-o})^{-1}} & \dfrac{G^i_{co}}{1 + \dfrac{Z_L}{(Y^i_{o-o})^{-1}}} \end{bmatrix}$$

$$\times \begin{bmatrix} \hat{u}_{in} \\ \hat{e}_o \\ \hat{c} \end{bmatrix} \tag{2.37}$$

The load-affected current-output model (2.37) can also be given by using the transfer functions of the corresponding voltage-output converter, which would give much more information due to the usually well-known transfer functions applying the information given in Section 2.2.2 (Eq. (2.12)) [2]:

$$\begin{bmatrix} \hat{i}_{in} \\ \hat{i}_o \end{bmatrix} = \begin{bmatrix} Y_{o-o} + \dfrac{G_{io-o} T_{oi-o}}{Z_L + Z_{o-o}} & -\dfrac{T_{oi-o}}{Z_L + Z_{o-o}} & G_{ci} + \dfrac{G_{co} T_{oi-o}}{Z_L + Z_{o-o}} \\ \dfrac{\dfrac{G_{io-o}}{Z_{o-o}}}{1 + \dfrac{Z_L}{Z_{o-o}}} & -\dfrac{1}{Z_L + Z_{o-o}} & \dfrac{\dfrac{G_{co}}{Z_{o-o}}}{1 + \dfrac{Z_L}{Z_{o-o}}} \end{bmatrix}$$

$$\times \begin{bmatrix} \hat{u}_{in} \\ \hat{e}_o \\ \hat{c} \end{bmatrix} \tag{2.38}$$

The source effect on the converter dynamics can be found by computing $\hat{u}_{in}$ from Figure 2.10b, which yields

$$\hat{u}_{in} = \frac{1}{1 + Z_s Y^i_{in-o}} \cdot \hat{u}_{ins} - \frac{Z_s}{1 + Z_s Y^i_{in-o}} \cdot (T^i_{oi-o} \cdot \hat{u}_o + G^i_{ci} \cdot \hat{c}) \tag{2.39}$$

2 Basis for Dynamic Analysis and Control Dynamics

Substituting it in (2.35) yields the source-affected dynamical models as follows:

$$\begin{bmatrix} \hat{i}_{in} \\ \hat{i}_o \end{bmatrix} = \begin{bmatrix} \dfrac{Y^i_{in-o}}{1+Z_s Y^i_{in-o}} & \dfrac{T^i_{oi-o}}{1+Z_s Y^i_{in-o}} & \dfrac{G^i_{ci}}{1+Z_s Y^i_{in-o}} \\ \dfrac{G^i_{io-o}}{1+Z_s Y^i_{in-o}} & -\left(Y^i_{o-o} + \dfrac{Z_s G^i_{io-o} T^i_{oi-o}}{1+Z_s Y^i_{in-o}}\right) & G^i_{co} - \dfrac{Z_s G^i_{io-o} G^i_{ci}}{1+Z_s Y^i_{in-o}} \end{bmatrix}$$
$$\times \begin{bmatrix} \hat{u}_{ins} \\ \hat{u}_o \\ \hat{c} \end{bmatrix} \quad (2.40)$$

The models in (2.40) can be transformed to the formulation described in [30] yielding

$$\begin{bmatrix} \hat{i}_{in} \\ \hat{i}_o \end{bmatrix} = \begin{bmatrix} \dfrac{Y^i_{in-o}}{1+Z_s Y^i_{in-o}} & \dfrac{T^i_{oi-o}}{1+Z_s Y^i_{in-o}} & \dfrac{G^i_{ci}}{1+Z_s Y^i_{in-o}} \\ \dfrac{G^i_{io-o}}{1+Z_s Y^i_{in-o}} & -\dfrac{1+Z_s Y^i_{in-oc}}{1+Z_s Y^i_{in-o}} \cdot Y^i_{o-o} & \dfrac{1+Z_s Y^i_{in-\infty}}{1+Z_s Y^i_{in-o}} \cdot G^i_{co} \end{bmatrix}$$
$$\times \begin{bmatrix} \hat{u}_{ins} \\ \hat{u}_o \\ \hat{c} \end{bmatrix} \quad (2.41)$$

where the special admittances $Y^i_{in-\infty}$ (ideal input admittance) and Y^i_{in-oc} (open-circuit input admittance) are defined as follows.

$$Y^i_{in-\infty} = Y^i_{in-o} - \dfrac{G^i_{io-o} G^i_{ci}}{G^i_{co}}$$
$$Y^i_{in-oc} = Y^i_{in-o} + \dfrac{G^i_{io-o} T^i_{oi-o}}{Y^i_{o-o}} \quad (2.42)$$

The source-affected current-output model (2.41) can also be given by using the transfer functions of the corresponding voltage-output converter, which may be more informative compared to (2.41) due to the usually well-known transfer functions, applying the information given in Section 2.2.2 (Eq. 2.12):

$$\begin{bmatrix} \hat{i}_{in} \\ \hat{i}_o \end{bmatrix} = \begin{bmatrix} \dfrac{Y^i_{in-o}}{1+Z_s Y^i_{in-o}} & -\dfrac{\frac{T_{oi-o}}{Z_{o-o}}}{1+Z_s Y^i_{in-o}} & \dfrac{G_{ci}+\frac{G_{co}T_{oi-o}}{Z_{o-o}}}{1+Z_s Y^i_{in-o}} \\ \dfrac{\frac{G_{io-o}}{Z_{o-o}}}{1+Z_s Y^i_{in-o}} & -\dfrac{1+Z_s Y_{in-o}}{1+Z_s Y^i_{in-o}} \cdot \dfrac{1}{Z_{o-o}} & \dfrac{1+Z_s Y_{in-\infty}}{1+Z_s Y^i_{in-o}} \cdot \dfrac{G_{co}}{Z_{o-o}} \end{bmatrix}$$
$$\times \begin{bmatrix} \hat{u}_{ins} \\ \hat{u}_o \\ \hat{c} \end{bmatrix} \quad (2.43)$$

2.5 An Example LC Circuit

where the ideal input admittance $Y_{in-\infty}$ is defined in (2.34). The ideal current-output input admittance $Y_{in-\infty}^i$ equals the corresponding ideal input admittance $Y_{in-\infty}$ of the voltage-output converter. The open-circuit input admittance (Y_{in-oc}^i) equals the internal open-loop input admittance (Y_{in-o}) of the corresponding voltage-output converter, respectively.

2.5 An Example LC Circuit

We derive the state space for the system comprising of a *LC* circuit both at voltage-output (Figure 2.11a) and current-output (Figure 2.11b) modes, and we solve its input-to-output ($\mathbf{G}(s)$) and input-to-state ($\mathbf{\Phi}(s)$) descriptions as an example to illustrate the use of the theoretical formulations defined in the previous sections.

According to Section 2.2, the state variables of the system are usually the inductor current (i_L) and the capacitor voltage (u_C), and their derivatives have to be evaluated as a function of the state variables and the input variables. The input variables of the voltage-output circuit, Figure 2.11a, are the input voltage (u_{in}) and the output current (i_o). The input variables of the current-output circuit, Figure 2.11b, are the input voltage (u_{in}) and the output voltage (u_o). In addition to the derivatives of the state variables, the output variables have to be defined as a function of the state and input variables. The output voltage (u_o) and the input current (i_{in}) are the output variables of the voltage-output circuit, and the output current (i_o) and the input current (i_{in}) are the output variables of the current-output circuit, respectively. According to the circuit theory [19], the voltage across the inductor can be given by $u_L = L\frac{di_L}{dt}$ and the current through the capacitor by $i_C = C\frac{du_C}{dt}$, which explicitly define the desired derivatives of the state variables.

2.5.1 Voltage-Output Circuit

In order to construct the state-space representation for the system of Figure 2.11a, we have to solve u_L and u_o applying *Kirchhoff*'s voltage law

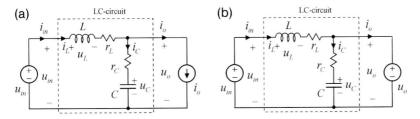

Figure 2.11 An example *LC* circuit: (a) voltage-output mode and (b) current-output mode.

yielding

$$\begin{aligned} u_L &= u_{in} - r_L i_L - u_o \\ u_o &= u_C + r_C i_C \end{aligned} \tag{2.44}$$

and i_C and i_{in} applying *Kirchhoff*'s current law yielding

$$\begin{aligned} i_C &= i_L - i_o \\ i_{in} &= i_L \end{aligned} \tag{2.45}$$

According to (2.44) and (2.45), the state space can be solved to be

$$\begin{aligned} \frac{di_L}{dt} &= -\frac{r_L + r_C}{L} \cdot i_L - \frac{1}{L} \cdot u_C + \frac{1}{L} \cdot u_{in} + \frac{r_C}{L} \cdot i_o \\ \frac{du_C}{dt} &= \frac{1}{C} \cdot i_L - \frac{1}{C} \cdot i_o \\ i_{in} &= i_L \\ u_o &= r_C i_L + u_C - r_C i_o \end{aligned} \tag{2.46}$$

The *LC* circuit is a linear system. Therefore, the state space (2.46) equally represents also the small-signal state space without any additional actions, and the desired transfer matrices can be solved directly applying *Laplace* transforms on it. We transform the state space (2.46) to the usual matrix form according to (2.3) yielding

$$\begin{aligned} \begin{bmatrix} \frac{d\hat{i}_L}{dt} \\ \frac{d\hat{u}_C}{dt} \end{bmatrix} &= \begin{bmatrix} -\frac{r_L + r_C}{L} & -\frac{1}{L} \\ \frac{1}{C} & 0 \end{bmatrix} \begin{bmatrix} \hat{i}_L \\ \hat{u}_C \end{bmatrix} + \begin{bmatrix} \frac{1}{L} & \frac{r_C}{L} \\ 0 & -\frac{1}{C} \end{bmatrix} \begin{bmatrix} \hat{u}_{in} \\ \hat{i}_o \end{bmatrix} \\ \begin{bmatrix} \hat{i}_{in} \\ \hat{u}_o \end{bmatrix} &= \begin{bmatrix} 1 & 0 \\ r_C & 1 \end{bmatrix} \begin{bmatrix} \hat{i}_L \\ \hat{u}_C \end{bmatrix} + \begin{bmatrix} 0 & 0 \\ 0 & -r_C \end{bmatrix} \begin{bmatrix} \hat{u}_{in} \\ \hat{i}_o \end{bmatrix} \end{aligned} \tag{2.47}$$

In order to solve the state-transition and transfer matrices, the *Laplace* transform is applied to the small-signal state space (2.47) yielding

$$\begin{aligned} s \begin{bmatrix} \hat{i}_L(s) \\ \hat{u}_C(s) \end{bmatrix} &= \begin{bmatrix} -\frac{r_L + r_C}{L} & -\frac{1}{L} \\ \frac{1}{C} & 0 \end{bmatrix} \begin{bmatrix} \hat{i}_L(s) \\ \hat{u}_C(s) \end{bmatrix} + \begin{bmatrix} \frac{1}{L} & \frac{r_C}{L} \\ 0 & -\frac{1}{C} \end{bmatrix} \begin{bmatrix} \hat{u}_{in}(s) \\ \hat{i}_o(s) \end{bmatrix} \\ \begin{bmatrix} \hat{i}_{in}(s) \\ \hat{u}_o(s) \end{bmatrix} &= \begin{bmatrix} 1 & 0 \\ r_C & 1 \end{bmatrix} \begin{bmatrix} \hat{i}_L(s) \\ \hat{u}_C(s) \end{bmatrix} + \begin{bmatrix} 0 & 0 \\ 0 & -r_C \end{bmatrix} \begin{bmatrix} \hat{u}_{in}(s) \\ \hat{i}_o(s) \end{bmatrix} \end{aligned} \tag{2.48}$$

Applying matrix manipulation techniques to (2.48) yields the state-transition matrix $\boldsymbol{\Phi}(s)$

$$\boldsymbol{\Phi}(s) = \frac{\begin{bmatrix} s & -\dfrac{1}{L} \\ \dfrac{1}{C} & s + \dfrac{r_L + r_C}{L} \end{bmatrix}}{s^2 + s \cdot \dfrac{r_L + r_C}{L} + \dfrac{1}{LC}} \tag{2.49}$$

the transfer matrix $\mathbf{G}(s)$

$$\mathbf{G}(s) = \begin{bmatrix} Y_{in} & T_{oi} \\ G_{io} & -Z_o \end{bmatrix} = \frac{\begin{bmatrix} \dfrac{s}{L} & \dfrac{1 + sr_C C}{LC} \\ \dfrac{1 + sr_C C}{LC} & -\dfrac{(r_L + sL)(1 + sr_C C)}{LC} \end{bmatrix}}{s^2 + s \cdot \dfrac{r_L + r_C}{L} + \dfrac{1}{LC}} \tag{2.50}$$

and the characteristic equation $\boldsymbol{\Delta}(s)$

$$\boldsymbol{\Delta}(s) = s^2 + s \cdot \frac{r_L + r_C}{L} + \frac{1}{LC} \tag{2.51}$$

The equivalent series resistances (ESR) r_L and r_C have to be small for an effective filter, and therefore, the roots of the characteristic polynomial are complex. As a consequence, the LC-circuit system, Figure 2.11a, would exhibit resonant-type behavior.

The transfer functions in (2.49) and (2.50) can be easily derived directly from the circuit without the state-space methods by applying superposition principle.

2.5.2
Current-Output Circuit

In order to construct the state-space representation for the system of Figure 2.11b, we have to solve u_L and i_C by applying *Kirchhoff*'s voltage law yielding

$$\begin{aligned} u_L &= u_{in} - r_L i_L - u_o \\ i_C &= \frac{u_o - u_C}{r_C} \end{aligned} \tag{2.52}$$

and i_o and i_{in} by applying *Kirchhoff*'s current law yielding

$$\begin{aligned} i_o &= i_L - i_C \\ i_{in} &= i_L \end{aligned} \tag{2.53}$$

According to (2.52) and (2.53), the state space can be solved to be

$$\frac{di_L}{dt} = -\frac{r_L}{L} \cdot i_L + \frac{1}{L} \cdot u_{in} - \frac{1}{L} \cdot u_o$$

$$\frac{du_C}{dt} = -\frac{1}{r_C C} \cdot u_C + \frac{1}{r_C C} \cdot u_o$$

$$i_{in} = i_L$$

$$i_o = i_L + \frac{1}{r_C} u_C - \frac{1}{r_C} u_o$$
(2.54)

We transform the state space (2.54) to the usual matrix form according to (2.3) yielding

$$\begin{bmatrix} \dfrac{d\hat{i}_L}{dt} \\ \dfrac{d\hat{u}_C}{dt} \end{bmatrix} = \begin{bmatrix} -\dfrac{r_L}{L} & 0 \\ 0 & -\dfrac{1}{r_C C} \end{bmatrix} \begin{bmatrix} \hat{i}_L \\ \hat{u}_C \end{bmatrix} + \begin{bmatrix} \dfrac{1}{L} & -\dfrac{1}{L} \\ 0 & \dfrac{1}{r_C C} \end{bmatrix} \begin{bmatrix} \hat{u}_{in} \\ \hat{u}_o \end{bmatrix}$$

$$\begin{bmatrix} \hat{i}_{in} \\ \hat{i}_o \end{bmatrix} = \begin{bmatrix} 1 & 0 \\ 1 & \dfrac{1}{r_C} \end{bmatrix} \begin{bmatrix} \hat{i}_L \\ \hat{u}_C \end{bmatrix} + \begin{bmatrix} 0 & 0 \\ 0 & -\dfrac{1}{r_C} \end{bmatrix} \begin{bmatrix} \hat{u}_{in} \\ \hat{u}_o \end{bmatrix}$$
(2.55)

In order to solve the state-transition and transfer matrices, the *Laplace* transform is applied to the small-signal state space (2.55) yielding

$$s \begin{bmatrix} \hat{i}_L(s) \\ \hat{u}_C(s) \end{bmatrix} = \begin{bmatrix} -\dfrac{r_L}{L} & 0 \\ 0 & -\dfrac{1}{r_C C} \end{bmatrix} \begin{bmatrix} \hat{i}_L(s) \\ \hat{u}_C(s) \end{bmatrix} + \begin{bmatrix} \dfrac{1}{L} & -\dfrac{1}{L} \\ 0 & \dfrac{1}{r_C C} \end{bmatrix} \begin{bmatrix} \hat{u}_{in}(s) \\ \hat{u}_o(s) \end{bmatrix}$$

$$\begin{bmatrix} \hat{i}_{in}(s) \\ \hat{u}_o(s) \end{bmatrix} = \begin{bmatrix} 1 & 0 \\ 1 & \dfrac{1}{r_C} \end{bmatrix} \begin{bmatrix} \hat{i}_L(s) \\ \hat{u}_C(s) \end{bmatrix} + \begin{bmatrix} 0 & 0 \\ 0 & -\dfrac{1}{r_C} \end{bmatrix} \begin{bmatrix} \hat{u}_{in}(s) \\ \hat{u}_o(s) \end{bmatrix}$$
(2.56)

Applying matrix manipulation techniques to (2.56) yields the state-transition matrix $\mathbf{\Phi}(s)$,

$$\mathbf{\Phi}(s) = \frac{\begin{bmatrix} 1 + s r_C C & -(1 + s r_C C) \\ 0 & r_L + sL \end{bmatrix}}{(r_L + sL)(1 + s r_C C)}$$
(2.57)

the transfer matrix $\mathbf{G}(s)$

$$\mathbf{G}(s) = \begin{bmatrix} Y_{in}^i & T_{oi}^i \\ G_{io}^i & -Y_o^i \end{bmatrix} = \frac{\begin{bmatrix} 1 + s r_C C & -(1 + s r_C C) \\ 1 + s r_C C & -(s^2 + s \dfrac{r_L + r_C}{L} + \dfrac{1}{LC}) \end{bmatrix}}{(r_L + sL)(1 + s r_C C)}$$
(2.58)

and the characteristic equation $\mathbf{\Delta}(s)$

$$\mathbf{\Delta}(s) = (r_L + sL)(1 + sr_C C) \tag{2.59}$$

The roots of the characteristic equation (2.59) are real, and all the other transfer functions except the output admittance (Y^i_{o-o}) are clearly first-order transfer functions. The output admittance would, however, exhibit resonant behavior similar to that of the output impedance (Z_{o-o}) of the voltage-output circuit, because $Y^i_o = 1/Z_{o-o}$.

According to Section 2.2.2, the current-output transfer functions can be derived from the corresponding transfer functions of the voltage-output circuit by applying (2.60)

$$G^i(s) = \begin{bmatrix} Y^i_{in-o} & T^i_{oi-o} \\ G^i_{io-o} & -Y^i_{o-o} \end{bmatrix} = \begin{bmatrix} Y_{in-o} + \dfrac{G_{io-o}T_{oi-o}}{Z_{o-o}} & -\dfrac{T_{oi-o}}{Z_{o-o}} \\ \dfrac{G_{io-o}}{Z_{o-o}} & -\dfrac{1}{Z_{o-o}} \end{bmatrix} \tag{2.60}$$

which yields

$$G^i(s) = \dfrac{\begin{bmatrix} 1 + sr_C C & -(1 + sr_C C) \\ 1 + sr_C C & -\left(s^2 + s\dfrac{r_L + r_C}{L} + \dfrac{1}{LC}\right) \end{bmatrix}}{(r_L + sL)(1 + sr_C C)} \tag{2.61}$$

The transfer functions in (2.61) are obviously equal to (2.58). It may be obvious that the change of mode from voltage output to current output would significantly change the dynamics of the associated system, and would be induced by the change of load type.

2.6
Review of Basic Mathematical Tools

In this subsection, we will review some basic mathematical and control-engineering concepts, which are vital to produce dynamical models and to understand the dynamics associated with a converter.

2.6.1
Linearization

The models representing the converter dynamics are constructed by linearizing the averaged behavior around the defined operating point. In power electronics, the recommended method is to substitute the averaged values by means of the sum of a DC value and small perturbation [14]. This method works well when the averaged model is only slightly nonlinear as in the case of continuous mode of operation, but usually fails when the average model is highly nonlinear as in the case of discontinuous mode of operation. In control engineering [31], the linearization is typically carried out computing the *Jacobian* matrix of the function $y = f(t, x)$, where x contains the state and input variables. The *Jacobian* matrix represents the partial derivatives of the function with respect to all the variables. In the one-dimensional case, the linearized function $\hat{y}$ can be given as follows:

$$\hat{y} = \frac{\partial y}{\partial x_1} \cdot \hat{x}_1 + \cdots + \frac{\partial y}{\partial x_n} \cdot \hat{x}_n$$

$$\hat{y} = \begin{bmatrix} \frac{\partial f}{\partial x_1}(t, X) & \cdots & \frac{\partial f}{\partial x_n}(t, X) \end{bmatrix} \begin{bmatrix} \hat{x}_1 \\ \vdots \\ \hat{x}_n \end{bmatrix} \quad (2.62)$$

where X in the *Jacobian* matrix contains the steady-state values of the corresponding variables at the operating point. As an example, we consider the function $y = \frac{u_{in}^2 i_L}{u_C}$, which is highly nonlinear. We consider that the steady-state values of the variables u_{in}, i_L, and u_C are U_{in}, I_L, and U_C, respectively. According to these assumptions, the linearized function $\hat{y}$ can be given by

$$\hat{y} = \frac{U_{in}^2}{U_C} \cdot \hat{i}_L - \frac{U_{in}^2 I_L}{U_C^2} \cdot \hat{u}_C + \frac{2 U_{in} I_L}{U_C} \cdot \hat{u}_{in}$$

In practice, this means that we treat each variable at a time and consider the other variables to be constant when developing the required derivative according to the basic mathematics.

2.6.2
Transfer Functions

A transfer function is usually given as a ratio of two polynomials in s (2.63), where s is the *Laplace* variable. The roots of the numerator polynomial ω_{zi} are called *zeros*, and the roots of the denominator polynomial ω_{pi} *poles*. The zeros and poles may be real or complex numbers and they are given with respect to angular frequency ω (rad/s). The frequency (f) in Hz and the angular frequency (ω) in rad/s are related by $f = \omega/2\pi$:

$$G(s) = \frac{a_n s^n + a_{n-1} s^{n-1} + \cdots + a_o}{b_m s^m + b_{m-1} s^{m-1} + \cdots + b_o}$$

$$G(s) = K \cdot \frac{(s - \omega_{z1})(s - \omega_{z2}) \cdots (s - \omega_{zn})}{(s - \omega_{p1})(s - \omega_{p2}) \cdots (s - \omega_{pm})} \quad (2.63)$$

The magnitude of the transfer function ($|G(s)|$) is commonly expressed in dB, that is, $|G(s)|_{dB} = 20\log_{10}(|G(s)|)$, and the phase ($\angle G(s)$) in degrees. The logarithmic magnitude means that the combined effect of zeros and poles can be found adding together the dB-values of the zeros and subtracting the dB-values of the poles, respectively. The phase of the transfer function can be found similarly adding together the phase contributions of the zeros and subtracting the phase contributions of the poles. The zeros and poles locating closest to the origin (i.e., zero frequency) are called dominant zeros and poles having the strongest effect on the time-domain behaviour of the corresponding system.

2.6.2.1 Single Zero

Let us consider the single-zero transfer function $G(s) = 1 + s/\omega_z$. Its magnitude (i.e., $|G(j\omega)| = \sqrt{1 + \frac{\omega^2}{\omega_z^2}}$), and phase (i.e., $\angle G(j\omega) = \arctan\left(\frac{\omega}{\omega_z}\right)$), may be given at certain interesting angular-frequency points as follows:

$$\omega = \frac{\omega_z}{10}$$

$$|G(j\omega)| = \sqrt{1.01} \triangleq 20\log_{10}(\sqrt{1.01}) = 0.04 \text{ dB}$$

$$\angle G(j\omega) = \arctan(0.1) = 5.7°$$

$$\omega = \omega_z$$

$$|G(j\omega)| = \sqrt{2} \triangleq 20\log_{10}(\sqrt{2}) = 3 \text{ dB}$$

$$\angle G(j\omega) = \arctan(1) = 45°$$

$$\omega = 10 \cdot \omega_z$$

$$|G(j\omega)| = 10 \triangleq 20\log_{10}(10) = 20 \text{ dB}$$

$$\angle G(j\omega) = \arctan(10) = 84.3°$$

As a consequence, the behavior of the single-zero transfer functions is typically considered to be such that its magnitude is unity (i.e., 0 dB) up to $\omega = \omega_z$, and starts increasing at a slope of +20 dB/decade (i.e., +20 dB for a 10 times increase in frequency) from $\omega = \omega_z$. Its phase is zero up to $\omega = \omega_z/10$, and starts increasing at a slope of +45°/decade up to $10\omega_z$, and stays constant at +90° after that. The phase is exactly equal to +45° at $\omega = \omega_z$. If $G(s) = 1 - s/\omega_z$, the magnitude behavior is same but the phase has the negative sign compared to $G(s) = 1 + s/\omega_z$. $s = -\omega_z$ is called left-half-plane (LHP) zero and $s = \omega_z$ right-half-plane (RHP) zero, based on their locations in the complex plane. The RHP zero would have a profound impact on the control loop design, and would exist also in the certain-type of the switched-mode converters.

2.6.2.2 Single Pole

Lets consider the single-pole transfer function $G(s) = (1 + s/\omega_p)^{-1}$. Its magnitude (i.e., $|G(j\omega)| = 1/\sqrt{1 + \frac{\omega^2}{\omega_z^2}}$), and phase (i.e., $\angle G(j\omega) = -\arctan\left(\frac{\omega}{\omega_z}\right)$), may be given at certain interesting angular-frequency points as follows:

$$\omega = \frac{\omega_z}{10}$$

$$|G(j\omega)| = 1/\sqrt{1.01} \triangleq -20\log_{10}(\sqrt{1.01}) = -0.04 \text{ dB}$$

$$\angle G(j\omega) = -\arctan(0.1) = -5.7°$$

$$\omega = \omega_z$$

$$|G(j\omega)| = 1/\sqrt{2} \triangleq -20\log_{10}(\sqrt{2}) = -3 \text{ dB}$$

$$\angle G(j\omega) = -\arctan(1) = -45°$$

$$\omega = 10 \cdot \omega_z$$

$$|G(j\omega)| = 1/10 \triangleq -20\log_{10}(10) = -20 \text{ dB}$$

$$\angle G(j\omega) = -\arctan(10) = -84.3°$$

As a consequence, the behavior of the single-pole transfer functions is typically considered to be such that its magnitude is unity (i.e., 0 dB) up to $\omega = \omega_z$, and starts decreasing at a slope of -20 dB/decade from $\omega = \omega_z$. Its phase is zero up to $\omega = \omega_z/10$, and starts decreasing at a slope of $-45°$/decade up to $10\omega_z$, and stays constant at $-90°$ after that. The phase is exactly equal to $-45°$ at $\omega = \omega_z$. If $G(s) = (1 - s/\omega_p)^{-1}$, the magnitude behavior is same but the phase has positive sign compared to $G(s) = (1 + s/\omega_p)^{-1}$. $s = -\omega_p$ is called left-half-plane (LHP) pole and $s = \omega_p$ right-half-plane (RHP) pole based on their location in the complex plane. An RHP pole means that the corresponding system is unstable.

Most important is to recognize that the zeros and poles in the transfer functions are given in the angular frequency and $\omega \approx 6.28 \cdot f$; the phase of the transfer function starts changing already 10 times earlier than the location of the corresponding zero or pole, and keep on changing up to 10 times higher frequencies than the location of the zero or pole. These facts are forgotten very often.

2.6.2.3 Second-Order Transfer Function

The second-order polynomial is typically expressed as $s^2 + s \cdot 2\zeta\omega_n + \omega_n^2$ or as $s^2 + s \cdot \frac{\omega_n}{Q} + \omega_n^2$ comprising either zeros or poles in the corresponding transfer function, where ζ is called *damping factor*, ω_n *undamped natural frequency*, and Q *quality factor*. The second-order polynomials are common in power electronics, and will be recognized because of the special features they will bring to the control design, the load and source interactions, and so on.

2.6 Review of Basic Mathematical Tools

The roots of the second-order polynomial can be expressed as $s_{1,2} = -\zeta\omega_n \pm \omega_n\sqrt{\zeta^2 - 1}$. The system characterized by it (i.e., the second-order polynomial is the denominator) may be classified according to the value of the damping factor as (1) underdamped, when $0 < \zeta < 1$, (2) critically damped, when $\zeta = 1$, (3) overdamped, when $\zeta > 1$, and (4) oscillatory, when $\zeta = 0$. The examples in Figure 2.12 correspond to the case where the transfer function is of the form $\omega_n^2/(s^2 + s \cdot 2\zeta\omega_n + \omega_n^2)$.

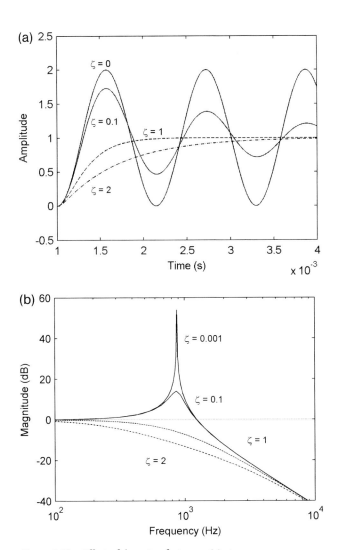

Figure 2.12 Effect of damping factor on (a) step response, (b) magnitude of the frequency response, and (c) phase of the frequency response of a second-order transfer function of the form $\omega_n^2/(s^2 + s \cdot 2\zeta\omega_n + \omega_n^2)$.

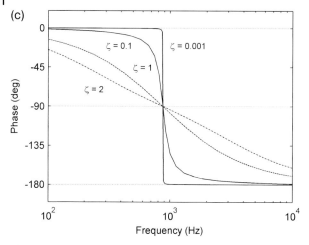

Figure 2.12 (Continued).

1. *Underdamped Case*: $0 < \zeta < 1$

The roots of the second-order function are complex conjugates of each other, which can be expressed by $s_{1,2} = -\zeta\omega_n \pm j\omega_d$, where $\omega_d = \omega_n\sqrt{1-\zeta^2}$ is called *damped natural frequency*. Step response of such a system applied to the reference input includes decaying oscillatory behavior, where the oscillation frequency is ω_d (Figure 2.12, $\zeta = 0.1$).

2. *Critically Damped Case*: $\zeta = 1$

The second-order function has a real double root, which can be expressed by $s_{1,2} = -\zeta\omega_n$. The step response resembles an exponential response but is faster (Figure 2.12, $\zeta = 1$).

3. *Overdamped Case*: $\zeta > 1$

The second-order function has two real roots, which can be expressed by $s_{1,2} = -\zeta\omega_n \pm \omega_n\sqrt{\zeta^2-1}$. The greater the damping factor, the more the roots are separated from each other. The step response of such a system contains only exponential behavior (Figure 2.12, $\zeta = 2$).

4. *Oscillatory Case*: $\zeta = 0$

The second-order function has two pure imaginary roots, which can be expressed by $s_{1,2} = \pm j\omega_n$. The system will oscillate at the underdamped natural frequency ω_n (Figure 2.12, $\zeta = 0$).

The damping factor ζ affects the system step response (Figure 2.12a) and the behavior of the corresponding transfer function as shown in Figures 2.12b and c, where the second-order system is assumed to be of the form $G(s) = \omega_n^2/(s^2 + s \cdot 2\zeta\omega_n + \omega_n^2)$. The damping factor $\zeta = 0$ would produce an infinite value in the magnitude of the transfer function, and therefore, $\zeta = 0.001$ is used in Figures 2.12b and c to demonstrate the effect of the decrease in the damping factor instead of zero. Characteristic to the magnitude of the second-order transfer function is that its high-frequency-magnitude slope is -40 dB/decade and the overall change in the phase is $\mp 180°$ depending on whether the poles are LHP or RHP, respectively. If the second-order polynomial is a numerator polynomial in the corresponding transfer function, then its effect is 40 dB/decade and $\pm 180°$, respectively. If the zeros or poles are complex, then there exists either dipping or peaking in the magnitude response (Figure 2.12b) and the phase changes rapidly $\pm 180°$. Such a behavior is called resonant behavior. The resonant frequency (ω_o) (i.e., the undamped natural frequency ω_n) can be found at a frequency where the phase has changed exactly $\pm 90°$ (Figure 2.12c). The maximum peak/dip value of the magnitude of the resonant-type transfer function corresponds to the quality factor $Q = 1/2\zeta$ (i.e., the difference between 0 dB and the peak/dip value) at $\omega = \sqrt{1 - 2\zeta^2} \cdot \omega_n$. This means that the maximum peak/dip value does not locate at the resonant frequency but at a slightly lower frequency (i.e., ω_n versus $\sqrt{1 - 2\zeta^2} \cdot \omega_n$).

2.6.2.4 Example

The practical transfer functions associated with the power electronic converters are typically a combination of first-order and second-order polynomials. The internal output impedance (Z_{o-o}) of the converter, shown in Figure 2.2, can be given symbolically by

$$Z_{o-o} = \frac{\frac{(r_L + Dr_{ds1} + D'r_{ds2})}{LC}\left(1 + s\frac{L}{r_L + Dr_{ds1} + D'r_{ds2}}\right)(1 + sr_C C)}{s^2 + s\frac{r_L + r_C + Dr_{ds1} + D'r_{ds2}}{L} + \frac{1}{LC}}$$

(2.64)

A practical converter, Figure 2.2, operating at the switching frequency of 100 kHz, the output power of 25 W, and the duty ratio (D) of 0.5, may have the circuit element values such as follows: $L = 100$ µH, $C = 330$ µF, $r_L = 20$ mΩ, $r_C = 33$ mΩ, and $r_{ds1,2} = 200$ mΩ, where r_L, r_C, and $r_{ds1,2}$ are called equivalent series resistances (ESR) of the associated circuit element, and $r_{ds1,2}$ includes also the switching losses of the associated semiconductor switches. Substituting the symbolical element values in (2.64) with the defined physical values yields

$$Z_{o-o} = \frac{6.667 \times 10^6 (1 + s \cdot 4.546 \times 10^{-4})(1 + s \cdot 1.089 \times 10^{-5})}{s^2 + s \cdot 2.53 \times 10^3 + 3.03 \times 10^7}$$

(2.65)

According to (2.64), we can conclude that the low-frequency value of $|Z_{o-o}|$ equals $r_L + Dr_{ds1} + D'r_{ds2}$ by setting $s = 0$. The corresponding high-frequency value can be found by letting s to approach infinity (i.e., $s = \infty$) equaling r_C, respectively. The damping factor (ζ) and the resonant frequency (ω_o) or the

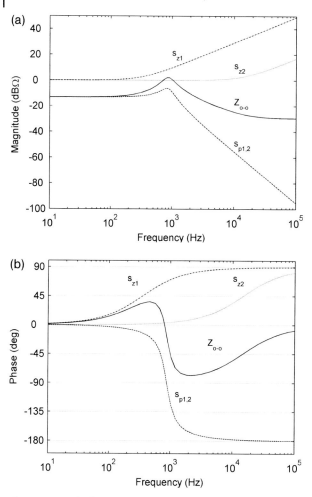

Figure 2.13 The frequency responses of the zeros and poles of Z_{o-o} (a) magnitude and (b) phase.

undamped natural frequency (ω_n) of the physical circuit can be computed to be 0.23 and 5.5045 krad/s. This means that the system poles are complex (i.e., $\zeta < 1$), and they can be given according to the denominator of (2.65) as $s_{p1,2} = (-1.265 \times 10^3 \pm j5.3572 \times 10^3)$ rad/s. The magnitude of $s_{p1,2}$ (i.e., $|s_{p1,2}|$) equals $\omega_n = 5.5045$ krad/s (i.e., 876 Hz). The system comprises also two zeros $s_{z1} = -2.1997 \times 10^3$ rad/s (i.e., 350 Hz) and $s_{z2} = -9.1827 \times 10^4$ rad/s (i.e., 14.615 kHz). The frequency responses of the defined transfer functions and their combination or Z_{o-o} are shown in Figure 2.13. The location of the poles and zeros can be found from Figure 2.13b according to the phase behavior, that is, the pole location, where the phase equals $-90°$, and the zero location, where the phase equals $45°$, respectively.

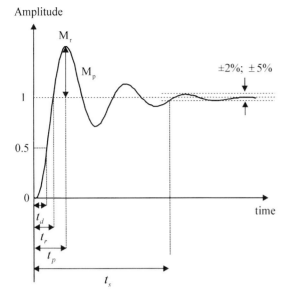

Figure 2.14 Typical parameters specifying the transient response in the control engineering textbooks.

2.6.3
Stability and Performance

The control engineering textbooks [1, 32] usually characterize the transient responses assuming a second-order system behavior (Figure 2.14) by means of parameters (2.66) such as rise time (t_r), peak time (t_p), settling time (t_s), maximum overshoot (M_p), and maximum peak value (M_r) in order to relate the time-domain behavior of the system to the frequency-domain loop-gain behavior. Similar attempts have also been pursued in [33].

$$t_r = \frac{1}{\omega_d} \cdot \arctan\left(-\frac{\omega_d}{\zeta \omega_n}\right)$$

$$t_p = \frac{\pi}{\omega_d}$$

$$t_s = \frac{4}{\zeta \omega_d}(\pm 2.5\%); \quad \frac{3}{\zeta \omega_d}(\pm 5\%) \tag{2.66}$$

$$M_p = e^{-\left(\zeta/\sqrt{1-\zeta^2}\right)\pi} \times 100\%$$

$$M_r = \frac{1}{2\zeta\sqrt{1-\zeta^2}}$$

Close correlation between the loop characteristics and the time-domain transient behavior can be obtained, however, only for the responses excited through the reference input, which is the most usual case in control engineering. In the switched-mode converters, the voltage or current reference is usually not available for transient testing, and consequently, the transients have to be injected through the input voltage or load current. Therefore, the transients recorded at the output voltage also contain the effect of the corresponding internal transfer functions associated with the input voltage and output current as well as the loop gain as defined in Section 2.3. The internal transfer functions tend to dominate in the responses, and therefore, the time-domain transients do not provide accurate information on the loop behavior [34–37].

2.6.3.1 Stability

The closed-loop output dynamics of the voltage-output converter are characterized (Section 2.2.1) by

$$\hat{u}_o = \frac{G_{io-o}}{1+L_v(s)} \cdot \hat{u}_{in} - \frac{Z_{o-o}}{1+L_v(s)} \cdot \hat{i}_o + \frac{1}{G_{se}} \cdot \frac{L_v(s)}{1+L_v(s)} \cdot \hat{u}_r \quad (2.67)$$

where $L_v(s)$ is the voltage-loop gain. The denominator term $1+L_v(s)$ is called the closed-loop system's characteristic polynomial. The inverse of the characteristic polynomial is called sensitivity function $S(s)$, and $L_v(s) \cdot S(s)$ the complementary sensitivity function $T(s)$ [1, 31, 32]. It is obvious that $S(s) + T(s) = 1$. Consequently, the output dynamics can be defined by

$$\hat{u}_o = G_{io-o} \cdot S(s) \cdot \hat{u}_{in} - Z_{o-o} \cdot S(s) \cdot \hat{i}_o + \frac{1}{G_{se}} \cdot T(s) \cdot \hat{u}_r \quad (2.68)$$

For a stable system, the roots of the characteristic polynomial $(1 + L_v(s))$ have to be located in the open left-half plane of the complex plane. A system having roots in the imaginary axis is considered to be *marginally stable* in control engineering, but in power electronics, a system with pure imaginary roots is deemed to be unstable.

If an accurate analytical expression for the loop gain $L(s)$ is available, then the study of the location of the roots of the characteristic polynomial can be made. In practice, the frequency response of the loop gain may be available only, from which the poles and zeros cannot be reliably extracted. Therefore, other methods based directly on the loop frequency response have to be used. The usual visualization methods of the loop frequency behavior are polar and *Bode* plots. The polar plot (Figure 2.15a) is constructed by plotting the locus of the magnitude of the loop gain in the complex plane with the x axis containing the real part of the loop gain and the y axis containing the imaginary part. Usually the locus tends to zero when the frequency approaches infinity. The frequency is not explicitly shown in the polar plot but only the direction of the increasing frequency. In order to study the stability, the polar plot is constructed both for the positive (Figure 2.15a, solid line) and negative (Figure 2.15a,

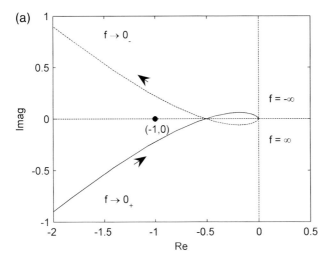

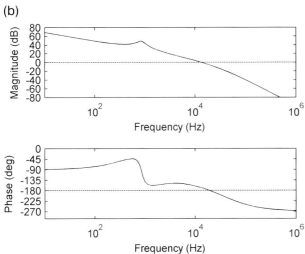

Figure 2.15 Visualization methods: (a) polar plot and (b) Bode plot.

dashed line) frequencies. In practice, this means that the imaginary part is as it is in the original loop gain for the positive frequencies, and its sign is changed negative for the negative frequencies producing a mirror effect with respect to the x axis. The *Bode* plot (Figure 2.15b) is constructed by plotting the magnitude in dB and the phase in degrees usually in the separate subplots in respect to frequency, where the x axis is the frequency in the logarithmic scale and the y axis the magnitude and phase in the linear scale. The polar and *Bode* plots in Figure 2.15 are drawn for the same loop gain, but the polar plot shows only the frequencies higher than 10 kHz when the *Bode* plot shows a much higher frequency range.

The stability of the closed-loop system can be studied by applying the *Nyquist* stability criterion [1, 32] to the loop gain $L(s)$ by means of a polar plot constructed both for the positive and negative frequencies as shown in Figure 2.15a. Such a polar plot is called *Nyquist plot*. According to the *Nyquist* stability criterion, the system is stable if the locus of the loop gain does not pass through the point $(-1, 0)$ or encircle it in the clockwise direction when both of the loci are considered. The situation may be much more complicated for concluding the state of stability as explained in detail in [32] (pp. 521–542).

Consider the following loop gain:

$$L(s) = \frac{K}{s(1 + s \times 10^{-3})(1 + s \times 10^{-2})} \tag{2.69}$$

where K is a constant. According to Figure 2.16a, the closed-loop system having the loop gain according to (2.69) would become unstable for high enough values of K because they either pass through the point $(-1, 0)$ or encircle it in the clockwise direction.

The closed-loop system may also be *conditionally stable* as shown in Figure 2.17, where the encirclement of the point $(-1, 0)$ takes place counterclockwise indicating stability, but the system may become unstable when the gain is slightly increased or decreased. Such a condition may take place, for instance, when the converter starts up due to a reduction of a gain in the associated control circuitry.

2.6.3.2 Loop-Gain-Related Dynamic Indices

The robustness of the stability is typically related to gain (GM) and phase (PM) margins and the dynamical transient performance to the control bandwidth. The closed-loop systems usually employ negative feedback, which is also taken into account when constructing the characteristic polynomial (i.e., $1 + L(s)$): if the loop gain is physically measured, the overall phase change for unstable operation is $360°$, which is also the typical reading (i.e., $0°$) in the frequency response analyzers. For complying with the control engineering domain, $180°$ is subtracted from the reading or data produced by the analyzers. According to this, the PM (Figure 2.18a) is defined by

$$\text{PM} = \angle L(s) + 180° \tag{2.70}$$

at the frequency where $|L(s)| = 1$ (i.e., the loop-gain crossover frequency, ω_{gco} or f_{gco}). Similarly, the GM (Figure 2.18a) is defined by

$$\text{GM} = \frac{1}{|L(s)|} \tag{2.71}$$

at the frequency where $\angle L(s) = -180°$ (i.e., the phase-crossover frequency, ω_{phco} or f_{phco}). Figure 2.18b shows the same definitions using the *Bode* plot.

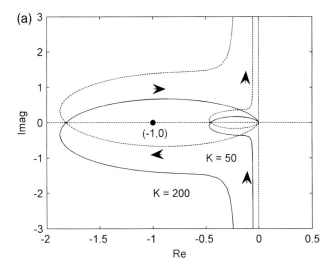
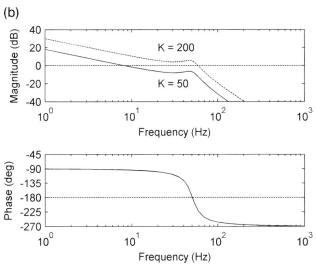

Figure 2.16 Stable ($K = 50$) and unstable ($K = 200$) systems: (a) *Nyquist* plot and (b) *Bode* plot.

In power electronics, the loop crossover frequency (f_{gco}), Figure 2.19, is typically called control bandwidth, but the theoretical control bandwidth is defined to be the frequency range from zero to the frequency at which the sensitivity function ($S(s)$) equals -3 dB (i.e., $1/\sqrt{2}$). The loop crossover frequency is naturally slightly higher than the control bandwidth (i.e., Figure 2.19; $f_{\text{gco}} = 2.36$ kHz and $f_{s-3\text{ dB}} = 1.79$ kHz). Another definition can also be found based on the corner frequency of the complementary sensitivity function ($T(s)$).

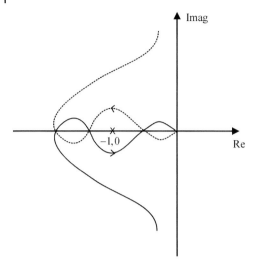

Figure 2.17 Conditionally stable system.

2.6.3.3 Right-Half-Plane Zero and Pole

Certain switched-mode converters contain an RHP zero in their control-to-output transfer function (G_{co}) [38–40]. Such a converter is also known as a nonminimum-phase system. Characteristic to the RHP zero is that it will increase the loop gain and degrease the loop phase, which would make the corresponding converters more susceptible to instability without proper control design actions. The effect of RHP zero cannot be removed [41–43] even if such claims have been presented [38–40]. The RHP zero would limit the achievable maximum control bandwidth at the frequency of the RHP zero, and even lower if proper phase and gain margins are to be obtained [41–43]. Sometimes, the internal dynamics of a switched-mode converter may contain an RHP pole or it may be imposed by the load impedance [44]. The RHP pole would determine the minimum control bandwidth for stability to exist [41]. As a consequence, the control bandwidth should be designed as high as possible to minimize the possible load-imposed stability problems.

Figure 2.20 shows a control-to-output transfer function containing an RHP zero: the flat portion of the magnitude at higher frequencies and a phase less than −180° implies that the compensator gain and, consequently, the control bandwidth would be limited. It was claimed in [40] that the application of peak-current-mode control would remove the effect of RHP zero, but in reality it would not do that.

2.6.4
Matrix Algebra

Manipulation of the simultaneous linear equations constituting the small-signal state space is carried out using a matrix technique. Such a group

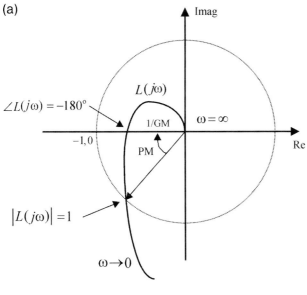
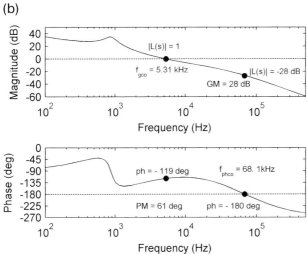

Figure 2.18 Gain and phase margins: (a) *Nyquist* plot and (b) *Bode* plot.

of linear equations can be viewed as the transformation of one vector into another. Consider, for example, the *n* simultaneous linear equations:

$$y_1 = a_{11}x_1 + a_{12}x_2 + \cdots + a_{1n}x_n$$
$$y_2 = a_{21}x_1 + a_{22}x_2 + \cdots + a_{2n}x_n$$
$$\vdots$$
$$y_n = a_{n1}x_1 + a_{n2}x_2 + \cdots + a_{nn}x_n$$

(2.72)

52 | *2 Basis for Dynamic Analysis and Control Dynamics*

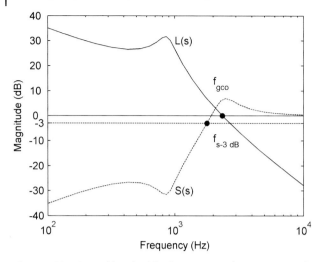

Figure 2.19 Control bandwidth ($f_{s-3\,dB}$) versus loop crossover frequency (f_{gco}).

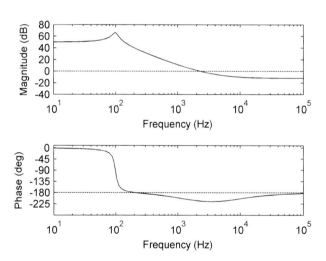

Figure 2.20 Typical effect of RHP zero on the control-to-output transfer function.

If we define two column vectors **x** and **y** by

$$\mathbf{x} = \begin{bmatrix} x_1 \\ x_2 \\ . \\ x_n \end{bmatrix}, \mathbf{y} = \begin{bmatrix} y_1 \\ y_2 \\ . \\ y_n \end{bmatrix} \quad (2.73)$$

then the linear transformation in (2.72) can be given by

$$\begin{bmatrix} y_1 \\ y_2 \\ \cdot \\ y_n \end{bmatrix} = \begin{bmatrix} a_{11} & a_{12} & \cdot & a_{1n} \\ a_{21} & a_{22} & \cdot & a_{2n} \\ \cdot & \cdot & \cdot & \cdot \\ a_{n1} & a_{n2} & \cdot & a_{nn} \end{bmatrix} \begin{bmatrix} x_1 \\ x_2 \\ \cdot \\ x_n \end{bmatrix} \quad (2.74)$$

or $\mathbf{y} = \mathbf{A} \cdot \mathbf{x}$, where the operator $\cdot$ stands for multiplication and will be defined later.

An identity matrix $\mathbf{I}$ is defined such that its diagonal elements or entries (i.e., a_{ii}) equals 1, and all the other elements are zeros as shown in (2.75).

$$\mathbf{I} = \begin{bmatrix} 1 & 0 & 0 & 0 \\ 0 & 1 & 0 & 0 \\ 0 & 0 & 1 & 0 \\ 0 & 0 & 0 & 1 \end{bmatrix} \quad (2.75)$$

Transpose of a matrix $\mathbf{B} = (b_{ij})_{m \times n}$ denoted by $\mathbf{B}^T = (b_{ji})_{m \times n}$ is such that the diagonal elements of the original matrix and its transpose are the same but the upper and lower triangular elements are interchanged as illustrated in (2.76).

$$\mathbf{B} = \begin{bmatrix} b_{11} & b_{12} & b_{13} \\ b_{21} & b_{22} & b_{23} \\ b_{31} & b_{32} & b_{33} \end{bmatrix} \quad \mathbf{B}^T = \begin{bmatrix} b_{11} & b_{21} & b_{31} \\ b_{12} & b_{22} & b_{32} \\ b_{13} & b_{23} & b_{33} \end{bmatrix} \quad (2.76)$$

2.6.4.1 Addition of Matrices
If we have two matrices $\mathbf{A}$ and $\mathbf{B}$, both of the same order (i.e., $m \times n$)

$$\mathbf{A} = \begin{bmatrix} a_{11} & a_{12} & a_{13} \\ a_{21} & a_{22} & a_{23} \\ a_{31} & a_{32} & a_{33} \end{bmatrix} \quad \mathbf{B} = \begin{bmatrix} b_{11} & b_{12} & b_{13} \\ b_{21} & b_{22} & b_{23} \\ b_{31} & b_{32} & b_{33} \end{bmatrix} \quad (2.77)$$

then the sum $\mathbf{A} \pm \mathbf{B}$ is defined by

$$\mathbf{A} \pm \mathbf{B} = \begin{bmatrix} a_{11} \pm b_{11} & a_{12} \pm b_{12} & a_{13} \pm b_{13} \\ a_{21} \pm b_{21} & a_{22} \pm b_{22} & a_{23} \pm b_{23} \\ a_{31} \pm b_{31} & a_{32} \pm b_{32} & a_{33} \pm b_{33} \end{bmatrix} \quad (2.78)$$

2.6.4.2 Multiplication by Scalar
The multiplication of a matrix by a scalar is defined by

$$c\mathbf{A} = \begin{bmatrix} ca_{11} & ca_{12} & ca_{13} \\ ca_{21} & ca_{22} & ca_{23} \\ ca_{31} & ca_{32} & ca_{33} \end{bmatrix} \quad (2.79)$$

2.6.4.3 Matrix Multiplication

If the product **AB** is defined by **C** = **AB**, where c_{ij}, the element of **C** in the ith row and jth column, is found by adding the products of the elements of **A** in the ith row with the corresponding elements of **B** in the jth column. Thus

$$c_{ij} = a_{i1}b_{1j} + a_{i2}b_{2j} + \cdots + a_{in}b_{nj} \tag{2.80}$$

It will be noted that the number of rows in **A** has to be the same as the number of columns in **B**. The order of the multiplication is also important because usually **AB** $\neq$ **BA**.

For example, if

$$\mathbf{A} = \begin{bmatrix} a_{11} & a_{12} \\ a_{21} & a_{22} \end{bmatrix} \quad \mathbf{B} = \begin{bmatrix} b_{11} & b_{12} \\ b_{21} & b_{22} \end{bmatrix} \tag{2.81}$$

then the product

$$\mathbf{AB} = \begin{bmatrix} a_{11}b_{11} + a_{12}b_{21} & a_{11}b_{12} + a_{12}b_{22} \\ a_{21}b_{11} + a_{22}b_{21} & a_{21}b_{12} + a_{22}b_{22} \end{bmatrix} \tag{2.82}$$

2.6.4.4 Matrix Determinant

The determinant of the matrix **A** denoted by det **A** or |**A**| can be found as follows:

$$\det \mathbf{A} = \sum_{i=1}^{n} a_{ij}(-1)^{i+j} \cdot c_{ij} \tag{2.83}$$

where c_{ij} is the cofactor of the element a_{ij} in the matrix **A**. The cofactor c_{ij} can be found by eliminating the ith row and jth column associated with the element a_{ij}. The summation is carried out for example for the chosen ith row. Matrix determinant exists only for a square matrix, that is, the matrix order has to be $n \times n$.

For example, if

$$\mathbf{A} = \begin{bmatrix} a_{11} & a_{12} & a_{13} \\ a_{21} & a_{22} & a_{23} \\ a_{31} & a_{32} & a_{33} \end{bmatrix} \tag{2.84}$$

then the determinant can be given by

$$\det \mathbf{A} = (-1)^2 a_{11} \cdot \begin{vmatrix} a_{22} & a_{23} \\ a_{32} & a_{33} \end{vmatrix} + (-1)^3 a_{21} \cdot \begin{vmatrix} a_{21} & a_{23} \\ a_{31} & a_{33} \end{vmatrix}$$

$$+ (-1)^4 a_{31} \cdot \begin{vmatrix} a_{21} & a_{22} \\ a_{31} & a_{32} \end{vmatrix} \tag{2.85}$$

$$\det \mathbf{A} = a_{11}(a_{22}a_{33} - a_{23}a_{33}) - a_{21}(a_{21}a_{33} - a_{23}a_{31}) + a_{32}(a_{21}a_{32} - a_{22}a_{31})$$

2.6.4.5 Matrix Inversion

The inverse of the matrix $\mathbf{A}$ is denoted by $\mathbf{A}^{-1}$. The inverse exists only when $\det \mathbf{A} \neq 0$. The inverse can be found as follows:

$$\mathbf{A}^{-1} = \frac{\text{adj } \mathbf{A}}{\det \mathbf{A}} \tag{2.86}$$

where adj $\mathbf{A}$ is the adjugate of $\mathbf{A}$. adj $\mathbf{A}$ is a matrix where all the elements of $\mathbf{A}$ are first replaced by their corresponding cofactors c_{ij} and the resulting matrix is transposed.

For example, if

$$\mathbf{A} = \begin{bmatrix} a_{11} & a_{12} & a_{13} \\ a_{21} & a_{22} & a_{23} \\ a_{31} & a_{32} & a_{33} \end{bmatrix} \tag{2.87}$$

then $\det \mathbf{A}$ is as defined in (2.85), and

$$\text{adj } \mathbf{A} = \begin{bmatrix} (-1)^2 \begin{vmatrix} a_{22} & a_{23} \\ a_{32} & a_{33} \end{vmatrix} & (-1)^3 \begin{vmatrix} a_{21} & a_{23} \\ a_{31} & a_{33} \end{vmatrix} & (-1)^4 \begin{vmatrix} a_{21} & a_{22} \\ a_{31} & a_{32} \end{vmatrix} \\ (-1)^3 \begin{vmatrix} a_{12} & a_{13} \\ a_{32} & a_{33} \end{vmatrix} & (-1)^4 \begin{vmatrix} a_{11} & a_{23} \\ a_{31} & a_{33} \end{vmatrix} & (-1)^5 \begin{vmatrix} a_{11} & a_{12} \\ a_{31} & a_{32} \end{vmatrix} \\ (-1)^4 \begin{vmatrix} a_{12} & a_{13} \\ a_{22} & a_{23} \end{vmatrix} & (-1)^5 \begin{vmatrix} a_{11} & a_{13} \\ a_{21} & a_{23} \end{vmatrix} & (-1)^6 \begin{vmatrix} a_{11} & a_{12} \\ a_{21} & a_{22} \end{vmatrix} \end{bmatrix}^T \tag{2.88}$$

as well as the inverse of $\mathbf{A}$

$$\mathbf{A}^{-1} = \frac{\begin{bmatrix} a_{22}a_{33} - a_{23}a_{32} & -a_{12}a_{33} + a_{13}a_{32} & a_{12}a_{33} - a_{13}a_{22} \\ -a_{21}a_{33} + a_{23}a_{31} & a_{11}a_{33} - a_{23}a_{31} & -a_{11}a_{23} + a_{13}a_{21} \\ a_{21}a_{32} - a_{22}a_{31} & -a_{11}a_{32} + a_{12}a_{31} & a_{11}a_{22} - a_{12}a_{22} \end{bmatrix}}{a_{11}(a_{22}a_{33} - a_{23}a_{32}) - a_{12}(a_{21}a_{33} - a_{23}a_{31}) + a_{13}(a_{21}a_{32} - a_{22}a_{31})} \tag{2.89}$$

2.7 Operational and Control Modes

The operation of a converter is classified as continuous, boundary, or discontinuous based on the behavior of the inductor current during a switching cycle (T_s), where the cycle time is an inverse of the switching frequency (f_s).

The operation mode is typically defined to be the continuous conduction mode (CCM) [17] when the inductor during the switching cycle is always greater than zero (Figure 2.21, CCM-1). This definition has become obsolete because the synchronous rectification (Figure 2.2) has replaced the diodes usually used earlier to implement part of the PWM switch, and consequently, the inductor current can have values less than zero (Figure 2.2, CCM-2). As

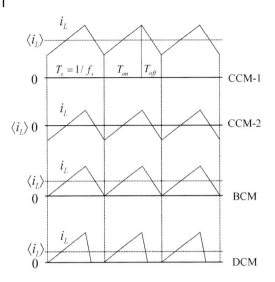

Figure 2.21 Converter operational modes.

a consequence, the operation is continuous if the inductor current has exactly two different slopes within a switching cycle. Even this definition is not covering all the cases but defines well most of them.

The operation mode is defined to be the boundary conduction mode (BCM, Figure 2.2) when the inductor current momentarily touches the zero level but does not cross it. This operation mode is also called the critical or transition mode.

The operation mode is defined to be discontinuous when the inductor current is equal to zero during a part of the switching cycle. Replacing the low-side FET switch in Figure 2.2 with a diode would limit the inductor current to zero by effectively disconnecting the current carrying path [14].

The way of producing the control of the pulsewidth (i.e., the length of the on-time (T_{on}), Figure 2.21) during which the main switch (i.e., the high-side switch in Figure 2.2) is turned on would define the control mode. The basic control mode is the *direct-on-time* or *voltage-mode control* [11], which is also called the *direct-duty-ratio control* under fixed-switching-frequency operation [9]. The *direct-on-time control* is implemented in such a way that the length of the on-time is controlled based on the constant-slope ramp signal. The *current-mode control* is such that the length of the on-time is directly or partly generated using the instantaneous inductor current [6, 7]. The *hysteretic control* is such that the length of the on-time and the off-time are controlled based on either inductor current or output voltage by providing upper and lower limits for the associated variable. The *self-oscillation or boundary-mode control* is a class of hysteretic control, where the upper limit is controlled and the lower limit is zero, that is, the converter is forced to operate in BCM. The voltage-mode control is usually misinterpreted to mean the constant-voltage feedback control but is related to the way of generating the pulsewidth as described above.

References

1. R.C. Dorf and R.H. Bishop, *Modern Control Systems*, Addison-Wesley, Menlo Park, CA, USA, **1998**, 8th Edition.
2. M. Hankaniemi and T. Suntio, 'Small-signal models for constant-current regulated converters,' in *Proc. IEEE Industrial Electronics Society Annual Conf.*, **2006**, pp. 2037–2042.
3. M. Hankaniemi, M. Sippola, and T. Suntio, 'Analysis of load interactions in constant-current-controlled buck converter,' in *Proc. IEEE International Telecommunications Energy Conf.*, **2006**, pp. 343–348.
4. T. Suntio, I. Gadoura, J. Lempinen, and K. Zenger, 'Practical design issues of multiloop controller for a telecom rectifier,' in *Proc. IEEE Telecommunications Energy Special Conf.*, **2000**, pp. 197–201.
5. A. Tenno, R. Tenno, and T. Suntio, 'Battery impedance and its relation to battery characteristics,' in *Proc. IEEE International Telecommunications Energy Conf.*, **2002**, pp. 176–183.
6. C.W. Deisch, 'Simple switching control method changes power converter into a current source,' in *Proc. IEEE Power Electronics Specialists Conf.*, **1978**, pp. 300–306.
7. L. Dixon, 'Average current mode control,' in *Proc. Unitrode Power Supply Design Seminar*, **1991**, pp. C1-1–C1-14.
8. M. Karppanen, T. Suntio, and M. Sippola, 'Dynamical characterization of input-voltage-feedforward-controlled buck converter,' *IEEE Trans. Indust. Electron.*, vol. 54, no. 2, **2007**, pp. 1005–1013.
9. R.D. Middlebrook and S. Cuk, 'A general unified approach to modeling switching-converter power stages,' *Int. J. Electron.*, vol. 42, no. 6, **1977**, pp. 521–550.
10. J. Sun, D.M. Mitchell, M.F. Greuel, P.T. Krein, and R.M. Bass, 'Average modeling of PWM converters in discontinuous modes,' *IEEE Trans. Power Electron.*, vol. 16, no. 4, **2001**, pp. 482–492.
11. T. Suntio, 'Unified average and small-signal modeling of direct-on-time control,' *IEEE Trans. Indust. Electron.*, vol. 53, no. 1, **2006**, pp. 287–295.
12. A.S. Kislovski, R. Redl, and N.O. Sokal, *Dynamic Analysis of Switching-Mode DC/DC Converters*, Van Nostrand Reinhold, New York, USA, **1991**.
13. D.M. Mitchell, *DC–DC Switching Regulator Analysis*, DMMitchell Consultants, Cedar Rapids, IA, USA, **1992**.
14. R.W. Erickson and D. Maksimovic, *Fundamentals of Power Electronics*, Kluwer, Norwell, MA, USA, **2001**, 2nd Edition.
15. T. Suntio and I. Gadoura, 'Dynamic analysis of switched-mode converters using two-port modeling technique,' in *Proc. Power Conversion and Intelligent Motion Conf.*, **2002**, pp. 387–392.
16. B.H. Cho, 'Modeling and analysis of spacecraft power systems,' PhD Thesis, Virginia Polytechnic Institute and State University, **1985**, 181 pp.
17. M. Shoyama, Y. Hamafuku, N. Matsuzaki, and T. Ninomiya, 'Simplification of transfer function in switching converter with general load impedance,' in *Proc. IEEE Power Electronics and Drives Conf.*, **1995**, pp. 155–161.
18. C.T. Chen, *Linear System Theory and Design*, Oxford University Press, New York, USA, **1999**, 3rd Edition.
19. C.K. Tse, *Linear Circuit Analysis*, Addison-Wesley Longman, Harlow, UK, **1998**.
20. R. Mammano, 'Isolating the control loop,' in *Proc. Unitrode Power Supply Seminar*, SEM-1000, **1994**, pp. C2–1–C2–15.
21. M.P. Sayani, R.V. White, D.N. Nason, and W.A. Taylor, 'Isolated feedback for off-line switching power supplies with primary-side control,' in *Proc. IEEE Applied Power Electronics Conf.*, **1988**, pp. 203–211.
22. Y. Panov and M. Jovanovic, 'Small-signal analysis and control design of isolated power supplies with

22. optocoupler feedback,' *IEEE Trans. Power Electron.*, vol. 20, no. 4, **2005**, pp. 823–832.
23. T. Tepsa and T. Suntio, 'Adjustable shunt regulator based control systems,' *IEEE Power Electron. Lett.*, vol. 1, no. 4, **2003**, pp. 93–96.
24. R. Kollman and J. Betten, 'Closing the loop with a popular shunt regulator,' *Power Electronics Technology*, September **2003**, pp. 30–36.
25. C. Basso, 'Biasing the TL431 for improved output impedance,' *Power Electronics Technology*, January **2005**, pp. 56–57.
26. Y. Panov and M. Jovanovic, 'Small-signal measurement techniques in switching power supplies,' in *Proc. IEEE Applied Power Electronics Conf.*, **2004**, pp. 770–776.
27. T. Suntio and M. Hankaniemi, 'Unified small-signal model for PCM control in CCM – Unterminated modeling approach,' *HIT J. Sci. Eng. B*, vol. 2, nos. 3–4, **2005**, pp. 452–475.
28. R.D. Middlebrook, 'Input filter considerations in design and application of switching regulators,' in *Proc. IEEE Industry Application Society Annual Conf.*, **1976**, pp. 366–382.
29. M. Hankaniemi, M. Karppanen, and T. Suntio, 'Load imposed instability and performance degradation in a regulated converter,' *IEE Proc. Electric Power Appl.*, vol. 153, no. 6, **2006**, pp. 905–910.
30. R.D. Middlebrook, 'Null double injection and the extra element theorem,' *IEEE Trans. Educ.*, vol. 32, no. 3, **1989**, pp. 167–180.
31. P.J. Antsaklis and A.N. Michel, *Linear Systems*, McGraw-Hill, New York, USA, **1997**.
32. K. Ogata, *Modern Control Engineering*, Prentice-Hall, Upper Saddle River, NJ, USA, **1997**.
33. W.H. Tutle, 'Relating converter transient response characteristics to feedback loop control design,' in *Proc. Powercon* 11, **1984**, pp. 10.1–10.12.
34. T. Suntio, M. Hankaniemi, and M. Karppanen, 'Analysing the dynamics of regulated converters,' *IEE Proc. Electric Power Appl.*, vol. 153, no. 6, **2006**, pp. 905–910.
35. B. Choi, 'Step load response of a current-mode-controlled DC-to-DC converter,' *IEEE Trans. Aerosp. Electron. Syst.*, vol. 33, no. 4, **1997**, pp. 1115–1121.
36. C. Gezgin, 'Predicting load transient response of output voltage in DC–DC converters,' in *Proc. IEEE Applied Power Electronics Conf.*, **2004**, pp. 1339–1343.
37. J. Betten and R. Kollman, 'Easy calculation yields load transient response,' *Power Electronics Technology*, **2005**, pp. 40–48.
38. D.M. Sable, B.H. Cho, and R.B. Ridley, 'Use of leading-edge modulation to transform boost and flyback converters into minimum-phase-zero systems,' *IEEE Trans. Power Electron.*, vol. 6, no. 4, **1991**, pp. 704–711.
39. W.C. Wu, R.M. Bass, and J.R. Yeargan, 'Eliminating the effects of the right-hand plane zero in fixed frequency boost converter,' in *Proc. IEEE Power Electron. Specialists Conf.*, **1988**, pp. 362–366.
40. D.M. Mitchell, 'Tricks of the trade: Understanding the right-hand-plane zero in the small-signal DC/DC converter models,' *IEEE Power Electron. Soc. Newsletter*, January **2001**, pp 5–6.
41. J.M. Freudenberg and D.P. Looze, 'Right half plane poles and zeros and design tradeoffs in feedback systems,' *IEEE Trans. Autom. Control*, vol. AC-30, no. 6, **1985**, pp. 555–565.
42. J.M. Maciejowski, *Multivariable Feedback Design*, Addison-Wesley, Reading, MA, USA, **1989**.
43. S. Skogestad and I. Postlethwaite, *Multivariable Feedback Control, Analysis and Design*, Wiley, New York, USA, **1996**.
44. V. Grigore, J. Hätönen, J. Kyyrä, and T. Suntio, 'Dynamics of a buck converter with a constant power load,' in *Proc. IEEE Power Electronics Specialists Conf.*, **1998**, pp. 72–78.

3
Average and Small-Signal Modeling of Direct-On-Time Controlled Converters

3.1
Introduction

The chapter provides the unified basis for the average and small-signal modeling of the voltage-output switching-mode converters under direct-on-time control in continuous (CCM) and discontinuous (DCM) conduction modes. Direct on-time control is commonly known as voltage-mode control (VMC) and also direct-duty-ratio control in the case of constant switching frequency [1]. It will be shown that the classical state-space-averaging (SSA) technique [2, 3] yields the same fixed-frequency dynamic models in CCM as the more general method introduced in [10, 11]. The classical SSA method [5] failed to produce accurate small-signal models when applied directly to the converters operating in DCM. The method of producing the more accurate DCM models was developed in the late 1990s [7–9] after 20 years of intensive efforts [6, 12]. The dynamic models associated with the direct-on-time control are important because they would provide the basis for modeling the other control modes such as current-mode control, self-oscillation control, hysteretic control, and so on.

The SSA method was developed without considering the dynamic processes inside the converter and, therefore, the good results in CCM were actually obtained by accident, which the failing in DCM proves: The time-averaged state variables where the averaging is done over one switching cycle would mainly contribute to the dynamics observed at the output or input of the converter [11]. In practice, this means that the real state variables are the time-averaged values of those variables which are also continuous signals of time regardless of the operation mode. If feedback or feedforward signals are applied either to implementing the usual feedback control or to producing different control modes such as current-mode control, the ripple components superimposed on the time-averaged signals may produce ripple effects on the converter dynamics [13–16], which may even lead to instability under certain conditions [17].

It has been observed that the pulsewidth modulation (PWM) process would corrupt the sinus-form excitation signals [20]. The results of the corruption

Dynamic Profile of Switched-Mode Converter. Teuvo Suntio
© 2009 WILEY-VCH Verlag GmbH & Co. KGaA, Weinheim
ISBN: 978-3-527-40708-8

start appearing clearly when the excitation signal approaches half the switching frequency. Typically, the phenomenon would cause an excess phase lag and slight increase in the magnitude compared to the predictions where the phenomenon is not considered [18, 19, 21, 22].

The modeling and the models are typically presented including the load impedance (i.e., usually a resistor) in them [1–8]. As a consequence, those models do not represent the internal dynamics of the associated converter, and they may even totally hide its real dynamic behavior. Such an approach has led to widespread misunderstanding. A good example is [23], where it is explicitly claimed that the resonant nature in a VMC buck converter would become more severe (i.e., damping decreases) when the load decreases. Actually the internal dynamics does not change if the other operational conditions are kept constant.

The concept of unterminated or internal load-and-source-independent models has been well known for a long time [24, 25] but their meaning has not been understood until [26]. The average- and small-signal models presented in this chapter are the real internal models. We derive and present the state spaces in general form applicable both to variable and fixed-frequency operation, but the transfer functions we derive are only applicable to fixed-frequency operation. The variable-frequency operation will be treated more in detail in Chapter 6. The converters we treat more in detail are buck, boost, and buck–boost converters. The introduced methods are also readily applicable to the more complicated converters.

3.2
Direct-on-Time Control

The direct on-time control is the basic control method of a switched-mode converter (Figure 3.1) where the length of the on-time (Figure 3.1b, t_{on}) is controlled to keep the output voltage or current constant at the predefined level [1]. During the on-time, the main switch (Figure. 3.1a, the high-side switch) is conducting and the other switch (Figure 3.1a, the low-side switch) is turned off. During the off-time (Figure 3.1b, t_{off}), the main switch is turned off, and the other switch is conducting. This process is repeated in a sequential manner, that is, the operation is periodical. If the switching frequency (Figure 3.1a, f_s) is constant, then the on-time control equals direct-duty-ratio control, where the duty ratio (d) can be expressed by $\frac{t_{on}}{T_s}$, and its complement ($d' = 1 - d$) by $\frac{t_{off}}{T_s}$. The inverse of switching frequency (f_s) denoted by T_s (Figure 3.1b, $T_s = t_{on} + t_{off}$) is called cycle time. The fixed-frequency operation (i.e., constant cycle time) is the usual choice of controlling the converter, although other controlling methods do exist yielding variable switching frequency, for example, self-oscillation, constant-on-time, and hysteretic control [1]. In reality, the variable-frequency direct-on-time-control mode is fictive serving only the modeling purposes.

3.2 Direct-on-Time Control

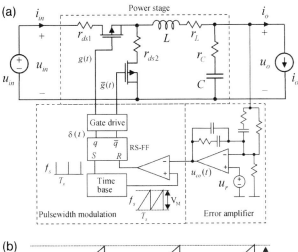

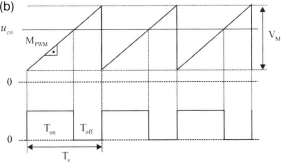

Figure 3.1 Direct-duty-ratio-controlled synchronous buck converter: (a) schematics and (b) PWM generation.

The on-time or PWM under the fixed-frequency VMC is implemented (Figure 3.1b) by comparing the voltage provided by the controller (i.e., u_{co}) to a constant-slope PWM ramp signal (i.e., $M_{PWM} = V_M/T_s$) provided by the modulator. The pulse is switched on in the beginning of the cycle by means of the pace signal applied to the set input (S) of the RS-flip-flop. The pace signal also determines the switching frequency (f_s). The pulse will be terminated when the PWM ramp exceeds the control signal (u_{co}) by turning the comparator output high and resetting the RS-flip-flop. The flip-flop controls the gate drivers, which in turn provide the gate-control signals of the associated switches.

When considering the situation from the dynamic point of view, the slope of the PWM ramp signal would affect the gain (i.e., G_a, Figure 3.2) of the dynamic actions through the modulator. This gain is usually called *modulator gain*. It can be found by constructing the comparator equations (Figure 3.1a), which define the length of the on-time, as follows:

$$u_{co} = \frac{V_M}{T_s} \cdot t_{on} \qquad (3.1)$$

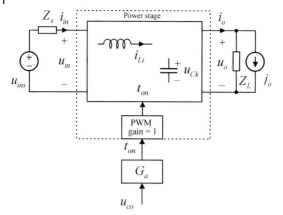

Figure 3.2 Direct-on-time-controlled converter at open loop with a PWM modulator.

and developing the partial derivatives of (3.1), which yields in general form

$$\hat{t}_{on} = \frac{T_s}{V_M} \cdot \hat{u}_{co} \qquad (3.2)$$

Under fixed-frequency operation, Eq. (3.2) becomes

$$\hat{d} = \frac{1}{V_M} \cdot \hat{u}_{co} \qquad (3.3)$$

and consequently, the modulator gain (G_a) equals V_M^{-1} [1]. Due to the PWM process, the modulator gain may also contain some frequency-dependent components [18–22] in addition to the constant gain defined above.

When defining the dynamic models for the direct-on-time-controlled power stages in the subsequent sections, we will assume that the modulator gain equals 1 as illustrated in Figure 3.2. The derived models are also the internal models, that is, the source impedance (i.e., Z_s, Figure 3.2) is assumed to be zero and the load impedance (i.e., Z_L, Figure 3.2) is assumed to be infinite when deriving the corresponding state spaces.

3.3
Generalized Modeling Technique

The switched-mode converters are nonlinear variable-structure systems where different structures are periodically switched on and off. The nonlinearity can be removed by averaging the converter operation within one switching cycle. Linearizing the obtained averaged model by developing the partial derivatives with respect to the state and input variables at a certain operating point would yield the required dynamic representation of the converter. It is obvious that the models are valid only up to half the switching frequency due to the nature

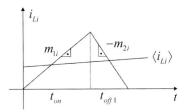

Figure 3.3 Waveforms of the inductor current at an arbitrary switching cycle.

of averaging. In order to construct the averaged state space, we have to consider the variables as the time-varying averages of the corresponding instantaneous values (x), which we denote by $\langle x \rangle$.

The behavior of the time-averaged inductor current $\langle i_L \rangle$ would determine most of the dynamics in a converter. The portion of the inductor current having positive up-slope (m_{1i}) and negative down-slope (m_{2i}) is presented in Figure 3.3 for an arbitrary cycle. The corresponding portions of the cycle time (t_s) are denoted by t_{on} and t_{off1}. It should be noted that the cycle time $t_s = t_{on} + t_{off1}$ in CCM but $t_s > t_{on} + t_{off1}$ in DCM (see Figure 2.21, Chapter 2). The slopes m_{1i}, m_{2i} are local averages within the on-time or off-time, respectively.

The time-averaged inductor current $\langle i_{Li} \rangle$ is a continuous signal of time regardless of the conduction mode but its charge potential within those subcycles is distributed as

$$\langle i_{Li} \rangle_{on} = \frac{t_{on}}{t_{on} + t_{off1}} \cdot \langle i_{Li} \rangle$$
$$\langle i_{Li} \rangle_{off} = \frac{t_{off1}}{t_{on} + t_{off1}} \cdot \langle i_{Li} \rangle \tag{3.4}$$

For developing the averaged state space, the derivatives of the time-averaged state variables have to be defined. The derivative of the time-averaged inductor current can be approximated according to the average slope of the instantaneous inductor current (Figure 3.3) by

$$\frac{d\langle i_{Li} \rangle}{dt} = \frac{t_{on}}{t_s} \cdot m_{1i} - \frac{t_{off1}}{t_s} \cdot m_{2i} \tag{3.5}$$

where $m_{1i} = \frac{\overline{u}_{Li-on}}{L_i}$ and $m_{2i} = \frac{-\overline{u}_{Li-off1}}{L_i}$ as well as $\overline{u}_{Li-on}$, the voltage across the inductor during the on-time (t_{on}), and $\overline{u}_{Li-off1}$, the voltage across the inductor during the part (t_{off1}) of the off-time (t_{off}) when the inductor current is nonzero. This relation may be obvious because $u_L = L\frac{di_L}{dt}$. The bar over the voltage denotes that the variables constituting the voltage are the corresponding local averages during the corresponding subcycle.

The derivative of the time-averaged capacitor voltage can be approximated (Figure 3.4) by

$$\frac{d\langle u_{Ci} \rangle}{dt} = \frac{\langle i_+ \rangle}{C_i} - \frac{\langle i_- \rangle}{C_i} \tag{3.6}$$

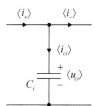

Figure 3.4 Capacitor charge balance.

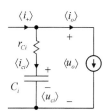

Figure 3.5 Converter output stage.

where $\langle i_+ \rangle$ is the time-averaged current charging the capacitor, and $\langle i_- \rangle$ is the time-averaged current discharging the capacitor within a cycle. This relation is obvious because $i_c = C \frac{du_C}{dt}$ and $i_C = i_+ - i_-$ according to *Kirchhoff*'s current law (Figure 3.4).

The output of the converter is most often provided with an output capacitor as illustrated in Figure 3.5. Therefore, the average output voltage can be given by

$$\langle u_o \rangle = \langle u_{Ci} \rangle + r_{Ci} \langle i_+ \rangle - r_{Ci} \langle i_o \rangle \tag{3.7}$$

or by

$$\langle u_o \rangle = \langle u_{Ci} \rangle + r_{Ci} C_i \frac{d\langle u_{Ci} \rangle}{dt} \tag{3.8}$$

which is a useful form in the final state space for the output voltage.

The time-varying averaged input current ($\langle i_{in} \rangle$) is usually the current of a certain circuit inductor or part of it, and therefore, its value can be found according to (3.4). The output voltage (u_o) may affect the slope of the inductor current. The slopes (m_{ij}) are the local averages taken over the corresponding subcycle, and therefore, the effect of the output voltage will be considered correspondingly: the inductor current shall be taken always as such if involved in the process not as an average over the whole cycle. This statement will be clarified when we will present the derived state spaces for the basic converters in the subsequent subsections.

3.3.1
Buck Converter

A buck or step-down converter (Figure 3.6) is a converter where the input voltage has to be higher than the output voltage for a proper operation. We

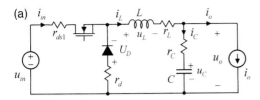
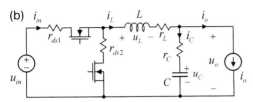

Figure 3.6 Buck converter using (a) diode switching and (b) synchronous switching.

consider the modeling of two type of buck converters differing from each other with respect to the implementation of the freewheeling or low-side switch either with a diode (Figure 3.6a) or with a MOSFET (Figure 3.6b). The use of the diode means that the converter may operate in both CCM and DCM. The use of MOSFET means that the converter operates only in CCM without any special control arrangements. The internal parasitic resistances (i.e., r_d, r_{ds1}, r_{ds2}) in the switching elements also include the switching losses. The high-side switch is the main switch, which is turned on during the on-time.

According to the generalized method, we divide the switching cycle into two subcycles: during the on-time both of the buck converters (Figure 3.6) have the same structure as shown in Figure 3.7a. During the off-time1 (i.e., during the

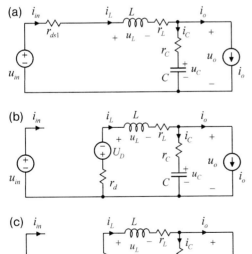

Figure 3.7 The subcircuit structures of the buck converter during the (a) on-time and off-time1 with (b) diode switching and (c) synchronous switching.

part of the off-time when the inductor current slope is negative (Figure 3.3)), the structures are basically the same except the loss components: Figure 3.7b, with the diode switching, and Figure 3.7c, with the synchronous switching.

According to Figure 3.7, we may conclude that the current $\langle i_+ \rangle$ charging the output capacitor equals the time-averaged inductor current $\langle i_L \rangle$ because the inductor current charges the output during both of the subcycles. For the same reason, the local average of the output voltage during the on- and off-times can be given by

$$\overline{u}_{o-\text{on/off}} = r_C \langle i_L \rangle + \langle u_C \rangle - r_C \langle i_o \rangle \tag{3.9}$$

The time-averaged input current ($\langle i_\text{in} \rangle$) equals the on-time inductor current, and therefore, the first equation in (3.4) applies.

According to these conclusions and applying Eqs. (3.4)–(3.8), the general averaged state space can be given for the buck converter by

$$\begin{aligned}
\frac{d\langle i_L \rangle}{dt} &= \frac{t_\text{on}}{t_s} \cdot m_1 - \frac{t_\text{off1}}{t_s} \cdot m_2 \\
\frac{d\langle u_C \rangle}{dt} &= \frac{\langle i_L \rangle}{C} - \frac{\langle i_o \rangle}{C} \\
\langle i_\text{in} \rangle &= \frac{t_\text{on}}{t_\text{on} + t_\text{off1}} \cdot \langle i_L \rangle \\
\langle u_o \rangle &= \langle u_C \rangle + r_C C \frac{d\langle u_C \rangle}{dt}
\end{aligned} \tag{3.10}$$

where the up-slope

$$m_1 = \frac{\langle u_\text{in} \rangle - (r_L + r_\text{ds1} + r_C)\langle i_L \rangle - \langle u_C \rangle + r_C \langle i_o \rangle}{L} \tag{3.11}$$

the down-slope of the diode-switching converter

$$m_2 = \frac{(r_L + r_d + r_C)\langle i_L \rangle + \langle u_C \rangle - r_C \langle i_o \rangle + U_D}{L} \tag{3.12}$$

and the down-slope of the synchronous-switching converter

$$m_2 = \frac{(r_L + r_\text{ds2} + r_C)\langle i_L \rangle + \langle u_C \rangle - r_C \langle i_o \rangle}{L} \tag{3.13}$$

3.3.2 Boost Converter

The boost or step-up converter is a converter where the input voltage has to be lower than the output voltage for a proper operation. We consider the modeling of both the diode-switching (Figure 3.8a) and the synchronous-switching

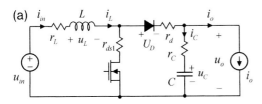
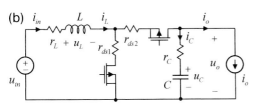

Figure 3.8 Boost converter using (a) diode switching and (b) synchronous switching.

(Figure 3.8b) boost converter. The parasitic loss resistances (r_d, r_{ds1}, r_{ds2}) also include the switching losses. The main switch is the low-side MOSFET. The converter may operate in both CCM and DCM when the high-side switch is a diode but only in CCM when the high-side switch is a MOSFET.

During the on-time both of the boost converters (Figure 3.8) have the same structure as shown in Figure 3.9a. During the off-time1, the structures are basically the same but the loss components are different as shown in Figures 3.9b (diode switching) and 3.9c (synchronous switching).

From Figure 3.9, we may conclude that the current $\langle i_+ \rangle$ charging the output capacitor equals the time-averaged off-time inductor current $\langle i_L \rangle_{\text{off}}$ defined in

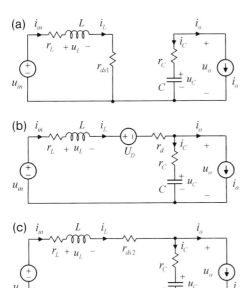

Figure 3.9 The subcircuit structures of the boost converter during the (a) on-time and off-time1 using (b) diode switching and (c) synchronous switching.

(3.4) because the inductor current charges the output only during the off-time. For the same reason, the local average of the output voltage during the on- and off-times can be given by

$$\bar{u}_{o-on} = \langle u_C \rangle - r_C \langle i_o \rangle$$
$$\bar{u}_{o-off} = r_C \langle i_L \rangle + \langle u_C \rangle - r_C \langle i_o \rangle \qquad (3.14)$$

The time-averaged input current ($\langle i_{in} \rangle$) equals the time-averaged inductor current.

According to these conclusions and applying Eqs. (3.4)–(3.8), the general averaged state space can be given for the boost converter by

$$\frac{d\langle i_L \rangle}{dt} = \frac{t_{on}}{t_s} \cdot m_1 - \frac{t_{off1}}{t_s} \cdot m_2$$
$$\frac{d\langle u_C \rangle}{dt} = \frac{t_{off1}}{t_{on}} + t_{off1} \cdot \frac{\langle i_L \rangle}{C} - \frac{\langle i_o \rangle}{C}$$
$$\langle i_{in} \rangle = \langle i_L \rangle$$
$$\langle u_o \rangle = \langle u_C \rangle + r_C C \frac{d\langle u_C \rangle}{dt}$$
(3.15)

where the up-slope

$$m_1 = \frac{\langle u_{in} \rangle - (r_L + r_{ds1})\langle i_L \rangle}{L} \qquad (3.16)$$

the down-slope of the diode-switching converter

$$m_2 = \frac{(r_L + r_d + r_C)\langle i_L \rangle + \langle u_C \rangle - r_C \langle i_o \rangle + U_D - \langle u_{in} \rangle}{L} \qquad (3.17)$$

and the down-slope of the synchronous-switching converter

$$m_2 = \frac{(r_L + r_{ds2} + r_C)\langle i_L \rangle + \langle u_C \rangle - r_C \langle i_o \rangle - \langle u_{in} \rangle}{L} \qquad (3.18)$$

3.3.3
Buck–Boost Converter

The buck–boost, step-up/down or flyback converter is a converter where the input voltage can be lower or higher than the output voltage and where the polarity of the output voltage is negative compared to the polarity of the input voltage. We consider the modeling of both the diode-switched (Figure 3.10a) and synchronous-switched (Figure 3.10b) versions. We have arranged the polarities of the output voltage and current so that the associated equations would have positive sign as can be concluded from Figure 3.10, when comparing to Figures. 3.6 and 3.8. The parasitic loss resistances (r_d, r_{ds1}, r_{ds2})

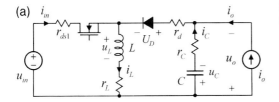
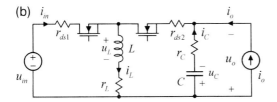

Figure 3.10 A buck–boost converter.

of the switching components also include the switching losses. The main switch is the high-side MOSFET. The converter can operate in both CCM and DCM when the diode switching is in use, but only in CCM when the synchronous switching is in use.

During the on-time, the converter has the structure shown in Figure 3.11a and during the off-time1 the structures are the same in principle but the loss components are different as shown in Figures 3.11b (diode switching) and Figure 3.11c (synchronous switching).

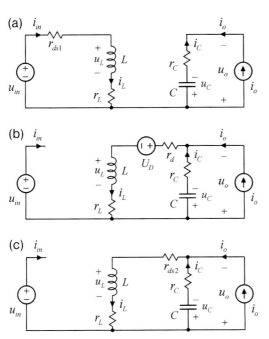

Figure 3.11 The subcircuit structures of the buck–boost converter during the (a) on-time and off-time1 using (b) diode switching and (c) synchronous switching.

From Figure 3.10, we may conclude that the current $\langle i_+ \rangle$ charging the output capacitor equals the time-averaged off-time inductor current $\langle i_L \rangle_{\text{off}}$ defined in (3.4), because the inductor current charges the output only during the off-time. For the same reason, the local average of the output voltage during the on- and off-times can be given by

$$\overline{u}_{o-\text{on}} = \langle u_C \rangle - r_C \langle i_o \rangle$$
$$\overline{u}_{o-\text{off}} = r_C \langle i_L \rangle + \langle u_C \rangle - r_C \langle i_o \rangle \tag{3.19}$$

The time-averaged input current ($\langle i_{\text{in}} \rangle$) equals the time-averaged on-time inductor current defined in (3.4). According to these conclusions and applying Eqs. (3.4)–(3.8), the general averaged state space can be given for the buck–boost converter by

$$\frac{d\langle i_L \rangle}{dt} = \frac{t_{\text{on}}}{t_s} \cdot m_1 - \frac{t_{\text{off1}}}{t_s} \cdot m_2$$
$$\frac{d\langle u_C \rangle}{dt} = \frac{t_{\text{off1}}}{t_{\text{on}} + t_{\text{off1}}} \cdot \frac{\langle i_L \rangle}{C} - \frac{\langle i_o \rangle}{C}$$
$$\langle i_{\text{in}} \rangle = \frac{t_{\text{on}}}{t_{\text{on}} + t_{\text{off1}}} \cdot \langle i_L \rangle \tag{3.20}$$
$$\langle u_o \rangle = \langle u_C \rangle + r_C C \frac{d\langle u_C \rangle}{dt}$$

where the up-slope

$$m_1 = \frac{\langle u_{\text{in}} \rangle - (r_L + r_{\text{ds1}})\langle i_L \rangle}{L} \tag{3.21}$$

the down-slope of the diode-switching converter

$$m_2 = \frac{(r_L + r_d + r_C)\langle i_L \rangle + \langle u_C \rangle - r_C \langle i_o \rangle + U_D}{L} \tag{3.22}$$

and the down-slope of the synchronous-switching converter

$$m_2 = \frac{(r_L + r_{\text{ds2}} + r_C)\langle i_L \rangle + \langle u_C \rangle - r_C \langle i_o \rangle}{L} \tag{3.23}$$

According to Sections 3.1 and 3.2, the buck–boost converter has inherited features from the buck and the boost converters as its name implies.

3.4
Fixed-Frequency Operation in CCM

The only practical voltage-mode-control principle is the fixed-frequency operation mode described in Section 3.2. The variable-frequency-operation modes are fictive in that respect. The generalized modeling method would give exactly the same models as the classical state-space averaging [4] in CCM because the time-varying averaged inductor current lies exactly in the middle of the inductor-current ripple [11], and consequently, its charge distribution would be directly related to the length of the on-time and off-time according to (3.4). In practice, this means that we can apply the circuit theory without necessity to considering the nature of the inductor current or the other variables more in detail. Consequently, the averaging can be done by multiplying the on-time equations with the duty ratio d and the off-time equations with its complement d' as instructed in [3, 4]. Naturally, the direct use of the circuit theory (i.e., applying *Kirchhoff*'s voltage and current laws) is the most convenient method to solve the required averaged state spaces, and also a recommended method. The steady-state values of the circuit variables can be solved from the averaged state spaces by setting the derivatives to zero. This method would give more accurate steady-state models compared for example to those given in [1], because the averaging methods presented in [1] cannot correctly model the steady-state behavior of the converter. Consequently, the effect of some important loss mechanisms such as the ESR of the circuit capacitors does not appear in the results. The synchronous buck converter will be treated more in detail in Section 3.4.1 as an example. The dynamical models for the other converters will be given in Section 3.4.2 including also the canonical small-signal and steady-state equivalent circuits introduced in [3, 4].

3.4.1
Synchronous Buck Converter

We apply the state-space averaging technique to the synchronous buck converter shown in Figure 3.12. The converter can operate only in CCM. Therefore, two subcircuit structures have to be studied. In order to construct the state space, the derivatives of the inductor current and the capacitor voltage have to be defined as well as the equations for the output variables (i_{in}, u_o) utilizing basic circuit theory and *Kirchhoff*'s voltage and current laws.

The on-time subcircuit is shown in Figure 3.13a and the off-time subcircuit in Figure. 3.13b. We have to solve the equations for u_L, i_C, i_{in}, and u_o during both of the subcycles as follows:

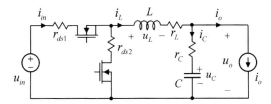

Figure 3.12 Synchronous buck converter.

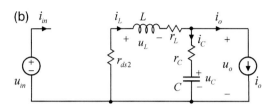

Figure 3.13 Subcircuit structures for (a) on-time and (b) off-time.

On-time:

$$u_L = u_{in} - (r_L + r_{ds1})i_L - u_o$$

$$i_C = i_L - i_o$$

$$i_{in} = i_L$$

$$u_o = u_C + r_C i_C$$

Off-time:

$$u_L = -(r_L + r_{ds2})i_L - u_o$$

$$i_C = i_L - i_o$$

$$i_{in} = i_L$$

$$u_o = u_C + r_C i_C$$

When arranging the on-time and off-time equations as required by the state-space representation and applying the identities $\frac{di_L}{dt} = \frac{u_L}{L}$ and $\frac{du_C}{dt} = \frac{i_C}{C}$, we get

On-time:

$$\frac{di_L}{dt} = -\frac{(r_L + r_{ds1} + r_C)}{L}i_L - \frac{1}{L}u_C + \frac{1}{L}u_{in} + \frac{r_C}{L}i_o$$

$$\frac{du_C}{dt} = \frac{i_L}{C} - \frac{i_o}{C}$$

$$i_{in} = i_L$$

$$u_o = r_C i_L + u_C - r_C i_o$$

Off-time:

$$\frac{di_L}{dt} = -\frac{(r_L + r_{ds2} + r_C)}{L}i_L - \frac{1}{L}u_C + \frac{r_C}{L}i_o$$

$$\frac{du_C}{dt} = \frac{i_L}{C} - \frac{i_o}{C}$$
$$i_{in} = i_L$$
$$u_o = r_C i_L + u_C - r_C i_o$$

The averaging is done by multiplying the on-time equations with d and the off-time equations with d', adding them together, and applying the identity $d + d' = 1$ if applicable. This procedure yields

$$\frac{di_L}{dt} = -\frac{(r_L + dr_{ds1} + d'r_{ds2} + r_C)}{L}i_L - \frac{1}{L}u_C + \frac{d}{L}u_{in} + \frac{r_C}{L}i_o$$
$$\frac{du_C}{dt} = \frac{i_L}{C} - \frac{i_o}{C} \quad (3.24)$$
$$i_{in} = di_L$$
$$u_o = r_C i_L + u_C - r_C i_o$$

The averaged state space (3.24) is nonlinear due to the product of two variables (i.e., du_{in} and di_L), and needs to be linearized by developing the partial derivatives as instructed in Chapter 2 (Section 2.6.1): this procedure yields

$$\frac{d\hat{i}_L}{dt} = -\frac{(r_L + Dr_{ds1} + D'r_{ds2} + r_C)}{L}\hat{i}_L - \frac{1}{L}\hat{u}_C + \frac{D}{L}\hat{u}_{in} + \frac{r_C}{L}\hat{i}_o$$
$$+ \frac{U_{in} + (r_{ds2} - r_{ds1})I_L}{L}\hat{d}$$
$$\frac{d\hat{u}_C}{dt} = \frac{\hat{i}_L}{C} - \frac{\hat{i}_o}{C} \quad (3.25)$$
$$\hat{i}_{in} = D\hat{i}_L + I_L\hat{d}$$
$$\hat{u}_o = r_C\hat{i}_L + \hat{u}_C - r_C\hat{i}_o$$

As discussed earlier, the output-voltage equation in (3.25) would be most convenient to be substituted with $\hat{u}_C\left(1 + r_C C \frac{d}{dt}\right)$ when solving the state space in frequency domain (i.e., applying *Laplace* transformation). This yields as the final time-domain state space

$$\frac{d\hat{i}_L}{dt} = -\frac{(r_L + Dr_{ds1} + D'r_{ds2} + r_C)}{L}\hat{i}_L - \frac{1}{L}\hat{u}_C + \frac{D}{L}\hat{u}_{in} + \frac{r_C}{L}\hat{i}_o$$
$$+ \frac{U_{in} + (r_{ds2} - r_{ds1})I_L}{L}\hat{d}$$
$$\frac{d\hat{u}_C}{dt} = \frac{\hat{i}_L}{C} - \frac{\hat{i}_o}{C} \quad (3.26)$$

3 Average and Small-Signal Modeling of Direct-On-Time Controlled Converters

$$\hat{i}_{in} = D\hat{i}_L + I_L\hat{d}$$

$$\hat{u}_o = \left(1 + r_C C \frac{d}{dt}\right)\hat{u}_C$$

The set of equations in (3.26) can be presented in matrix form as

$$\begin{bmatrix} \dfrac{d\hat{i}_L}{dt} \\ \dfrac{d\hat{u}_C}{dt} \end{bmatrix} = \begin{bmatrix} -\dfrac{r_L + Dr_{ds1} + D'r_{ds2} + r_C}{L} & -\dfrac{1}{L} \\ \dfrac{1}{C} & 0 \end{bmatrix} \begin{bmatrix} \hat{i}_L \\ \hat{u}_C \end{bmatrix}$$

$$+ \begin{bmatrix} \dfrac{D}{L} & \dfrac{r_C}{L} & \dfrac{U_{in} + (r_{ds2} - r_{ds1})I_L}{L} \\ 0 & -\dfrac{1}{C} & 0 \end{bmatrix} \begin{bmatrix} \hat{u}_{in} \\ \hat{i}_o \\ \hat{d} \end{bmatrix} \quad (3.27)$$

$$\begin{bmatrix} \hat{i}_{in} \\ \hat{u}_o \end{bmatrix} = \begin{bmatrix} D & 0 \\ 0 & 1 + r_C C \dfrac{d}{dt} \end{bmatrix} \begin{bmatrix} \hat{i}_L \\ \hat{u}_C \end{bmatrix} + \begin{bmatrix} 0 & 0 & I_L \\ 0 & 0 & 0 \end{bmatrix} \begin{bmatrix} \hat{u}_{in} \\ \hat{i}_o \\ \hat{d} \end{bmatrix}$$

Applying *Laplace* transform to (3.26) yields

$$s\begin{bmatrix} \hat{i}_L(s) \\ \hat{u}_C(s) \end{bmatrix} = \begin{bmatrix} -\dfrac{r_L + Dr_{ds1} + D'r_{ds2} + r_C}{L} & -\dfrac{1}{L} \\ \dfrac{1}{C} & 0 \end{bmatrix} \begin{bmatrix} \hat{i}_L(s) \\ \hat{u}_C(s) \end{bmatrix}$$

$$+ \begin{bmatrix} \dfrac{D}{L} & \dfrac{r_C}{L} & \dfrac{U_{in} + (r_{ds2} - r_{ds1})I_L}{L} \\ 0 & -\dfrac{1}{C} & 0 \end{bmatrix} \begin{bmatrix} \hat{u}_{in}(s) \\ \hat{i}_o(s) \\ \hat{d}(s) \end{bmatrix} \quad (3.28)$$

$$\begin{bmatrix} \hat{i}_{in}(s) \\ \hat{u}_o(s) \end{bmatrix} = \begin{bmatrix} D & 0 \\ 0 & 1 + sr_C C \end{bmatrix} \begin{bmatrix} \hat{i}_L(s) \\ \hat{u}_C(s) \end{bmatrix} + \begin{bmatrix} 0 & 0 & I_L \\ 0 & 0 & 0 \end{bmatrix} \begin{bmatrix} \hat{u}_{in}(s) \\ \hat{i}_o(s) \\ \hat{d}(s) \end{bmatrix}$$

and

$$\begin{bmatrix} \hat{i}_L(s) \\ \hat{u}_C(s) \end{bmatrix} = \begin{bmatrix} s + \dfrac{r_L + Dr_{ds1} + D'r_{ds2} + r_C}{L} & \dfrac{1}{L} \\ -\dfrac{1}{C} & s \end{bmatrix}^{-1} \begin{bmatrix} \hat{i}_L(s) \\ \hat{u}_C(s) \end{bmatrix}$$

$$+ \begin{bmatrix} \dfrac{D}{L} & \dfrac{r_C}{L} & \dfrac{U_{in} + (r_{ds2} - r_{ds1})I_L}{L} \\ 0 & -\dfrac{1}{C} & 0 \end{bmatrix} \begin{bmatrix} \hat{u}_{in}(s) \\ \hat{i}_o(s) \\ \hat{d}(s) \end{bmatrix} \quad (3.29)$$

$$\begin{bmatrix} \hat{i}_{in}(s) \\ \hat{u}_o(s) \end{bmatrix} = \begin{bmatrix} D & 0 \\ 0 & 1 + sr_C C \end{bmatrix} \begin{bmatrix} \hat{i}_L(s) \\ \hat{u}_C(s) \end{bmatrix} + \begin{bmatrix} 0 & 0 & I_L \\ 0 & 0 & 0 \end{bmatrix} \begin{bmatrix} \hat{u}_{in}(s) \\ \hat{i}_o(s) \\ \hat{d}(s) \end{bmatrix}$$

3.4 Fixed-Frequency Operation in CCM

The determinant of the inverse matrix (i.e., $[\ ]^{-1}$) determines the characteristic polynomial of the transfer functions to be

$$\Delta = s^2 + s\frac{r_L + Dr_{ds1} + D'r_{ds2} + r_C}{L} + \frac{1}{LC} \quad (3.30)$$

The transfer functions from the input variables to the state variables (i.e., $[s\mathbf{I} - \mathbf{A}]^{-1}\mathbf{B}$, Chapter 2, Section 2.2) can be computed to be

$$\begin{bmatrix} G_{iL-o} & G_{oL-o} \\ G_{iC-o} & -G_{oC-o} \end{bmatrix} = \frac{\begin{bmatrix} \dfrac{Ds}{L} & \dfrac{1 + sr_C C}{LC} \\ \dfrac{D}{LC} & -\dfrac{r_L + Dr_{ds1} + D'r_{ds2} + sL}{LC} \end{bmatrix}}{s^2 + s\dfrac{r_L + Dr_{ds1} + D'r_{ds2} + r_C}{L} + \dfrac{1}{LC}}$$

$$\begin{bmatrix} G_{cL} \\ G_{cC} \end{bmatrix} = \frac{\begin{bmatrix} \dfrac{(U_{in} + (r_{ds2} - r_{ds1})I_L)s}{L} \\ \dfrac{U_{in} + (r_{ds2} - r_{ds1})I_L}{LC} \end{bmatrix}}{s^2 + s\dfrac{r_L + Dr_{ds1} + D'r_{ds2} + r_C}{L} + \dfrac{1}{LC}} \quad (3.31)$$

where
G_{iL-o} = transfer function from input voltage to inductor current
G_{oL-o} = transfer function from output current to inductor current
G_{iC-o} = transfer function from input voltage to output capacitor voltage
G_{oC-o} = transfer function from output current to output capacitor voltage
G_{cL} = transfer function from control to inductor current
G_{cC} = transfer function from control to inductor current.

The control-to-inductor-current transfer function (G_{cL}) of the set of the input-to-state transfer functions (3.31) may be used to construct the dynamical models for the current-mode controls introduced in Chapters 4 and 5.

The input-to-output description (i.e., $\mathbf{C}[s\mathbf{I} - \mathbf{A}]^{-1}\mathbf{B} + \mathbf{D}$, Chapter 2, Section 2.2) can be found to be

$$\begin{bmatrix} Y_{in-o} & T_{oi-o} \\ G_{io-o} & -Z_{o-o} \end{bmatrix} = \frac{\begin{bmatrix} \dfrac{D^2 s}{L} & \dfrac{D(1 + sr_C C)}{LC} \\ \dfrac{D(1 + sr_C C)}{LC} & -\dfrac{(r_L + Dr_{ds1} + D'r_{ds2} + sL)(1 + sr_C C)}{LC} \end{bmatrix}}{s^2 + s\dfrac{r_L + Dr_{ds1} + D'r_{ds2} + r_C}{L} + \dfrac{1}{LC}}$$

$$\begin{bmatrix} G_{ci} \\ G_{co} \end{bmatrix} = \frac{\begin{bmatrix} \dfrac{D(U_{in} + (r_{ds2} - r_{ds1})I_L)s}{L} \\ \dfrac{(U_{in} + (r_{ds2} - r_{ds1})I_L)(1 + sr_C C)}{LC} \end{bmatrix}}{s^2 + s\dfrac{r_L + Dr_{ds1} + D'r_{ds2} + r_C}{L} + \dfrac{1}{LC}} + \begin{bmatrix} I_L \\ 0 \end{bmatrix} \quad (3.32)$$

The steady-state values of the variables can be found setting the derivatives $\frac{di_L}{dt}$ and $\frac{du_C}{dt}$ in (3.24) to zero. This procedure yields

$$I_L = I_o$$
$$I_{in} = DI_o$$
$$U_o = U_C \quad (3.33)$$
$$U_o = DU_{in} - (r_L + Dr_{ds1} + D'r_{ds2})I_o$$
$$D = \frac{U_o + (r_L + r_{ds2})I_o}{U_{in} + (r_{ds2} - r_{ds1})I_o}$$

The transfer functions in (3.32) would represent the open-loop internal dynamics of a synchronous buck converter at a certain operating point defined in (3.33). The operating-point equations (3.33) show that the ESR of the output capacitor does not affect the operating point. It may be obvious that the internal dynamics does not change if the operating point is kept constant. The parasitic resistances (r_{ds1} and r_{ds2}) are dependent on the level of output current and input voltage as well as the gate-driver implementation and the dead time applied to the switching control, which would affect the damping factor of the resonant behavior as discussed in [23].

The ideal input admittance ($Y_{in-\infty} = Y_{in-o} - G_{io-o}G_{ci}/G_{co}$) has a significant role in the source interactions, and therefore, it has to be known: it can be given by

$$Y_{in-\infty} = -\frac{DI_L}{U_{in} + U_D + (r_{ds2} - r_{ds1})I_L} \quad (3.34)$$

Similarly, the short-circuit input admittance ($Y_{in-sc} = Y_{in-o} + G_{io-o}T_{oi-o}/Z_{o-o}$) has a significant role in the source interactions, and therefore, it has to be known also: It can be given by

$$Y_{in-sc} = \frac{D^2}{r_L + Dr_{ds1} + D'r_{ds2} + sL} \quad (3.35)$$

3.4.2
Dynamic Descriptions of Buck, Boost, and Buck–Boost Converters

3.4.2.1 Diode-Switched Buck (Figure 3.6a)

$$\frac{d\hat{i}_L}{dt} = -\frac{(r_L + Dr_{ds1} + D'r_d + r_C)}{L}\hat{i}_L - \frac{1}{L}\hat{u}_C + \frac{D}{L}\hat{u}_{in} + \frac{r_C}{L}\hat{i}_o$$
$$+ \frac{U_{in} + U_D + (r_d - r_{ds1})I_L}{L}\hat{d}$$
$$\frac{d\hat{u}_C}{dt} = \frac{\hat{i}_L}{C} - \frac{\hat{i}_o}{C}$$

$$\hat{i}_{\text{in}} = D\hat{i}_L + I_L\hat{d}$$
$$\hat{u}_o = \left(1 + r_C C \frac{d}{dt}\right)\hat{u}_C \qquad (3.36)$$

$$I_L = I_o$$
$$I_{\text{in}} = DI_o$$
$$U_o = U_C \qquad (3.37)$$
$$U_o = DU_{\text{in}} - D'U_D - (r_L + Dr_{\text{ds1}} + D'r_d)I_o$$
$$D = \frac{U_o + U_D + (r_L + r_d)I_o}{U_{\text{in}} + U_D + (r_d - r_{\text{ds1}})I_o}$$

$$\begin{bmatrix} Y_{\text{in}-o} & T_{\text{oi}-o} \\ \\ G_{\text{io}-o} & -Z_{o-o} \end{bmatrix} = \frac{\begin{bmatrix} \dfrac{D^2 s}{L} & \dfrac{D(1+sr_C C)}{LC} \\ \dfrac{D(1+sr_C C)}{LC} & -\dfrac{(r_L + Dr_{\text{ds1}} + D'r_d + sL)(1+sr_C C)}{LC} \end{bmatrix}}{s^2 + s\dfrac{r_L + Dr_{\text{ds1}} + D'r_d + r_C}{L} + \dfrac{1}{LC}}$$

$$\begin{bmatrix} G_{\text{ci}} \\ G_{\text{co}} \end{bmatrix} = \frac{\begin{bmatrix} \dfrac{D(U_{\text{in}} + U_D + (r_d - r_{\text{ds1}})I_L)s}{L} \\ \dfrac{(U_{\text{in}} + U_D + (r_d - r_{\text{ds1}})I_L)(1+sr_C C)}{LC} \end{bmatrix}}{s^2 + s\dfrac{r_L + Dr_{\text{ds1}} + D'r_d + r_C}{L} + \dfrac{1}{LC}} + \begin{bmatrix} I_L \\ 0 \end{bmatrix}$$

$$G_{cL} = \frac{\dfrac{(U_{\text{in}} + U_D + (r_d - r_{\text{ds1}})I_L)s}{L}}{s^2 + s\dfrac{r_L + Dr_{\text{ds1}} + D'r_d + r_C}{L} + \dfrac{1}{LC}} \qquad (3.38)$$

$$Y_{\text{in}-\infty} = -\frac{DI_L}{U_{\text{in}} + U_D + (r_d - r_{\text{ds1}})I_L} \qquad (3.39)$$
$$Y_{\text{in}-\text{sc}} = \frac{D^2}{r_L + Dr_{\text{ds1}} + D'r_d + sL}$$

3.4.2.2 Diode-Switched Boost (Figure 3.8a)

$$\frac{d\hat{i}_L}{dt} = -\frac{(r_L + Dr_{\text{ds1}} + D'r_d + D'r_C)}{L}\cdot\hat{i}_L - \frac{D'}{L}\cdot\hat{u}_C + \frac{1}{L}\cdot\hat{u}_{\text{in}} + \frac{D'r_C}{L}\cdot\hat{i}_o$$
$$+ \frac{U_o + U_D + (r_d + Dr_C - r_{\text{ds1}})I_L}{L}\cdot\hat{d}$$

$$\frac{d\hat{u}_C}{dt} = \frac{D'}{C} \cdot \hat{i}_L - \frac{1}{C} \cdot \hat{i}_o - \frac{I_L}{C} \cdot \hat{d}$$

$$\hat{i}_{in} = \hat{i}_L \qquad (3.40)$$

$$\hat{u}_o = \left(1 + r_C C \frac{d}{dt}\right) \cdot \hat{u}_C$$

$$I_L = \frac{I_o}{D'}$$

$$I_{in} = \frac{I_o}{D'}$$

$$U_o = U_C \qquad (3.41)$$

$$U_o = \frac{U_{in}}{D'} - U_D - \frac{(r_L + Dr_{ds1} + D'r_d + DD'r_C)}{D'^2} \cdot I_o$$

$$(U_o + U_D - r_C I_o)D'^2 - (U_{in} - (r_d - r_{ds1} + r_C)I_o)D'$$
$$+ (r_L + r_{ds1})I_o = 0$$

$$\begin{bmatrix} Y_{in-o} & T_{oi-o} \\ G_{io-o} & -Z_{o-o} \end{bmatrix} = \frac{\begin{bmatrix} \dfrac{s}{L} & \dfrac{D'(1 + sr_C C)}{LC} \\ \dfrac{D'(1 + sr_C C)}{LC} & -\dfrac{(r_L + Dr_{ds1} + D'r_d + DD'r_C + sL)(1 + sr_C C)}{LC} \end{bmatrix}}{s^2 + s\dfrac{r_L + Dr_{ds1} + D'(r_d + r_C)}{L} + \dfrac{D'^2}{LC}}$$

$$\begin{bmatrix} G_{ci} \\ G_{co} \end{bmatrix} = \frac{\begin{bmatrix} \dfrac{D'I_L}{LC}\left(1 + s\left(\dfrac{U_o + U_D}{D'I_L} + \dfrac{r_d + Dr_C - r_{ds1}}{D'}\right)C\right) \\ \dfrac{(D'(U_o + U_D) - (r_L + r_{ds1} + D'^2 r_C)I_L - sLI_L)(1 + sr_C C)}{LC} \end{bmatrix}}{s^2 + s\dfrac{r_L + Dr_{ds1} + D'(r_d + r_C)}{L} + \dfrac{D'^2}{LC}}$$

$$G_{cL} = \frac{\dfrac{D'I_L}{LC}\left(1 + s\left(\dfrac{U_o + U_D}{D'I_L} + \dfrac{r_d + Dr_C - r_{ds1}}{D'}\right)C\right)}{s^2 + s\dfrac{r_L + Dr_{ds1} + D'(r_d + r_C)}{L} + \dfrac{D'^2}{LC}} \qquad (3.42)$$

$$Y_{in-\infty} = -\frac{I_L}{D'(U_o + U_D) - (r_L + r_{ds1} + D'^2 r_C)I_L - sLI_L}$$

$$Y_{in-sc} = \frac{1}{r_L + Dr_{ds1} + D'r_d + DD'r_C + sL} \qquad (3.43)$$

$$\cdot \hat{u}_C$$

$$\underline{ - r_{ds1})I_L} \cdot \hat{d}$$

(3.44)

$$U_o = U_C$$ (3.45)

$$U_o = \frac{U_{in}}{D'} - \frac{(r_L + Dr_{ds1} + D'r_{ds2} + DD'r_C)}{D'^2} \cdot I_o$$

$$(U_o - r_C I_o)D'^2 - (U_{in} - (r_{ds2} - r_{ds1} + r_C)I_o)D' + (r_L + r_{ds1})I_o = 0$$

$$\begin{bmatrix} Y_{in-o} & T_{oi-o} \\ G_{io-o} & -Z_{o-o} \end{bmatrix} = \frac{\begin{bmatrix} \dfrac{s}{L} & \dfrac{D'(1 + sr_C C)}{LC} \\ \dfrac{D'(1 + sr_C C)}{LC} & -\dfrac{(r_L + Dr_{ds1} + D'r_{ds2} + DD'r_C + sL)(1 + sr_C C)}{LC} \end{bmatrix}}{s^2 + s\dfrac{r_L + Dr_{ds1} + D'(r_{ds2} + r_C)}{L} + \dfrac{D'^2}{LC}}$$

$$\begin{bmatrix} G_{ci} \\ G_{co} \end{bmatrix} = \frac{\begin{bmatrix} \dfrac{D'I_L}{LC}\left(1 + s\left(\dfrac{U_o}{D'I_L} + \dfrac{r_{ds2} + Dr_C - r_{ds1}}{D'}\right)C\right) \\ \dfrac{(D'U_o - (r_L + r_{ds1} + D'^2 r_C)I_L - sLI_L)(1 + sr_C C)}{LC} \end{bmatrix}}{s^2 + s\dfrac{r_L + Dr_{ds1} + D'(r_{ds2} + r_C)}{L} + \dfrac{D'^2}{LC}}$$

$$G_{cL} = \frac{\dfrac{D'I_L}{LC}\left(1 + s\left(\dfrac{U_o}{D'I_L} + \dfrac{r_{ds2} + Dr_C - r_{ds1}}{D'}\right)C\right)}{s^2 + s\dfrac{r_L + Dr_{ds1} + D'(r_{ds2} + r_C)}{L} + \dfrac{D'^2}{LC}}$$ (3.46)

$$Y_{\text{in}-\infty} = -\frac{I_L}{D'U_o - (r_L + r_{ds1} + D'^2 r_C)I_L - sLI_L}$$

$$Y_{\text{in}-sc} = \frac{1}{r_L + Dr_{ds1} + D'r_{ds2} + DD'r_C + sL} \quad (3.47)$$

3.4.2.4 Diode-Switched Buck–Boost (Figure 3.10a)

$$\frac{d\hat{i}_L}{dt} = -\frac{(r_L + Dr_{ds1} + D'r_d + D'r_C)}{L} \cdot \hat{i}_L - \frac{D'}{L} \cdot \hat{u}_C + \frac{D}{L} \cdot \hat{u}_{\text{in}}$$
$$\quad + \frac{D'r_C}{L} \cdot \hat{i}_o + \frac{U_{\text{in}} + U_o + U_D + (r_d + Dr_C - r_{ds1})I_L}{L} \cdot \hat{d}$$

$$\frac{d\hat{u}_C}{dt} = \frac{D'}{C} \cdot \hat{i}_L - \frac{1}{C} \cdot \hat{i}_o - \frac{I_L}{C} \cdot \hat{d} \quad (3.48)$$

$$\hat{i}_{\text{in}} = D \cdot \hat{i}_L + I_L \cdot \hat{d}$$

$$\hat{u}_o = \left(1 + r_C C \frac{d}{dt}\right) \cdot \hat{u}_C$$

$$I_L = \frac{I_o}{D'}$$

$$I_{\text{in}} = \frac{D}{D'} I_o$$

$$U_o = U_C \quad (3.49)$$

$$U_o = \frac{DU_{\text{in}}}{D'} - U_D - \frac{(r_L + Dr_{ds1} + D'r_d + DD'r_C)}{D'^2} \cdot I_o$$

$$(U_{\text{in}} + U_o + U_D - r_C I_o)D'^2 - (U_{\text{in}} - (r_d - r_{ds1} + r_C)I_o)D'$$
$$+ (r_L + r_{ds1})I_o = 0$$

$$\begin{bmatrix} Y_{\text{in}-o} & T_{oi-o} \\ G_{io-o} & -Z_{o-o} \end{bmatrix} = \frac{\begin{bmatrix} \dfrac{D^2 s}{L} & \dfrac{DD'(1 + sr_C C)}{LC} \\ \dfrac{DD'(1 + sr_C C)}{LC} & -\dfrac{(r_L + Dr_{ds1} + D'r_d + DD'r_C + sL)(1 + sr_C C)}{LC} \end{bmatrix}}{s^2 + s\dfrac{r_L + Dr_{ds1} + D'(r_d + r_C)}{L} + \dfrac{D'^2}{LC}}$$

$$\begin{bmatrix} G_{ci} \\ G_{co} \end{bmatrix} = \frac{\begin{bmatrix} \dfrac{DD'I_L}{LC}\left(1 + s\left(\dfrac{U_{\text{in}} + U_o + U_D}{D'I_L} + \dfrac{r_d + Dr_C - r_{ds1}}{D'}\right)C\right) \\ \dfrac{(D'(U_{\text{in}} + U_o + U_D) - (r_L + r_{ds1} + D'^2 r_C)I_L - sLI_L)(1 + sr_C C)}{LC} \end{bmatrix}}{s^2 + s\dfrac{r_L + Dr_{ds1} + D'(r_d + r_C)}{L} + \dfrac{D'^2}{LC}} + \begin{bmatrix} I_L \\ 0 \end{bmatrix}$$

$$G_{cL} = \cfrac{\cfrac{D'I_L}{LC}\left(1 + s\left(\cfrac{U_{in} + U_o + U_D}{D'I_L} + \cfrac{r_d + Dr_C - r_{ds1}}{D'}\right)\right)}{s^2 + s\cfrac{r_L + Dr_{ds1} + D'(r_d + r_C)}{L} + \cfrac{D'^2}{LC}}$$

(3.50)

$$Y_{in-\infty} = -\cfrac{DI_L}{D'(U_{in} + U_o + U_D) - (r_L + r_{ds1} + D'^2 r_C)I_L - sLI_L}$$

$$Y_{in-sc} = \cfrac{D^2}{r_L + Dr_{ds1} + D'r_d + DD'r_C + sL}$$

(3.51)

3.4.2.5 Synchronous Buck–Boost (Figure 3.10b)

$$\begin{aligned}
\frac{d\hat{i}_L}{dt} &= -\frac{(r_L + Dr_{ds1} + D'r_{ds2} + D'r_C)}{L}\cdot\hat{i}_L - \frac{D'}{L}\cdot\hat{u}_C + \frac{D}{L}\cdot\hat{u}_{in} \\
&\quad + \frac{D'r_C}{L}\cdot\hat{i}_o + \frac{U_{in} + U_o + (r_{ds2} + Dr_C - r_{ds1})I_L}{L}\cdot\hat{d} \\
\frac{d\hat{u}_C}{dt} &= \frac{D'}{C}\cdot\hat{i}_L - \frac{1}{C}\cdot\hat{i}_o - \frac{I_L}{C}\cdot\hat{d} \\
\hat{i}_{in} &= D\cdot\hat{i}_L + I_L\cdot\hat{d} \\
\hat{u}_o &= \left(1 + r_C C\frac{d}{dt}\right)\cdot\hat{u}_C
\end{aligned}$$

(3.52)

$$\begin{aligned}
I_L &= \frac{I_o}{D'} \\
I_{in} &= \frac{D}{D'}I_o \\
U_o &= U_C \\
U_o &= \frac{DU_{in}}{D'} - \frac{(r_L + Dr_{ds1} + D'r_{ds2} + DD'r_C)}{D'^2}\cdot I_o \\
(U_{in} &+ U_o - r_C I_o)D'^2 - (U_{in} - (r_{ds2} - r_{ds1} + r_C)I_o)D' \\
&+ (r_L + r_{ds1})I_o = 0
\end{aligned}$$

(3.53)

$$\begin{bmatrix} Y_{in-o} & T_{oi-o} \\ G_{io-o} & -Z_{o-o} \end{bmatrix} = \cfrac{\begin{bmatrix} \cfrac{D^2 s}{L} & \cfrac{DD'(1 + sr_C C)}{LC} \\ \cfrac{DD'(1 + sr_C C)}{LC} & -\cfrac{(r_L + Dr_{ds1} + D'r_{ds2} + DD'r_C + sL)(1 + sr_C C)}{LC} \end{bmatrix}}{s^2 + s\cfrac{r_L + Dr_{ds1} + D'(r_{ds2} + r_C)}{L} + \cfrac{D'^2}{LC}}$$

$$\begin{bmatrix} G_{ci} \\ G_{co} \end{bmatrix} = \frac{\begin{bmatrix} \dfrac{DD'I_L}{LC}\left(1+s\left(\dfrac{U_{in}+U_o}{D'I_L}+\dfrac{r_{ds2}+Dr_C-r_{ds1}}{D'}\right)C\right) \\ \dfrac{(D'(U_{in}+U_o)-(r_L+r_{ds1}+D'^2r_C)I_L-sLI_L)(1+sr_CC)}{LC} \end{bmatrix}}{s^2+s\dfrac{r_L+Dr_{ds1}+D'(r_{ds2}+r_C)}{L}+\dfrac{D'^2}{LC}} + \begin{bmatrix} I_L \\ 0 \end{bmatrix}$$

$$G_{cL} = \frac{\dfrac{D'I_L}{LC}\left(1+s\left(\dfrac{U_{in}+U_o}{D'I_L}+\dfrac{r_{ds2}+Dr_C-r_{ds1}}{D'}\right)\right)}{s^2+s\dfrac{r_L+Dr_{ds1}+D'(r_{ds2}+r_C)}{L}+\dfrac{D'^2}{LC}} \tag{3.54}$$

$$Y_{in-\infty} = -\frac{DI_L}{D'(U_{in}+U_o)-(r_L+r_{ds1}+D'^2r_C)I_L-sLI_L}$$

$$Y_{in-sc} = \frac{D^2}{r_L+Dr_{ds1}+D'r_{ds2}+DD'r_C+sL} \tag{3.55}$$

The operating point equations (3.41), (3.45), (3.49), and (3.53) show that the ESR of the output capacitor affects the steady-state operation of the boost and buck–boost converters: The effect is reflected as a series resistance (Dr_C/D') at the output of the converter, which increases the required duty ratio for a given output voltage and has to be considered in the design.

3.4.3
Steady-State and Small-Signal Equivalent Circuits

The steady-sate operation point of the converter can be solved using an equivalent circuit shown in Figure 3.14 in CCM, when the source and/or load conditions are changed. The parameter values are given in Table 3.1 for the buck, boost, and buck–boost converters treated in Sections 3.4.1 and 3.4.2 based on (3.33), (3.37), (3.41), (3.45), (3.49) and (3.53).

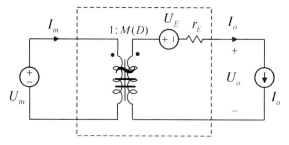

Figure 3.14 Steady-state equivalent circuit.

Table 3.1 Steady-state equivalent-circuit parameters for buck, boost and buck–boost converters.

Converter	M(D)	U_E	r_E
Buck	D		
Diode		$D'U_D$	$r_L + Dr_{ds1} + D'r_d$
Synch		0	$r_L + Dr_{ds1} + D'r_{ds2}$
Boost	$\dfrac{1}{D'}$		
Diode		U_D	$\dfrac{r_L}{D'^2} + \dfrac{Dr_{ds1}}{D'^2} + \dfrac{r_d}{D'} + \dfrac{Dr_C}{D'}$
Synch		0	$\dfrac{r_L}{D'^2} + \dfrac{Dr_{ds1}}{D'^2} + \dfrac{r_{ds2}}{D'} + \dfrac{Dr_C}{D'}$
Buck–boost	$-\dfrac{D}{D'}$		
Diode		U_D	$\dfrac{r_L}{D'^2} + \dfrac{Dr_{ds1}}{D'^2} + \dfrac{r_d}{D'} + \dfrac{Dr_C}{D'}$
Synch		0	$\dfrac{r_L}{D'^2} + \dfrac{Dr_{ds1}}{D'^2} + \dfrac{r_{ds2}}{D'} + \dfrac{Dr_C}{D'}$

The small-signal dynamical behavior of the basic second-order converters can be given as an equivalent circuit shown in Figure 3.15 derived originally in [3, 4]. The equivalent circuit presented in [3, 4] is not correctly constructed, because the effect of the load resistor is added in the parameters defining the internal model. The equivalent-circuit parameters for the basic converters treated in Sections 3.4.1 and 3.4.2 are given in Tables 3.2 and 3.3.

The equivalent circuit in Figure 3.15 can be utilized to develop a general set of transfer functions for the corresponding converters applying the circuit theory. This procedure yields

$$\begin{bmatrix} Y_{\text{in}-o} & T_{oi-o} \\ G_{io-o} & -Z_{o-o} \end{bmatrix} = \dfrac{\begin{bmatrix} \dfrac{M(D)^2 s}{L_e} & \dfrac{M(D)(1+sr_CC)}{L_eC} \\ \dfrac{M(D)(1+sr_CC)}{L_eC} & -\dfrac{(r_e+sL_e)(1+sr_CC)}{L_eC} \end{bmatrix}}{s^2 + s\dfrac{r_e+r_C}{L_e} + \dfrac{1}{L_eC}}$$

$$\begin{bmatrix} G_{ci} \\ G_{co} \end{bmatrix} = \dfrac{\begin{bmatrix} \dfrac{M(D)^2 \hat{e}(s) s}{L_e} \\ \dfrac{M(D)\hat{e}(s)(1+sr_CC)}{L_eC} \end{bmatrix}}{s^2 + s\dfrac{r_e+r_C}{L_e} + \dfrac{1}{L_eC}} + \begin{bmatrix} \hat{j}(s) \\ 0 \end{bmatrix}$$

(3.56)

The negative sign of s in the voltage-source-type parameter ($\hat{e}(s)$, Table 3.3) reflects the existence of a RHP zero in the control-to-output transfer function

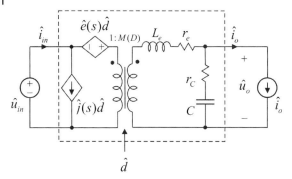

Figure 3.15 Small-signal equivalent circuit.

Table 3.2 Small-signal-equivalent-circuit output-side parameters for buck, boost and buck–boost converters.

Converter	M(D)	L_e	r_e
Buck	D	L	
Diode			$r_L + Dr_{ds1} + D'r_d$
Synch			$r_L + Dr_{ds1} + D'r_{ds2}$
Boost	$\dfrac{1}{D'}$	$\dfrac{1}{D'^2}$	
Diode			$\dfrac{r_L}{D'^2} + \dfrac{Dr_{ds1}}{D'^2} + \dfrac{r_d}{D'} + \dfrac{Dr_C}{D'}$
Synch			$\dfrac{r_L}{D'^2} + \dfrac{Dr_{ds1}}{D'^2} + \dfrac{r_{ds2}}{D'} + \dfrac{Dr_C}{D'}$
Buck–boost	$-\dfrac{D}{D'}$	$\dfrac{L}{D'^2}$	
Diode			$\dfrac{r_L}{D'^2} + \dfrac{Dr_{ds1}}{D'^2} + \dfrac{r_d}{D'} + \dfrac{Dr_C}{D'}$
Synch			$\dfrac{r_L}{D'^2} + \dfrac{Dr_{ds1}}{D'^2} + \dfrac{r_{ds2}}{D'} + \dfrac{Dr_C}{D'}$

(G_{co}) and consequently, a nonminimum-phase system with the associated control-bandwidth limitation as discussed in Chapter 2 (Section 2.6.3). The peak values of the open-loop output impedance and the input admittance can be given in general form for the basic converters by

$$|Z_{o-o}|_{\max} = \frac{R_{oe}^2 \sqrt{1 + \dfrac{r_e^2}{R_{oe}^2}} \cdot \sqrt{1 + \dfrac{r_C^2}{R_{oe}^2}}}{r_e + r_C}$$

$$|Y_{in-o}|_{\max} = \frac{M(D)^2}{r_e + sL_e} \quad (3.57)$$

Table 3.3 Small-signal-equivalent-circuit output-side parameters for buck, boost and buck–boost converters.

Converter	$\hat{e}(s)$	$\hat{j}(s)$
Buck		I_L
Diode	$\dfrac{U_m + U_D + (r_d - r_{ds1})I_L}{D}$	
Synch	$\dfrac{U_{in} + (r_{ds2} - r_{ds1})I_L}{D}$	
Boost		$\dfrac{I_L}{D'}$
Diode	$U_o + U_D - (r_L + r_{ds1} + D'^2 r_C)\dfrac{I_L}{D'} - s\dfrac{LI_L}{D'}$	
Synch	$U_o - (r_L + r_{ds1} + D'^2 r_C)\dfrac{I_L}{D'} - s\dfrac{LI_L}{D'}$	
Buck–boost		$\dfrac{I_L}{D'}$
Diode	$\dfrac{U_m + U_o + U_D - (r_L + r_{ds1} + D'^2 r_C)\dfrac{I_L}{D'} - s\dfrac{LI_L}{D'}}{D}$	
Synch	$\dfrac{U_m + U_o - (r_L + r_{ds1} + D'^2 r_C)\dfrac{I_L}{D'} - s\dfrac{LI_L}{D'}}{D}$	

where $R_{oe} = \sqrt{\dfrac{L_e}{C}}$ is the equivalent characteristic impedance of the averaging filter. When the source effects on the converter dynamics in Chapter 2 (Section 2.4) were introduced, the ideal input admittance ($Y_{in-\infty}$) and the short-circuit input admittance (Y_{in-sc}) were defined as

$$Y_{in-\infty} = Y_{in-o} - \dfrac{G_{io-o} G_{ci}}{G_{co}}$$
$$Y_{in-sc} = Y_{in-o} + \dfrac{G_{io-o} T_{oi-o}}{Z_{o-o}} \qquad (3.58)$$

As a consequence, the admittances can be given for the basic converters by

$$Y_{in-\infty} = -\dfrac{\hat{j}(s)}{\hat{e}(s)}$$
$$Y_{in-sc} = \dfrac{M(D)^2}{r_e + sL_e} \qquad (3.59)$$

3.5
Fixed-Frequency Operation in DCM

The diode-switched converters (i.e., Figures. 3.6a, 3.8a, and 3.10a) will enter into DCM when the inductor current tends to become negative: the diode in the

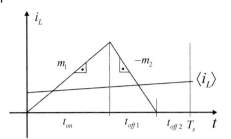

Figure 3.16 Inductor-current waveforms in DCM.

current-carrying path will become reverse biased, which effectively disconnects the path. As a consequence, the converters will have three different subcircuit structures during the switching cycle: one for the on-time, one for the off-time1, and one for the off-time2 (Figure 3.16). The inductor-current waveforms in DCM are shown in Figure 3.16. The time-averaged inductor current ($\langle i_L \rangle$) is naturally continuous within the switching cycle and less than half the peak inductor current.

When considering the general state-space equations we defined in Section 3.3, the only unknown variable is the length of the off-time1 (t_{off1}). The dynamics associated with t_{off1} can be recovered by computing its relation to $\langle i_L \rangle$[8] according to the waveforms of Figure 3.16, which gives

$$\langle i_L \rangle = \frac{1}{t_s} \int_0^{t_s} i_L(t) dt = \frac{1}{2t_s} \cdot m_1 t_{on}(t_{on} + t_{off1}) \tag{3.60}$$

and which under the fixed-frequency operation equals

$$\langle i_L \rangle = \frac{1}{2} \cdot m_1 d(d + d_1) T_s \tag{3.61}$$

According to (3.61), we find that

$$d_1 = \frac{2\langle i_L \rangle}{d m_1 T_s} - d \tag{3.62}$$

In order to find the fixed-frequency averaged models in DCM, t_{off1} or d_1 has to be replaced with (3.62) in the general averaged state-space equations given in Section 3.3. We treat more in detail the diode-switched buck converter in Section 3.5.1 and give the corresponding averaged state spaces and transfer functions for the boost and buck–boost converters without going into the details in Section 3.5.2.

3.5.1
Buck Converter

The set of equations defining the averaged state space for a buck converter in DCM can be given as follows (Section 3.3.1):

$$\frac{d\langle i_L \rangle}{dt} = dm_1 - d_1 m_2$$

$$\frac{d\langle u_C \rangle}{dt} = \frac{\langle i_L \rangle}{C} - \frac{\langle i_o \rangle}{C}$$

$$\langle i_{\text{in}} \rangle = \frac{d}{d + d_1} \cdot \langle i_L \rangle$$

$$\langle u_o \rangle = \langle u_C \rangle + r_C C \frac{d\langle u_C \rangle}{dt} \quad (3.63)$$

$$m_1 = \frac{\langle u_{\text{in}} \rangle - (r_L + r_{\text{ds1}} + r_C)\langle i_L \rangle - \langle u_C \rangle + r_C \langle i_o \rangle}{L}$$

$$m_2 = \frac{(r_L + r_d + r_C)\langle i_L \rangle + \langle u_C \rangle - r_C \langle i_o \rangle + U_D}{L}$$

$$d_1 = \frac{2\langle i_L \rangle}{d T_s m_1} - d$$

Substituting m_1 and m_2 in (3.63) with their defined formulas in (3.11) and (3.12) yields

$$\frac{d\langle i_L \rangle}{dt} = \frac{d(\langle u_{\text{in}} \rangle + (r_d - r_{\text{ds1}})\langle i_L \rangle + U_D)}{L}$$

$$- \frac{2\langle i_L \rangle}{dT_s} \cdot \frac{(r_L + r_d + r_C)\langle i_L \rangle + \langle u_C \rangle - r_C \langle i_o \rangle + U_D}{\langle u_{\text{in}} \rangle - (r_L + r_{\text{ds1}} + r_C)\langle i_L \rangle - \langle u_C \rangle + r_C \langle i_o \rangle}$$

$$\frac{d\langle u_C \rangle}{dt} = \frac{\langle i_L \rangle}{C} - \frac{\langle i_o \rangle}{C} \quad (3.64)$$

$$\langle i_{\text{in}} \rangle = \frac{d^2 T_s}{2L}(\langle u_{\text{in}} \rangle - (r_L + r_{\text{ds1}} + r_C)\langle i_L \rangle - \langle u_C \rangle + r_C \langle i_o \rangle)$$

$$\langle u_o \rangle = \langle u_C \rangle + r_C C \frac{d\langle u_C \rangle}{dt}$$

which is extremely nonlinear but can be naturally linearized using the methods introduced in Chapter 2 (Section 2.6.1). The parasitic losses do not, however, significantly affect the dynamical behavior except the zero from the output-capacitor ESR (r_C), and therefore, the parasitics may be neglected [11]: the DCM operation would create in small-signal sense a lossless resistor in series with the other resistive losses at the output, which is much greater than the other parasitic resistive losses together. As a consequence, the simplified averaged state space can be given by

$$\frac{d\langle i_L\rangle}{dt} = \frac{d\langle u_{\text{in}}\rangle}{L} - \frac{2\langle i_L\rangle\langle u_C\rangle}{dT_s(\langle u_{\text{in}}\rangle - \langle u_C\rangle)}$$

$$\frac{d\langle u_C\rangle}{dt} = \frac{\langle i_L\rangle}{C} - \frac{\langle i_o\rangle}{C}$$

$$\langle i_{\text{in}}\rangle = \frac{d^2 T_s(\langle u_{\text{in}}\rangle - \langle u_C\rangle)}{2L} \qquad (3.65)$$

$$\langle u_o\rangle = \langle u_C\rangle + r_C C\frac{d\langle u_C\rangle}{dt}$$

The parasitic elements may, however, affect the steady-state operating point, which can be solved from (3.64) by setting the derivatives to zero. The input–output relation (i.e., $M(D, K)$ in [1]) cannot be defined directly with the parasitic losses included. Usually the input and output voltages as well as the output current are known in the specifications, and we have to solve only the duty ratio D. As a consequence, the operation point can be given as follows:

$$I_L = I_o$$

$$I_{\text{in}} = \frac{((r_L + r_d)I_o + U_o + U_D)}{(U_{\text{in}} + (r_d - r_{\text{ds1}})I_o + U_D)} \cdot I_o$$

$$U_o = U_C$$

$$D = \sqrt{\frac{2LI_o}{T_s} \cdot \frac{(r_L + r_d)I_o + U_o + U_D}{(U_{\text{in}} - (r_L + r_{\text{ds1}})I_o - U_o)(U_{\text{in}} + (r_d - r_{\text{ds1}})I_o + U_D)}}.$$

(3.66)

The commonly available equations [1] describing the steady-state and small-signal behavior includes two notations M and K in order to simplify the final equations: the steady-state input–output relation (U_o/U_{in}) equals M, and the dimensionless value $K = \frac{2L}{T_s R_{\text{eq}}}$, where $R_{\text{eq}} = U_o/I_o$. (Note: commonly R_{eq} is denoted using R because of assuming resistive load. The same relation is valid, however, with arbitrary loads, and therefore, we use the general notation R_{eq}, which would also suite well for describing the internal dynamics.)

Applying the above-defined notations and (3.66), the operating point can be given by

$$I_L = I_o$$

$$I_{\text{in}} = \frac{\left(1 + \frac{(r_L + r_d)}{R_{\text{eq}}} + \frac{U_D}{U_o}\right)}{\left(1 + \frac{(r_d - r_{\text{ds1}})}{R_{\text{eq}}} M + \frac{U_D}{U_{\text{in}}}\right)} \cdot MI_o$$

$$U_o = U_C$$

$$D = M\sqrt{K \cdot \frac{1 + \frac{(r_L + r_d)}{R_{eq}} + \frac{U_D}{U_o}}{\left(1 - \left(1 + \frac{(r_L + r_{ds1})}{R_{eq}}\right)M\right)\left(1 + \frac{(r_d - r_{ds1})}{R_{eq}}M + \frac{U_D}{U_{in}}\right)}}$$

$$D_1 = \frac{KM}{D\left(1 - M - \frac{r_L + r_{ds1}}{R_{eq}}\right)} - D.$$

(3.67)

According to (3.67), it may be obvious that the losses do not significantly affect the operating point. Therefore, it may be justified to neglect them, which yields the commonly given formulation [1] as follows:

$$\begin{aligned}
I_L &= I_o \\
I_{in} &= M I_o \\
U_o &= U_C \\
M &= \frac{2}{1 + \sqrt{1 + \frac{4K}{D^2}}} \\
D &= M\sqrt{\frac{K}{1 - M}} \\
D_1 &= \sqrt{K(1 - M)}
\end{aligned}$$

(3.68)

We derive the small-signal state space without the losses based on the simplified averaged state space given in (3.65). The linearization yields

$$\begin{aligned}
\frac{d\hat{i}_L}{dt} &= -\frac{2U_C}{DT_s(U_{in} - U_C)} \cdot \hat{i}_L - \frac{2U_{in}I_L}{DT_s(U_{in} - U_C)^2} \cdot \hat{u}_C \\
&\quad + \left(\frac{D}{L} + \frac{2U_C I_L}{DT_s(U_{in} - U_C)^2}\right) \cdot \hat{u}_{in} + \left(\frac{U_{in}}{L} + \frac{2U_C I_L}{D^2 T_s(U_{in} - U_C)}\right) \cdot \hat{d} \\
\frac{d\hat{u}_C}{dt} &= \frac{\hat{i}_L}{C} - \frac{\hat{i}_o}{C} \\
\hat{i}_{in} &= -\frac{D^2 T_s}{2L} \cdot \hat{u}_C + \frac{D^2 T_s}{2L} \cdot \hat{u}_{in} + \frac{DT_s(U_{in} - U_C)}{2L} \cdot \hat{d} \\
\hat{u}_o &= \hat{u}_C + r_C C \frac{d\hat{u}_C}{dt}
\end{aligned}$$

(3.69)

Applying the operation-point definitions (3.68) and the notations M and K yields

$$\frac{d\hat{i}_L}{dt} = -\frac{R_{eq}}{L}\sqrt{\frac{K}{1-M}} \cdot \hat{i}_L - \frac{1}{L(1-M)}\sqrt{\frac{K}{1-M}} \cdot \hat{u}_C$$
$$+ \frac{(2-M)M}{L(1-M)}\sqrt{\frac{K}{1-M}} \cdot \hat{u}_{in} + \frac{2U_{in}}{L} \cdot \hat{d}$$
$$\frac{d\hat{u}_C}{dt} = \frac{\hat{i}_L}{C} - \frac{\hat{i}_o}{C} \qquad (3.70)$$
$$\hat{i}_{in} = -\frac{M^2}{R_{eq}(1-M)} \cdot \hat{u}_C + \frac{M^2}{R_{eq}(1-M)} \cdot \hat{u}_{in} + \frac{2U_o}{R_{eq}}\sqrt{\frac{1-M}{K}} \cdot \hat{d}$$
$$\hat{u}_o = \hat{u}_C + r_C C \frac{d\hat{u}_C}{dt}$$

The set of equations in (3.70) can be given in matrix form as follows:

$$\begin{bmatrix} \frac{d\hat{i}_L}{dt} \\ \frac{d\hat{u}_C}{dt} \end{bmatrix} = \begin{bmatrix} -\frac{R_{eq}}{L}\sqrt{\frac{K}{1-M}} & -\frac{1}{L(1-M)}\sqrt{\frac{K}{1-M}} \\ \frac{1}{C} & 0 \end{bmatrix} \begin{bmatrix} \hat{i}_L \\ \hat{u}_C \end{bmatrix}$$
$$+ \begin{bmatrix} \frac{(2-M)M}{L(1-M)}\sqrt{\frac{K}{1-M}} & 0 & \frac{2U_{in}}{L} \\ 0 & -\frac{1}{C} & 0 \end{bmatrix} \begin{bmatrix} \hat{u}_{in} \\ \hat{i}_o \\ \hat{d} \end{bmatrix}$$
$$\begin{bmatrix} \hat{i}_{in} \\ \hat{u}_o \end{bmatrix} = \begin{bmatrix} 0 & -\frac{M^2}{R_{eq}(1-M)} \\ 0 & 1 + r_C C \frac{d}{dt} \end{bmatrix} \begin{bmatrix} \hat{i}_L \\ \hat{u}_C \end{bmatrix} \qquad (3.71)$$
$$+ \begin{bmatrix} \frac{M^2}{R_{eq}(1-M)} & 0 & \frac{2U_o}{R_{eq}}\sqrt{\frac{1-M}{K}} \\ 0 & 0 & 0 \end{bmatrix} \begin{bmatrix} \hat{u}_{in} \\ \hat{i}_o \\ \hat{d} \end{bmatrix}$$

Applying *Laplace* transform to (3.71) yields

$$\begin{bmatrix} \hat{i}_L(s) \\ \hat{u}_C(s) \end{bmatrix} = \begin{bmatrix} s + \frac{R_{eq}}{L}\sqrt{\frac{K}{1-M}} & \frac{1}{L(1-M)}\sqrt{\frac{K}{1-M}} \\ -\frac{1}{C} & s \end{bmatrix}^{-1}$$
$$\times \begin{bmatrix} \frac{(2-M)M}{L(1-M)}\sqrt{\frac{K}{1-M}} & 0 & \frac{2U_{in}}{L} \\ 0 & -\frac{1}{C} & 0 \end{bmatrix} \begin{bmatrix} \hat{u}_{in}(s) \\ \hat{i}_o(s) \\ \hat{d}(s) \end{bmatrix}$$

3.5 Fixed-Frequency Operation in DCM

$$\begin{bmatrix} \hat{i}_{in}(s) \\ \hat{u}_o(s) \end{bmatrix} = \begin{bmatrix} 0 & -\dfrac{M^2}{R_{eq}(1-M)} \\ 0 & 1+sr_CC \end{bmatrix} \begin{bmatrix} \hat{i}_L(s) \\ \hat{u}_C(s) \end{bmatrix}$$

$$+ \begin{bmatrix} \dfrac{M^2}{R_{eq}(1-M)} & 0 & \dfrac{2U_o}{R_{eq}}\sqrt{\dfrac{1-M}{K}} \\ 0 & 0 & 0 \end{bmatrix} \begin{bmatrix} \hat{u}_{in}(s) \\ \hat{i}_o(s) \\ \hat{d}(s) \end{bmatrix} \quad (3.72)$$

The determinant of the matrix inverse (i.e., $[\ ]^{-1}$) determines the characteristic polynomial as

$$\Delta = s^2 + s\cdot\frac{R_{eq}}{L}\sqrt{\frac{K}{1-M}} + \frac{1}{LC}\cdot\sqrt{\frac{K}{1-M}}\cdot\frac{1}{1-M} \quad (3.73)$$

The transfer functions from the input variables to the state variables (i.e., $[s\mathbf{I}-\mathbf{A}]^{-1}\mathbf{B}$, Chapter 2, Section 2.2) can be computed to be

$$\begin{bmatrix} G_{iL-o} & G_{oL-o} \\ G_{iC-o} & -G_{oC-o} \end{bmatrix} = \frac{\begin{bmatrix} \dfrac{M(2-M)}{L(1-M)}\sqrt{\dfrac{K}{1-M}}\cdot s & \dfrac{1}{LC(1-M)}\sqrt{\dfrac{K}{1-M}} \\ \dfrac{M(2-M)}{LC(1-M)}\sqrt{\dfrac{K}{1-M}} & -\dfrac{1}{LC}\left(R_{eq}\sqrt{\dfrac{K}{1-M}}+sL\right) \end{bmatrix}}{s^2+s\cdot\dfrac{R_{eq}}{L}\sqrt{\dfrac{K}{1-M}}+\dfrac{1}{LC}\cdot\sqrt{\dfrac{K}{1-M}}\cdot\dfrac{1}{1-M}}$$

$$\begin{bmatrix} G_{cL} \\ G_{cC} \end{bmatrix} = \frac{\begin{bmatrix} \dfrac{2U_{in}\cdot s}{L} \\ \dfrac{2U_{in}}{LC} \end{bmatrix}}{s^2+s\cdot\dfrac{R_{eq}}{L}\sqrt{\dfrac{K}{1-M}}+\dfrac{1}{LC}\cdot\sqrt{\dfrac{K}{1-M}}\cdot\dfrac{1}{1-M}}$$

(3.74)

The input-to-output description (i.e., $\mathbf{C}[s\mathbf{I}-\mathbf{A}]^{-1}\mathbf{B}+\mathbf{D}$, Chapter 2, Section 2.2) can be found to be

$$\begin{bmatrix} Y_{in-o} & T_{oi-o} \\ G_{io-o} & -Z_{o-o} \end{bmatrix}$$

$$= \frac{\begin{bmatrix} -\dfrac{M^3(2-M)}{LCR_{eq}(1-M)^2}\sqrt{\dfrac{K}{1-M}} & \dfrac{M^2}{CR_{eq}(1-M)}\cdot\left(s+\dfrac{R_{eq}}{L}\sqrt{\dfrac{K}{1-M}}\right) \\ \dfrac{M(2-M)}{LC(1-M)}\sqrt{\dfrac{K}{1-M}}\cdot(1+sr_CC) & -\dfrac{\left(sL+R_{eq}\sqrt{\dfrac{K}{1-M}}\right)(1+sr_CC)}{LC} \end{bmatrix}}{s^2+s\cdot\dfrac{R_{eq}}{L}\sqrt{\dfrac{K}{1-M}}+\dfrac{1}{LC}\cdot\sqrt{\dfrac{K}{1-M}}\cdot\dfrac{1}{1-M}}$$

$$+\begin{bmatrix} \dfrac{M^2}{R_{eq}(1-M)} & 0 \\ 0 & 0 \end{bmatrix}$$

$$\begin{bmatrix} G_{ci} \\ G_{co} \end{bmatrix} = \dfrac{\begin{bmatrix} \dfrac{2U_{in}M^2}{LCR_{eq}(1-M)} \\ \dfrac{2U_{in}(1+sr_CC)}{LC} \end{bmatrix}}{s^2 + s \cdot \dfrac{R_{eq}}{L}\sqrt{\dfrac{K}{1-M}} + \dfrac{1}{LC} \cdot \sqrt{\dfrac{K}{1-M}} \cdot \dfrac{1}{1-M}}$$

$$+\begin{bmatrix} \dfrac{2U_o}{R_{eq}}\sqrt{\dfrac{1-M}{K}} \\ 0 \end{bmatrix} \tag{3.75}$$

The transfer functions in (3.75) would represent the open-loop internal dynamics of the buck converter at a certain operating point defined in (3.68). According to the operating point definition, the ERS (r_C) of the output capacitor does not affect it. The lossless resistor providing the damping can be found to be $R_{eq}\sqrt{\dfrac{K}{1-M}}$ according to the output impedance (Z_{o-o}) shown in (3.75). The lossless resistor is so large that the roots of the characteristic polynomial (3.73) are typically real and well separated from each other .

3.5.2
Dynamic Models for Boost and Buck–Boost Converters

The averaged sate space with parasitic elements, the operation point as well as the transfer functions defining the internal dynamics for the boost and buck–boost converters will be provided in this subsection.

3.5.2.1 Boost Converter (Figure 3.8a)
Full-order averaged state space:

$$\dfrac{d\langle i_L\rangle}{dt} = \dfrac{d((r_d + r_C - r_{ds1})\langle i_L\rangle + \langle u_C\rangle - r_C\langle i_o\rangle + U_D)}{L}$$

$$- \dfrac{2\langle i_L\rangle}{dT_s} \cdot \dfrac{(r_L + r_d + r_C)\langle i_L\rangle + \langle u_C\rangle - r_C\langle i_o\rangle + U_D - \langle u_{in}\rangle}{\langle u_{in}\rangle - (r_L + r_{ds1})\langle i_L\rangle}$$

$$\dfrac{d\langle u_C\rangle}{dt} = \dfrac{\langle i_L\rangle}{C} - \dfrac{d^2T_s}{2LC}(\langle u_{in}\rangle - (r_L + r_{ds1})\langle i_L\rangle) - \dfrac{\langle i_o\rangle}{C} \tag{3.76}$$

$$\langle i_{in}\rangle = \langle i_L\rangle$$

$$\langle u_o\rangle = \langle u_C\rangle + r_CC\dfrac{d\langle u_C\rangle}{dt}$$

3.5 Fixed-Frequency Operation in DCM

Simplified state space:

$$\frac{d\langle i_L\rangle}{dt} = \frac{d\langle u_C\rangle}{L} - \frac{2\langle i_L\rangle(\langle u_C\rangle - \langle u_{in}\rangle)}{dT_s\langle u_{in}\rangle}$$

$$\frac{d\langle u_C\rangle}{dt} = \frac{\langle i_L\rangle}{C} - \frac{d^2 T_s\langle u_{in}\rangle}{2LC} - \frac{\langle i_o\rangle}{C} \quad (3.77)$$

$$\langle i_{in}\rangle = \langle i_L\rangle$$

$$\langle u_o\rangle = \langle u_C\rangle + r_C C \frac{d\langle u_C\rangle}{dt}$$

Simplified operating point:

$$I_L = MI_o$$

$$I_{in} = MI_o$$

$$U_o = U_C \quad (3.78)$$

$$M = \frac{1 + \sqrt{1 + \frac{4D^2}{K}}}{2}$$

$$D = \sqrt{KM(M-1)}$$

$$D_1 = \frac{1}{M-1}$$

Dynamic description:

$$\begin{bmatrix} Y_{in-o} & T_{oi-o} \\ G_{io-o} & -Z_{o-o} \end{bmatrix}$$

$$= \frac{\begin{bmatrix} \frac{M^2}{L}\sqrt{\frac{KM}{M-1}}\left(s + \frac{M-1}{MR_{eq}C}\right) & \frac{1}{LC}\sqrt{\frac{KM}{M-1}} \\ \left(\frac{2M-1}{LCM}\sqrt{\frac{KM}{M-1}} - s \cdot \frac{M(M-1)}{R_{eq}C}\right) \cdot (1+sr_CC) & -\frac{\left(sL + R_{eq}\sqrt{\frac{K(M-1)}{M}}\right)(1+sr_CC)}{LC} \end{bmatrix}}{s^2 + s \cdot \frac{R_{eq}}{L}\sqrt{\frac{K(M-1)}{M}} + \frac{1}{LC}\cdot\sqrt{\frac{KM}{M-1}}}$$

$$\begin{bmatrix} G_{ci} \\ G_{co} \end{bmatrix} = \frac{\begin{bmatrix} \frac{2U_o}{L}\left(s + \frac{1}{R_{eq}C}\right) \\ \frac{2U_{in}\left(1 - s\cdot\frac{L}{R_{eq}}\sqrt{\frac{M(M-1)}{K}}\right)(1+sr_CC)}{LC} \end{bmatrix}}{s^2 + s\cdot\frac{R_{eq}}{L}\sqrt{\frac{K(M-1)}{M}} + \frac{1}{LC}\cdot\sqrt{\frac{KM}{M-1}}}$$

$$G_{cL} = \cfrac{\dfrac{2U_o}{L}\left(s + \dfrac{1}{R_{eq}C}\right)}{s^2 + s \cdot \dfrac{R_{eq}}{L}\sqrt{\dfrac{K(M-1)}{M}} + \dfrac{1}{LC} \cdot \sqrt{\dfrac{KM}{M-1}}} \qquad (3.79)$$

The transfer functions in (3.79) would represent the open-loop internal dynamics of the boost converter at a certain operating point defined in (3.78). The lossless resistor providing the damping can be found to be $R_{eq}\sqrt{\dfrac{K(M-1)}{M}}$ according to the output impedance (Z_{o-o}) shown in (3.79). The lossless resistor is so large that the roots of the characteristic polynomial (i.e., the denominator in (3.79)) are typically real and well separated from each other.

3.5.2.2 Buck–Boost Converter (Figure 3.10a)

Full-order averaged state space:

$$\begin{aligned}
\dfrac{d\langle i_L\rangle}{dt} &= \dfrac{d((r_d + r_C - r_{ds1})\langle i_L\rangle + \langle u_C\rangle - r_C\langle i_o\rangle + U_D + \langle u_{in}\rangle)}{L} \\
&\quad - \dfrac{2\langle i_L\rangle}{dT_s} \cdot \dfrac{(r_L + r_d + r_C)\langle i_L\rangle + \langle u_C\rangle - r_C\langle i_o\rangle + U_D}{\langle u_{in}\rangle - (r_L + r_{ds1})\langle i_L\rangle} \\
\dfrac{d\langle u_C\rangle}{dt} &= \dfrac{\langle i_L\rangle}{C} - \dfrac{d^2 T_s}{2LC}(\langle u_{in}\rangle - (r_L + r_{ds1})\langle i_L\rangle) - \dfrac{\langle i_o\rangle}{C} \\
\langle i_{in}\rangle &= \dfrac{d^2 T_s}{2L}(\langle u_{in}\rangle - (r_L + r_{ds1})\langle i_L\rangle) \\
\langle u_o\rangle &= \langle u_C\rangle + r_C C \dfrac{d\langle u_C\rangle}{dt}
\end{aligned} \qquad (3.80)$$

Simplified averaged state space:

$$\begin{aligned}
\dfrac{d\langle i_L\rangle}{dt} &= \dfrac{d(\langle u_C\rangle + \langle u_{in}\rangle)}{L} - \dfrac{2\langle i_L\rangle\langle u_C\rangle}{dT_s\langle u_{in}\rangle} \\
\dfrac{d\langle u_C\rangle}{dt} &= \dfrac{\langle i_L\rangle}{C} - \dfrac{d^2 T_s\langle u_{in}\rangle}{2LC} - \dfrac{\langle i_o\rangle}{C} \\
\langle i_{in}\rangle &= \dfrac{d^2 T_s\langle u_{in}\rangle}{2L} \\
\langle u_o\rangle &= \langle u_C\rangle + r_C C \dfrac{d\langle u_C\rangle}{dt}
\end{aligned} \qquad (3.81)$$

Simplified operating point:

$$I_L = (1 + M)I_o$$
$$I_{in} = MI_o$$
$$U_o = U_C$$

$$M = \frac{D}{\sqrt{K}}$$

$$D = M\sqrt{K}$$

$$D_1 = \sqrt{K} \tag{3.82}$$

Dynamic description:

$$\begin{bmatrix} Y_{\text{in}-o} & T_{oi-o} \\ G_{io-o} & -Z_{o-o} \end{bmatrix} = \frac{\begin{bmatrix} \frac{2M\sqrt{K}}{LC}\left(1 - s \cdot \frac{LM}{2R_{eq}C\sqrt{K}}\right) \cdot (1 + sr_C C) & -\frac{(sL + R_{eq}\sqrt{K})(1 + sr_C C)}{LC} \\ 0 & 0 \end{bmatrix}}{s^2 + s \cdot \frac{R_{eq}\sqrt{K}}{L} + \frac{\sqrt{K}}{LC}} + \begin{bmatrix} \frac{M^2}{R_{eq}} & 0 \\ 0 & 0 \end{bmatrix}$$

$$\begin{bmatrix} G_{ci} \\ G_{co} \end{bmatrix} = \frac{\begin{bmatrix} \frac{2U_{\text{in}}}{LC}\left(1 - s \cdot \frac{LM}{R_{eq}\sqrt{K}}\right)(1 + sr_C C) \\ 0 \end{bmatrix}}{s^2 + s \cdot \frac{R_{eq}\sqrt{K}}{L} + \frac{\sqrt{K}}{LC}} + \begin{bmatrix} \frac{2U_o}{R_{eq}\sqrt{K}} \\ 0 \end{bmatrix}$$

$$G_{cL} = \frac{\frac{2U_o}{LC}\left(\frac{1}{R_{eq}} + s \cdot \frac{C}{M}\right)}{s^2 + s \cdot \frac{R_{eq}\sqrt{K}}{L} + \frac{\sqrt{K}}{LC}} \tag{3.83}$$

The transfer functions in (3.83) would represent the open-loop internal dynamics of the buck converter at a certain operating point defined in (3.82). The lossless resistor providing the damping can be found to equal $R_{eq}\sqrt{K}$ according to the output impedance (Z_{o-o}) shown in (3.83). The lossless resistor is so large that the roots of the characteristic polynomial (i.e., the denominator in (3.83)) are typically real and well separated from each other.

3.6 Dynamic Review

The buck and boost converters are dynamically reviewed. We compare the dynamic changes when the converters are designed to operate in CCM or DCM. Experimental frequency-response data are provided for a buck converter in both CCM and DCM to support the theoretical findings and to prove the accuracy of the modeling approach.

3.6.1
Buck Converter

The dynamic profile of the buck converter shown in Figure 3.17 [27] is analyzed in CCM and DCM at the output current of 2.5 A, and the input voltage of 20 V and 50 V, respectively. The power stage is same in both of the operation modes except the size of the inductor (i.e., 105 µH versus 5 µH). The experimental frequency responses have been measured using Venable Industries' frequency-response analyzer Model 3120 with an impedance measurement kit. The measurement data have been imported into Matlab™ for efficient figure handling.

The duty ratio (D) giving the desired output voltage (i.e., 10 V) can be computed according to (3.37) and (3.66) as follows: CCM, 0.53 (20 V), 0.21 (50); and DCM, 0.38 (20 V) and 0.116 (50 V). The CCM values are very accurate, and the DCM values within 2% of the exact values according to the simulation by means of the switching models. The value of K in DCM equals 0.25.

3.6.1.1 Control-to-Output Transfer Function

The control-to-output transfer functions are shown symbolically in (3.84). The magnitude of the DCM transfer function has near two times higher dependence on the input voltage than does the magnitude of the CCM transfer function has. It may be obvious that the controller should be designed for the maximum loop crossover frequency at the high input voltage in order to avoid the high-frequency ripple at the output voltage to cause instability [17]:

$$G_{co}^{CCM} = \frac{\frac{(U_{in} + U_D + (r_d - r_{ds1})I_L)(+sr_CC)}{L}}{s^2 + s\frac{r_L + Dr_{ds1} + D'r_d + r_C}{L} + \frac{1}{LC}}$$

$$G_{co}^{DCM} = \frac{\frac{2U_{in}(1 + sr_CC)}{LC}}{s^2 + s \cdot \frac{R_{eq}}{L}\sqrt{\frac{K}{1-M}} + \frac{1}{LC} \cdot \sqrt{\frac{K}{1-M}} \cdot \frac{1}{1-M}} \quad (3.84)$$

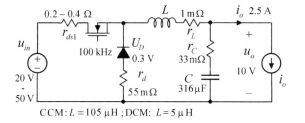

CCM: $L = 105\,\mu H$; DCM: $L = 5\,\mu H$

Figure 3.17 Experimental buck converter.

3.6 Dynamic Review

The numerical values of the transfer functions at 20 V are given by

$$G_{co}^{CCM} = \frac{s \cdot 6.2 \times 10^3 + 5.93 \times 10^8}{s^2 + s \cdot 2.08 \times 10^3 + 3 \times 10^7}$$
$$G_{co}^{DCM} = \frac{s \cdot 2.64 \times 10^5 + 2.53 \times 10^{10}}{s^2 + s \cdot 5.66 \times 10^5 + 8.95 \times 10^8}$$
(3.85)

and at 50 V by

$$G_{co}^{CCM} = \frac{s \cdot 1.6 \times 10^4 + 1.5 \times 10^9}{s^2 + s \cdot 1.33 \times 10^3 + 3 \times 10^7}$$
$$G_{co}^{DCM} = \frac{s \cdot 6.6 \times 10^5 + 6.33 \times 10^{10}}{s^2 + s \cdot 4.47 \times 10^5 + 4.42 \times 10^8}$$
(3.86)

The damping factor (ζ) (Chapter 2, Section 2.6.2) in the CCM converter varies from 0.19 to 0.12 and in the DCM converter from 9.5 to 10.6, where the first value corresponds to 20 V and the last to 50 V. This means that the CCM converter exhibits resonant behavior and, the DCM converter has real and well-separated roots of the characteristic polynomial (i.e., the denominator in (3.85) and (3.86)). The resonant frequency of the CCM converter is close to 872 Hz. The low-frequency pole of the DCM converter can be approximated to be close to $f_{low} \approx \frac{1}{2\pi R_{eq} C(1-M)}$, and the high-frequency pole close to $f_{high} \approx \frac{R_{eq}}{2\pi L} \sqrt{\frac{K}{1-M}}$. The high-frequency pole locates typically at the frequencies higher than the switching frequency yielding effectively first-order transfer functions.

The frequency responses of the transfer functions are shown in Figures 3.18a (CCM) and 3.18b (DCM) confirming the observations. The phase behavior of the CCM converter (Figure 3.18a) requires using a proportional-integral-derivative or PID-type compensator in order to ensure stability due to the resonant behavior. The phase behavior of the DCM converter depicts that a proportional-integral or PI-type compensator can be used. The flat high-frequency magnitude behavior (Figure 3.18b) may indicate difficulties to design proper control system for large input–output voltage ratios and high control bandwidth.

The predicted and measured CCM and DCM frequency responses are shown in Figure 3.19. The transfer functions also include the modulator gain ($G_a = 1/3$), and therefore, their magnitudes are about 10 dB lower than the corresponding magnitudes in Figure 3.18. The predictions and the measurements comply well with each other. In CCM, the losses are higher than those expected at the high input voltage yielding inaccuracy in the prediction. The high-frequency phase is clearly lower than the analytical predictions yield. The predictions could be corrected placing a zero at the switching frequency. The cause for the extra phase lag could be the modulation process [18–22] or simply a stray capacitor in the PWM modulator, which was a peak-current-mode modulator containing a Zener diode limiting the maximum instantaneous current signal approximately to 1 V.

98 | *3 Average and Small-Signal Modeling of Direct-On-Time Controlled Converters*

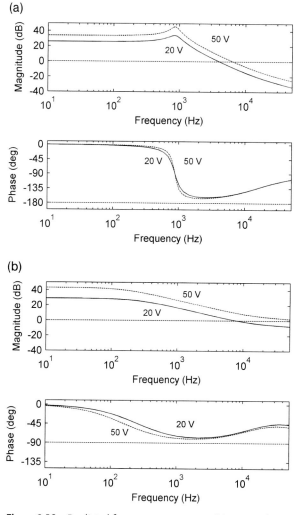

Figure 3.18 Predicted frequency responses of the control-to-output transfer functions: (a) CCM converter and (b) DCM converter.

3.6.1.2 Output Impedance

The open-loop output impedances are shown symbolically in (3.87) in CCM and DCM. The low-frequency value of the CCM output impedance equals $r_L + Dr_{ds1} + D'r_d$, and that of the DCM output impedance to $R_{eq}\sqrt{\frac{K}{1-M}}$. The high-frequency output impedance corresponds to r_C for both of the converters:

$$Z_{o-o}^{CCM} = \frac{\dfrac{(r_L + Dr_{ds1} + D'r_d + sL)(1 + sr_C C)}{LC}}{s^2 + s\dfrac{r_L + Dr_{ds1} + D'r_d + r_C}{L} + \dfrac{1}{LC}}$$

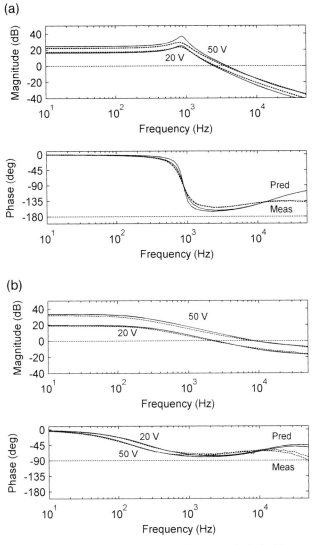

Figure 3.19 Predicted (solid lines) and measured (dashed lines) frequency responses of the control-to-output transfer functions including the modulator gain (i.e., 1/3) at the input voltages of 20 and 50 V: (a) CCM, and (b) DCM.

$$Z_{o-o}^{DCM} = \frac{\left(sL + R_{eq}\sqrt{\frac{K}{1-M}}\right)(1+sr_C C)}{LC} \Bigg/ \left(s^2 + s \cdot \frac{R_{eq}}{L}\sqrt{\frac{K}{1-M}} + \frac{1}{LC} \cdot \sqrt{\frac{K}{1-M}} \cdot \frac{1}{1-M}\right) \quad (3.87)$$

The CCM output impedance has the highest value close to the resonant frequency ($f_o = \frac{1}{2\pi\sqrt{LC}}$), which can be approximated by

$$|Z_{o-o}^{CCM}|_{max} = \frac{R_o^2 \sqrt{1 + \frac{r_e^2}{R_o^2}} \sqrt{1 + \frac{r_C^2}{R_o^2}}}{r_e + r_C} \approx \frac{R_o^2}{r_e + r_C} \quad (3.88)$$

where $r_e = r_L + Dr_{ds1} + D'r_d$ and $R_o = \sqrt{\frac{L}{C}}$ is the characteristic impedance of an LC circuit.

The numerical values of the output impedances are given at 20 V by

$$Z_{o-o}^{CCM} = \frac{s^2 \cdot 3.3 \times 10^{-2} + s \cdot 3.22 \times 10^3 + 5.6 \times 10^6}{s^2 + s \cdot 2.08 \times 10^3 + 3 \times 10^7}$$

$$Z_{o-o}^{DCM} = \frac{s^2 \cdot 3.3 \times 10^{-2} + s \cdot 2.18 \times 10^4 + 1.79 \times 10^9}{s^2 + s \cdot 5.66 \times 10^5 + 8.95 \times 10^8} \quad (3.89)$$

and at 50 V by

$$Z_{o-o}^{CCM} = \frac{s^2 \cdot 3.3 \times 10^{-2} + s \cdot 3.2 \times 10^3 + 3.23 \times 10^6}{s^2 + s \cdot 1.33 \times 10^3 + 3 \times 10^7}$$

$$Z_{o-o}^{DCM} = \frac{s^2 \cdot 3.3 \times 10^{-2} + s \cdot 1.8 \times 10^4 + 1.42 \times 10^9}{s^2 + s \cdot 4.47 \times 10^5 + 4.42 \times 10^8} \quad (3.90)$$

The predicted and measured frequency responses of the CCM output impedances are shown in Figure 3.20a. The measured output impedance does not change when the input voltage is changed, due to the increase in the switching losses in the high-side switch. The measured impedances are shown in the middle of the corresponding predicted values. The predicted (solid line) and measured (dashed line) frequency responses of the DCM output impedances are shown in Figure 3.20b. The first-order nature is obvious. The analytical predictions and the measurements have good compliance.

3.6.1.3 Input-to-Output Transfer Function

The open-loop input-to-output transfer functions are shown symbolically in (3.91),

$$G_{io-o}^{CCM} = \frac{\frac{D(1 + sr_C C)}{LC}}{s^2 + s\frac{r_L + Dr_{ds1} + D'r_d + r_C}{L} + \frac{1}{LC}}$$

$$G_{io-o}^{DCM} = \frac{\frac{M(2-M)}{LC(1-M)}\sqrt{\frac{K}{1-M}} \cdot (1 + sr_C C)}{s^2 + s \cdot \frac{R_{eq}}{L}\sqrt{\frac{K}{1-M}} + \frac{1}{LC} \cdot \sqrt{\frac{K}{1-M}} \cdot \frac{1}{1-M}} \quad (3.91)$$

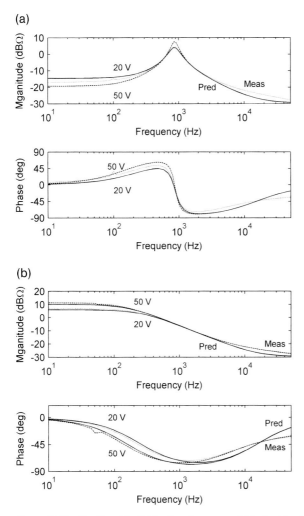

Figure 3.20 Predicted (solid line) and measured (dashed line) frequency responses of the output impedances at the input voltages of 20 and 50 V: (a) CCM and (b) DCM.

and the numerical values are given at 20 V by

$$G_{\text{io}-o}^{\text{CCM}} = \frac{s \cdot 167 + 1.6 \times 10^7}{s^2 + s \cdot 2.08 \times 10^3 + 3 \times 10^7}$$
$$G_{\text{io}-o}^{\text{DCM}} = \frac{s \cdot 7 \times 10^3 + 6.7 \times 10^8}{s^2 + s \cdot 5.66 \times 10^5 + 8.95 \times 10^8} \quad (3.92)$$

and at 50 V by

$$G_{\text{io}-o}^{\text{CCM}} = \frac{s \cdot 66 + 6.3 \times 10^6}{s^2 + s \cdot 1.33 \times 10^3 + 3 \times 10^7}$$

$$G_{io-o}^{DCM} = \frac{s \cdot 2 \times 10^3 + 1.9 \times 10^8}{s^2 + s \cdot 4.47 \times 10^5 + 4.42 \times 10^8} \qquad (3.93)$$

The predicted (solid line) and measured (dashed line) frequency responses of the CCM input-to-output transfer functions are shown in Figure 3.21a. The value of the duty ratio determines the low-frequency attenuation, which is poorest at the resonant frequency due to resonant peaking. The predicted (solid line) and measured (dashed line) frequency responses of the DCM input-to-output transfer functions are shown in Figure 3.21b. The low-frequency

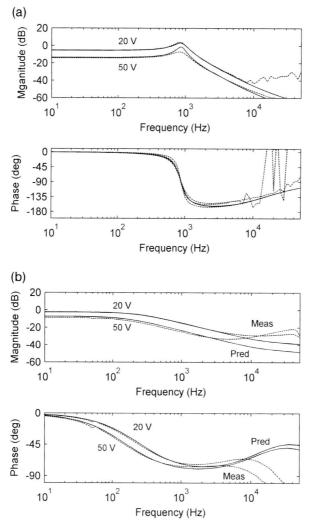

Figure 3.21 Measured and predicted frequency responses of the input-to-output transfer functions at the input voltages of 20 and 50 V: (a) CCM, and (b) DCM.

attenuation is lightly less than the corresponding attenuation in CCM. The reason for the high modeling inaccuracy at the high frequencies is the measurement problems and parasitic resonance.

3.6.1.4 Input Admittance
The open-loop input admittances are given symbolically by

$$Y_{in-o}^{CCM} = \frac{\dfrac{D^2 s}{L}}{s^2 + s\dfrac{r_L + Dr_{ds1} + D'r_d + r_C}{L} + \dfrac{1}{LC}}$$

$$Y_{in-o}^{DCM} = \frac{\dfrac{M^2}{R_{eq}(1-M)}\left(s^2 + s \cdot \dfrac{R_{eq}}{L}\sqrt{\dfrac{K}{1-M}} + \dfrac{1-M}{LC} \cdot \sqrt{\dfrac{K}{1-M}}\right)}{s^2 + s \cdot \dfrac{R_{eq}}{L}\sqrt{\dfrac{K}{1-M}} + \dfrac{1}{LC} \cdot \sqrt{\dfrac{K}{1-M}} \cdot \dfrac{1}{1-M}}.$$

(3.94)

The corresponding numerical values are given at 20 V by

$$Y_{in-o}^{CCM} = \frac{s \cdot 2.68 \times 10^3}{s^2 + s \cdot 2.08 \times 10^3 + 3 \times 10^7}$$

$$Y_{in-o}^{DCM} = 0.125 \cdot \frac{s^2 + s \cdot 5.66 \times 10^5 + 2.24 \times 10^8}{s^2 + s \cdot 5.66 \times 10^5 + 8.95 \times 10^8}$$

(3.95)

and at 50 V by

$$Y_{in-o}^{CCM} = \frac{s \cdot 4.2 \times 10^2}{s^2 + s \cdot 1.33 \times 10^3 + 3 \times 10^7}$$

$$Z_{o-o}^{DCM} = 0.0125 \cdot \frac{s^2 + s \cdot 4.47 \times 10^5 + 2.83 \times 10^8}{s^2 + s \cdot 4.47 \times 10^5 + 4.42 \times 10^8}$$

(3.96)

The predicted (solid line) and measured (dashed line) frequency responses of the CCM input admittances are shown in Figure 3.22a. The modeling accuracy is obvious. The reason for the high-frequency inaccuracy is the measurement problems and the parasitic resonance. The corresponding input impedances can be found by changing the sign of the magnitude and phase. The peak value of the admittance corresponds to

$$|Y_{in-o}^{CCM}|_{max} = \frac{D^2}{r_e + r_C}$$

(3.97)

where $r_e = r_L + Dr_{ds1} + D'r_d$. The resonant dip value of the corresponding input impedance is naturally the inverse of (3.97). The predicted (solid line) and measured (dashed line) frequency responses of the DCM input admittance

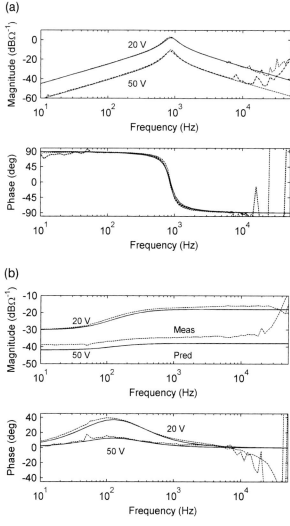

Figure 3.22 Predicted (solid line) and measured (dashed line) frequency responses of the open-loop input admittance at the input voltages of 20 and 50 V: (a) CCM and (b) DCM.

are shown in Figure 3.22b. The reason for the modeling inaccuracy is the measurement problems at low signal values.

3.6.1.5 Ideal Input Admittance
The ideal input admittances are symbolically given by

$$Y_{\text{in}-\infty}^{\text{CCM}} = -\frac{DI_L}{U_{\text{in}} + U_D + (r_d - r_{\text{ds}})I_L} \approx -\frac{I_{\text{in}}}{U_{\text{in}}}$$

$$Y_{\text{in}-\infty}^{\text{DCM}} = -\frac{M^2}{R_{\text{eq}}} \approx -\frac{I_{\text{in}}}{U_{\text{in}}} \tag{3.98}$$

The corresponding numerical values are $-6.25 \times 10^{-2}\,\Omega^{-1}$ at 20 V and $-10^{-2}\,\Omega^{-1}$ at 50 V. The predicted (solid line) and the measured (i.e., computed according to the measured data based on $Y_{\text{in}-\infty} = Y_{\text{in}-o} - \frac{G_{\text{io}-o}G_{\text{ci}}}{G_{\text{co}}}$) (dashed line) frequency responses of the CCM ideal admittances are shown in Figure 3.23a and those of the DCM admittances in Figure 3.23b. The corresponding

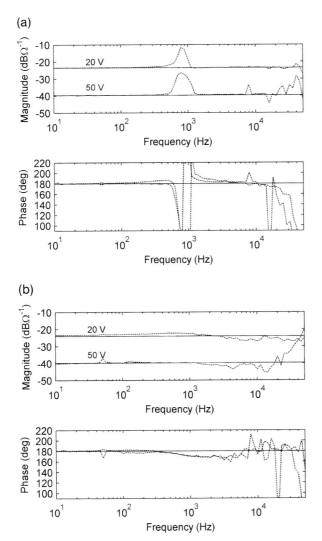

Figure 3.23 Predicted (solid line) and computed (dashed line) frequency responses of the ideal admittances at the input voltages of 20 and 50 V: (a) CCM and (b) DCM.

106 | *3 Average and Small-Signal Modeling of Direct-On-Time Controlled Converters*

impedances values can be obtained by changing the sign of the magnitude and phase. Figure 3.23 proves that the ideal admittance is a topology-based constant independent of operation mode but naturally dependent on the level of output power.

3.6.1.6 Short-Circuit Input Admittance

The short-circuit input admittances are given symbolically in (3.99) and the corresponding numerical values are given at 20 V in (3.100) and at 50 V in (3.101). The corresponding impedances are naturally inverses of the given admittances.

$$Y_{in-sc}^{CCM} = \frac{D^2}{sL + r_L + Dr_{ds1} + D'r_d}$$

$$Y_{in-sc}^{DCM} = \frac{M^2}{R_{eq}(1-M)} \tag{3.99}$$

$$Y_{in-sc}^{CCM} = \frac{0.28}{s \cdot 1.05 \times 10^{-4} + 0.186}$$

$$Y_{in-sc}^{DCM} = 0.125 \tag{3.100}$$

$$Y_{in-sc}^{CCM} = \frac{0.044}{s \cdot 1.05 \times 10^{-4} + 0.1075}$$

$$Y_{in-sc}^{DCM} = 0.0125 \tag{3.101}$$

The corresponding frequency responses in CCM are shown in Figure 3.24a, and those in DCM in Figure 3.24b. The experimental-data-based responses (i.e., those computed according to $Y_{in-sc} = Y_{in-o} + \frac{G_{io-o}T_{oi-o}}{Z_{o-o}}$) are shown with the dashed lines. The predictions and measurements comply well with each other.

3.6.2 Boost Converter

The dynamical profile of the boost converter shown in Figure 3.25 is analyzed in both CCM and DCM at the output current of 1.5 A, and the input voltage of 20 and 50 V. The power stage is the same in both of the operation modes except the size of the inductor (i.e., 350 μH versus 9 μH).

The duty ratio (D) giving the desired output voltage (i.e., 75 V) can be computed according to (3.41) in CCM to be 0.748 at 20 V and 0.339 at 50 V. The value of K in DCM equals 0.036, and $R_{eq} = 50\ \Omega$. The corresponding duty-ratio values (3.78) are 0.66 and 0.17.

3.6 Dynamic Review

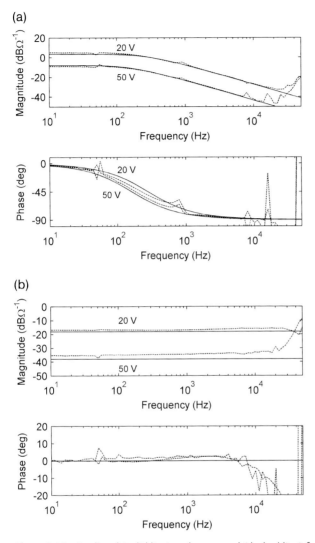

Figure 3.24 Predicted (solid line) and computed (dashed line) frequency responses of the short-circuit admittances at the input voltages of 20 and 50 V: (a) CCM and (b) DCM.

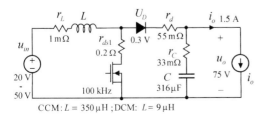

CCM: $L = 350\,\mu H$; DCM: $L = 9\,\mu H$

Figure 3.25 A boost converter.

3.6.2.1 Control-to-Output Transfer Function

The control-to-output transfer functions are shown symbolically in (3.102). The magnitude of the DCM transfer function has close to two times higher dependence on the input voltage than does the magnitude of the CCM transfer function:

$$G_{co}^{CCM} = \frac{\dfrac{(D'(U_o + U_D) - (r_L + r_{ds1} + D'^2 r_C)I_L - sLI_L)(1 + sr_C C)}{LC}}{s^2 + s\dfrac{r_L + Dr_{ds1} + D'(r_d + r_C)}{L} + \dfrac{D'^2}{LC}}$$

$$G_{co}^{DCM} = \frac{\dfrac{2U_{in}\left(1 - s\cdot\dfrac{L}{R_{eq}}\sqrt{\dfrac{M(M-1)}{K}}\right)(1 + sr_C C)}{LC}}{s^2 + s\cdot\dfrac{R_{eq}}{L}\sqrt{\dfrac{K(M-1)}{M}} + \dfrac{1}{LC}\cdot\sqrt{\dfrac{KM}{M-1}}}$$

(3.102)

According to (3.102), the RHP zero in CCM is approximately U_{in}/LI_L, which is closest to the origin when the input voltage is at minimum and the output current at maximum (i.e., $I_L = I_o/D'$ and $D' \approx U_{in}/U_o$). The RHP zero of the DCM converter can also be given by $DT_s/2$, which is closest to the origin when the input voltage is at minimum and the output current at maximum. This implies that the controllers should be designed under the same condition to give satisfactory performance due to the control-bandwidth limitation imposed by the RHP zero (Chapter 2, Section 2.6.3). It is also obvious that the resonant frequency in CCM (i.e., $f_o = D'/2\pi\sqrt{LC}$) would change according to the changes in the duty ratio, which would also affect the controller design. The low-frequency pole of the DCM converter can be approximated to be close to $f_{low} \approx \frac{1}{2\pi R_{eq} C}\cdot\frac{M}{M-1}$, and the high-frequency pole close to $f_{high} \approx \frac{R_{eq}}{2\pi L}\cdot\sqrt{\frac{K(M-1)}{M}}$. The high-frequency pole locates typically at the frequencies higher than the switching frequency yielding effectively first-order transfer functions. This means that the RHP zero may have more severe effect than would be expected.

The corresponding numerical values are given at 20 V by

$$G_{co}^{CCM} = \frac{-s^2 \cdot 0.2 - s\cdot 1.72 \times 10^4 + 1.61 \times 10^8}{s^2 + s\cdot 494 + 5.75 \times 10^5}$$

$$G_{co}^{DCM} = \frac{-s^2 \cdot 0.447 + s \times 1.039 \times 10^5 + 1.41 \times 10^{10}}{s^2 + s\cdot 9.027 \times 10^5 + 7.79 \times 10^7}$$

(3.103)

and at 50 V by

$$G_{co}^{CCM} = \frac{-s^2 \cdot 7.49 \times 10^{-2} - s\cdot 2.54 \times 10^3 + 4.45 \times 10^8}{s^2 + s\cdot 363 + 3.95 \times 10^6}$$

$$G_{co}^{DCM} = \frac{-s^2 \cdot 0.3 + s\cdot 3.38 \times 10^5 + 3.52 \times 10^{10}}{s^2 + s\cdot 6.086 \times 10^5 + 1.156 \times 10^8}$$

(3.104)

According to the numerical values, the damping factor (ζ) in CCM varies from 0.0326 to 0.0913, and in DCM from 51.1 to 28.3. This means that the roots of the characteristic polynomial in CCM are complex and in DCM real and well separated. The resonant frequency of the CCM converter is 120.7 Hz at 20 V and 316 Hz at 50 V. The RHP zero of the CCM converter is at 1.36 kHz at 20 V and at 9.86 kHz at 50 V. The low-frequency pole of the DCM converter is at 14.3 Hz at 20 V and at 30.2 Hz at 50 V. The corresponding high-frequency poles are at 143.7 and 96.8 kHz. The given approximated values of the DCM poles would predict quite close the same locations. The RHP zero of the DCM converter is at 52.2 kHz at 20 V and at 193.7 kHz at 50 V. As a consequence, the maximum control bandwidths would be limited approximately to the frequency corresponding to half the minimum RHP-zero location.

The frequency responses of the transfer functions are shown in Figures 3.26a (CCM) and 3.26b (DCM). The dots and squares represent the responses extracted from the switching model by means of simulation, which indicates that the analytical models are quite accurate. The flat high-frequency gain implies limitation on the achievable maximum crossover frequency for maintaining proper gain margin (i.e., typically 6 dB at least). Figure 3.26b implies that the controller of the DCM converter should be designed under the high-input voltage condition due to the highest high-frequency gain to avoid instability caused by the output-voltage ripple effects. This means that the lowest location of the RHP zero does not necessarily determine the worst case for the controller design, but the authentic frequency responses have to be studied carefully for the correct decision. In the case of the CCM converter (Figure 3.26a), the RHP zero clearly determines the worst case.

3.6.2.2 Output Impedance

The open-loop output impedances are symbolically shown in (3.105) in CCM and DCM. The low-frequency value of the CCM output impedance equals $\frac{r_L}{D'^2} + \frac{Dr_{ds1}}{D'^2} + \frac{r_d}{D'} + \frac{Dr_C}{D'}$, and the DCM output impedance to $\frac{M-1}{M} \cdot R_{eq}$, respectively. The high-frequency value corresponds to r_C for both of the converters.

$$Z_{o-o}^{CCM} = \frac{\frac{(r_L + Dr_{ds1} + D'r_d + DD'r_C + sL)(1 + sr_C C)}{LC}}{s^2 + s\frac{r_L + Dr_{ds1} + D'(r_d + r_C)}{L} + \frac{D'^2}{LC}}$$

$$Z_{o-o}^{DCM} = \frac{\frac{\left(sL + R_{eq}\sqrt{\frac{K(M-1)}{M}}\right)(1 + sr_C C)}{LC}}{s^2 + s \cdot \frac{R_{eq}}{L}\sqrt{\frac{K(M-1)}{M}} + \frac{1}{LC} \cdot \sqrt{\frac{KM}{M-1}}}$$

(3.105)

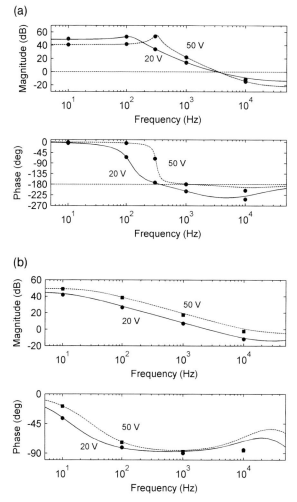

Figure 3.26 Predicted frequency responses of the control-to-output transfer functions: (a) CCM converter and (b) DCM converter. The dots and squares represent the switching-model-based frequency responses obtained by means of simulation.

The CCM output impedance has the highest value close to the resonant frequency $\left(f_o = \frac{D'}{\sqrt{LC}}\right)$, which can be approximated by

$$|Z_{o-o}^{CCM}|_{max} = \frac{R_{oe}^2\sqrt{1+\frac{r_e^2}{R_{oe}^2}}\sqrt{1+\frac{r_C^2}{R_{oe}^2}}}{r_e + r_C} \approx \frac{R_{oe}^2}{r_e + r_C} \qquad (3.106)$$

where $r_e = \frac{r_L}{D'^2} + \frac{Dr_{ds1}}{D'^2} + \frac{r_d}{D'} + \frac{Dr_C}{D'}$, and $R_{oe} = \frac{1}{D'}\sqrt{\frac{L}{C}}$ is the equivalent characteristic impedance of the LC circuit of the boost converter.

3.6 Dynamic Review

The numerical values of the output impedance at 20 V are

$$Z_{o-o}^{CCM} = \frac{s^2 \cdot 3.3 \times 10^{-2} + s \cdot 3.18 \times 10^3 + 1.54 \times 10^6}{s^2 + s \cdot 494 + 5.75 \times 10^5}$$

$$Z_{o-o}^{DCM} = \frac{s^2 \cdot 3.3 \times 10^{-2} + s \cdot 3.3 \times 10^4 + 2.86 \times 10^9}{s^2 + s \cdot 9.027 \times 10^5 + 7.79 \times 10^7}$$

(3.107)

and at 50 V

$$Z_{o-o}^{CCM} = \frac{s^2 \cdot 3.3 \times 10^{-2} + s \cdot 3.18 \times 10^3 + 1.02 \times 10^6}{s^2 + s \cdot 363 + 3.95 \times 10^6}$$

$$Z_{o-o}^{DCM} = \frac{s^2 \cdot 3.3 \times 10^{-2} + s \cdot 2.32 \times 10^4 + 1.93 \times 10^9}{s^2 + s \cdot 6.086 \times 10^5 + 1.156 \times 10^8}$$

(3.108)

The predicted frequency responses of the CCM output impedances are given in Figure 3.27a, where the effect of the varying duty cycle is observable only at low frequencies. The corresponding DCM output impedances are shown in Figure 3.27b, where the effect of the varying duty cycle is also observable at low frequencies but is much smaller than that in CCM.

3.6.2.3 Input-to-Output Transfer Functions

The input-to-output transfer functions are symbolically given in (3.109). The corresponding numerical values at 20 V are given in (3.110) and at 50 V in (3.111). The predicted frequency responses are given in Figure 3.28. Characteristic to the input-to-output transfer functions is that they do not provided input-noise attenuation but would amplify the noise:

$$G_{io-o}^{CCM} = \frac{\dfrac{D'(1 + sr_C C)}{LC}}{s^2 + s \dfrac{r_L + Dr_{ds1} + D'(r_d + r_C)}{L} + \dfrac{D'^2}{LC}}$$

$$G_{io-o}^{DCM} = \frac{\left(\dfrac{2M-1}{LCM}\sqrt{\dfrac{KM}{M-1}} - s \cdot \dfrac{M(M-1)}{R_{eq}C}\right)(1 + sr_C C)}{s^2 + s \cdot \dfrac{R_{eq}}{L}\sqrt{\dfrac{K(M-1)}{M}} + \dfrac{1}{LC}\cdot\sqrt{\dfrac{KM}{M-1}}}$$

(3.109)

$$G_{io-o}^{CCM} = \frac{s \cdot 23.76 + 2.28 \times 10^6}{s^2 + s \cdot 494 + 5.75 \times 10^5}$$

$$G_{io-o}^{DCM} = \frac{-s^2 \cdot 6.81 \times 10^{-3} + s \cdot 7.55 \times 10^2 + 1.35 \times 10^8}{s^2 + s \cdot 9.027 \times 10^5 + 7.79 \times 10^7}$$

(3.110)

$$G_{io-o}^{CCM} = \frac{s \cdot 62.32 + 5.98 \times 10^6}{s^2 + s \cdot 363 + 3.95 \times 10^6}$$

$$G_{io-o}^{DCM} = \frac{-s^2 \cdot 4.95 \times 10^{-4} + s \cdot 4.77 \times 10^3 + 4.62 \times 10^8}{s^2 + s \cdot 6.086 \times 10^5 + 1.156 \times 10^8}$$

(3.111)

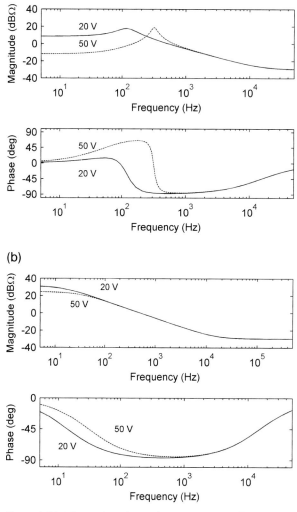

Figure 3.27 The predicted open-loop output impedances at 20 and 50 V: (a) CCM converter and (b) DCM converter.

3.6.2.4 Input Admittance

The open-loop input admittances are shown symbolically in (3.112). The resonant peak value in CCM can be given as shown in (3.113). The inverse of it corresponds to the resonant dip value of the open-loop input impedance:

$$Y_{\text{in}-o}^{\text{CCM}} = \frac{\dfrac{s}{L}}{s^2 + s\dfrac{r_L + Dr_{\text{ds1}} + D'(r_d + r_C)}{L} + \dfrac{D'^2}{LC}}$$

3.6 Dynamic Review

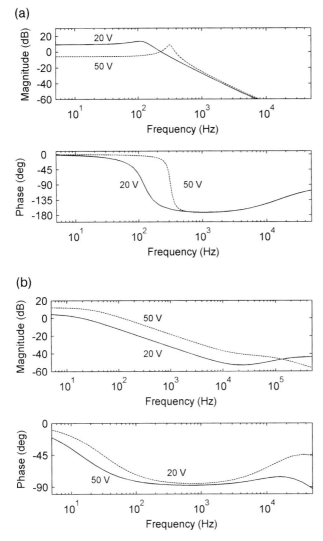

Figure 3.28 The predicted open-loop input-to-output transfer functions at 20 and 50 V: (a) CCM converter and (b) DCM converter.

$$Y_{\text{in}-o}^{\text{DCM}} = \frac{\dfrac{M^2}{L}\sqrt{\dfrac{KM}{M-1}}\left(s + \dfrac{M-1}{MR_{\text{eq}}C}\right)}{s^2 + s \cdot \dfrac{R_{\text{eq}}}{L}\sqrt{\dfrac{K(M-1)}{M}} + \dfrac{1}{LC}\cdot\sqrt{\dfrac{KM}{M-1}}} \qquad (3.112)$$

$$|Y_{\text{in}-o}^{\text{CCM}}|_{\max} = \frac{\dfrac{1}{D'^2}}{r_e + r_C} = \frac{1}{r_L + Dr_{\text{ds1}} + D'(r_d + r_C)} \qquad (3.113)$$

The numerical values of the input admittance at 20 V are given by

$$Y_{in-o}^{CCM} = \frac{s \cdot 2.86 \times 10^3}{s^2 + s \cdot 494 + 5.75 \times 10^5}$$

$$Y_{in-o}^{DCM} = \frac{s \cdot 3.46 \times 10^5 + 1.61 \times 10^7}{s^2 + s \cdot 9.027 \times 10^5 + 7.79 \times 10^7}$$

(3.114)

and at 50 V by

$$Y_{in-o}^{CCM} = \frac{s \cdot 2.86 \times 10^3}{s^2 + s \cdot 363 + 3.95 \times 10^6}$$

$$Y_{in-o}^{DCM} = \frac{s \cdot 8.22 \times 10^4 + 1.73 \times 10^6}{s^2 + s \cdot 6.086 \times 10^5 + 1.156 \times 10^8}$$

(3.115)

The corresponding frequency responses are shown in Figure 3.29. The resonant nature of the input admittance in CCM (Figure 3.29a) is obvious. The DCM input admittance implies close to resistive nature.

3.6.2.5 Ideal Input Admittance

The ideal input admittances are shown symbolically in (3.116). The corresponding numerical values are given at 20 V in (3.117) and at 50 V in (3.118). Theoretically, the numerators in (3.117) should be equal but the observed deviation is the result of omitting the effect of the parasitics in DCM (i.e., the numerator of (3.117) for an ideal converter would be 0.283). The corresponding predicted frequency responses are shown in Figure 3.30. The phase behavior of the ideal input admittance of the CCM converter (Figure 3.30a) implies increased sensitivity to source interactions:

$$Y_{in-\infty}^{CCM} = -\frac{\dfrac{I_L}{D'\left(U_o + U_D - (r_L + r_{ds1} + D'^2 r_C)\dfrac{I_L}{D'}\right)}}{1 - s \cdot \dfrac{LI_L}{D'\left(U_o + U_D - (r_L + r_{ds1} + D'^2 r_C)\dfrac{I_L}{D'}\right)}}$$

$$Y_{in-\infty}^{DCM} = -\frac{\dfrac{M^2}{R_{eq}}}{1 - s \cdot \dfrac{L}{R_{eq}} \cdot \sqrt{\dfrac{M(M-1)}{K}}}$$

(3.116)

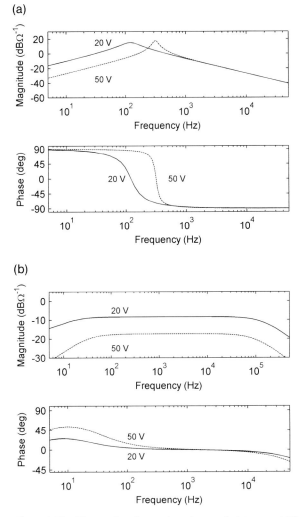

Figure 3.29 The predicted open-loop input admittances at 20 and 50 V: (a) CCM converter and (b) DCM converter.

$$Y_{\text{in}-\infty}^{\text{CCM}} = -\frac{0.3347}{1 - s \cdot 1.172 \times 10^{-4}} \cdot \Omega^{-1}$$
$$Y_{\text{in}-\infty}^{\text{DCM}} = -\frac{0.2813}{1 - s \cdot 3.047 \times 10^{-6}} \cdot \Omega^{-1}$$
(3.117)

$$Y_{\text{in}-\infty}^{\text{CCM}} = -\frac{0.046}{1 - s \cdot 1.61 \times 10^{-5}} \cdot \Omega^{-1}$$
$$Y_{\text{in}-\infty}^{\text{DCM}} = -\frac{0.045}{1 - s \cdot 8.216 \times 10^{-7}} \cdot \Omega^{-1}$$
(3.118)

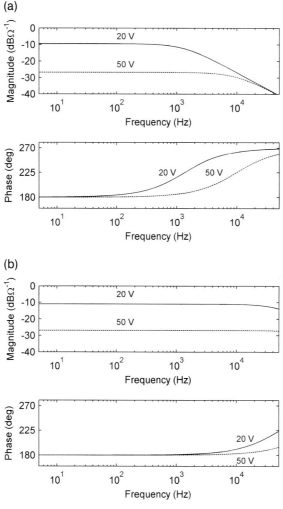

Figure 3.30 The predicted ideal input admittances at 20 and 50 V: (a) CCM converter and (b) DCM converter.

3.6.2.6 Short-Circuit Input Admittance

The short-circuit input admittances are shown symbolically in (3.119), and the corresponding numerical values at 20 V in (3.120) and at 50 V in (3.121). The predicted frequency responses are shown in Figure 3.31. The clear difference in CCM and DCM is obvious:

$$Y_{\text{in-sc}}^{\text{CCM}} = \dfrac{\dfrac{1}{r_L + Dr_{\text{ds1}} + D'(r_d + Dr_C)}}{1 + s \cdot \dfrac{L}{r_L + Dr_{\text{ds1}} + D'(r_d + Dr_C)}}$$

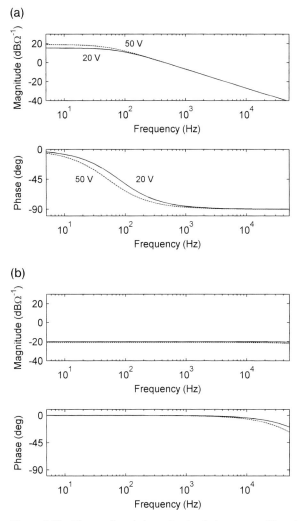

Figure 3.31 The predicted short-circuit admittances at 20 and 50 V: (a) CCM converter and (b) DCM converter.

$$Y_{\text{in-sc}}^{\text{DCM}} = \frac{\dfrac{M^2}{(M-1)R_{\text{eq}}}}{1 + s \cdot \dfrac{L}{R_{\text{eq}}} \cdot \sqrt{\dfrac{M}{K(M-1)}}} \tag{3.119}$$

$$Y_{\text{in-sc}}^{\text{CCM}} = \frac{5.86}{1 + s \cdot 2.05 \times 10^{-3}} \cdot \Omega^{-1}$$
$$Y_{\text{in-sc}}^{\text{DCM}} = \frac{0.1}{1 + s \cdot 1.1 \times 10^{-6}} \cdot \Omega^{-1} \tag{3.120}$$

$$Y_{\text{in-sc}}^{\text{CCM}} = \frac{8.9}{1 + s \cdot 3.11 \times 10^{-3}} \cdot \Omega^{-1}$$
$$Y_{\text{in-sc}}^{\text{DCM}} = \frac{0.09}{1 + s \cdot 1.64 \times 10^{-6}} \cdot \Omega^{-1} \tag{3.121}$$

References

1. R.W. Erickson and D. Maksimovic, *Fundamentals of Power Electronics*, Kluwer, Norwell, MA, USA, **2001**, 2nd Edition.
2. G.W. Wester and R.D. Middlebrook, 'Low-frequency characterization of switched-mode dc–dc converters,' *IEEE Trans. Aerosp. Electron. Syst.* vol. AES-9,, no. 3, **1973**, pp. 376–385.
3. R.D. Middlebrook and S. Cuk, 'A general unified approach to modeling switching-converter power stages,' in *Proc. IEEE Power Electronics Specialists Conf.*, **1976**, pp. 18–34.
4. R.D. Middlebrook and S. Cuk, 'A general unified approach to modeling switching-converter power stages,' *Int. J. Electron.*, vol. 42,, no. 6, **1977**, pp. 521–550.
5. S. Cuk and R.D. Middlebrook, 'A general unified approach to modeling switching DC-to-DC converters in discontinuous conduction mode,' in *Proc. IEEE Power Electron. Specialists Conf.*, **1977**, pp. 36–57.
6. D. Maksimovic and S. Cuk, 'A unified analysis of PWM converters in discontinuous modes,' *IEEE Trans. Power Electron.*, vol. 6,, no. 3, **1991**, pp. 476–490.
7. J. Sun, D.M. Mitchell, M.F. Greuel, P.T. Krein, and R.M. Bass, 'Modeling of PWM converters in discontinuous conduction mode- reexamination,' in *Proc. IEEE Power Electron. Specialists Conf.*, **1998**, pp. 615–622.
8. J. Sun, D.M. Mitchell, M.F. Greuel, P.T. Krein, and R.M. Bass, 'Averaged modeling of PWM converters operating in discontinuous conduction mode,' *IEEE Trans. Power Electron.*, vol. 16,, no. 4, **2001**, pp. 482–492.
9. D. Maksimovic, 'Computer-aided small-signal analysis based on the impulse response of DC/DC switching converters,' *IEEE Trans. Power Electron.*, vol. 15,, no. 6, **2000**, pp. 1183–1191.
10. T. Suntio, 'Small-signal modeling of switched-mode converters under direct-on-time control – A unified approach,' in *Proc. IEEE Indust. Electron. Society Annual Conf.*, **2002**, pp. 479–484.
11. T. Suntio, 'Unified average and small-signal modeling of direct-on-time control,' *IEEE Trans. Indust. Electron.*, vol. 53,, no. 1, **2006**, pp. 287–295.
12. D. Maksimovic, A.M. Stankovic, V.J. Thottuvelil, and G.C. Verghese, 'Modeling and simulation of power electronic converters,' *Proc. IEEE*, vol. 89,, no. 6, **2001**, pp. 898–912.
13. P.T. Krein, J. Bentsman, R.M. Bass, and B.L. Lesieutre, 'On the use of averaging for analysis of power electronic systems,' *IEEE Trans. Power Electron.*, vol. 5,, no. 2, **1990**, pp. 182–190.
14. P.T. Krein and R.M. Bass, 'A new approach to fast simulation of periodically switching power converters,' in *Proc. IEEE Industry Applications Society Annual Conf.*, **1990**, pp. 1185–1189.
15. B. Lehman and R.M. Bass, 'Switching frequency dependent averaged models for PWM DC–DC converters,' *IEEE Trans. Power Electron.*, vol. 11,, no 1, **1996**, pp. 89–98.
16. R.M. Bass and J. Sun, 'Large-signal averaging methods under large signal ripple conditions,' in *Proc. IEEE Power Electron. Specialists Conf.*, **1998**, pp. 630–632.

17. S. Banerje and G.C. Verghese, *Nonlinear Phenomena in Power Electronics – Attractors, Bifurcation, Chaos, and Control*, IEEE Press, New York, **2001**.
18. R.D. Middlebrook, 'Predicting modulator phase lag in PWM converter feedback loops,' in *Proc. Powercon 8*, **1981**, pp. H4.1–H4.6.
19. D.M. Mitchell, 'Pulsewidth modulator phase shift,' *IEEE Trans. Aerospace Electron. Syst.*, vol. AES-16,, no. 3, **1980**, pp. 272–278.
20. G.C. Verghese and V.J. Thottuvelil, 'Aliasing effects in PWM converters,' in *Proc. IEEE Power Electron. Specialists Conf.*, **1999**, pp. 1043–1049.
21. J. Sun, 'Small-signal modeling of variable-frequency pulsewidth modulators,' *IEEE Trans. Aerosp. Electron. Syst.*, vol. 38,, no. 3, **2002**, pp. 1104–1108.
22. Y. Qiu, M. Xu, K. Yao, J.J. Sun, and F.C. Lee, 'Multifrequency small-signal model for buck and multiphase buck converters,' *IEEE Trans. Power Electron.*, vol. 21,, no. 5, **2006**, pp. 1185–1192.
23. J.Y. Zhu, 'Interpreting small signal behavior of the synchronous buck converter at light load,' *IEEE Power Electron. Lett.*, vol. 3,, no. 4, **2005**, pp. 144–147.
24. B.H. Cho, 'Modeling and analysis of spacecraft power systems,' PhD Thesis, Virginia Polytechnic Institute and State University, USA, **1985**, pp. 181.
25. M. Shoyama, Y. Hamafuku, N. Matsuzaki, and T. Ninomiya, 'Simplification of transfer function in switching converter with general load impedance,' in *Proc. IEEE Power Electronics and Drives Systems Conf.*, **1995**, pp. 155–161.
26. T. Suntio, and I. Gadoura, 'Dynamic analysis of switched-mode converters using two-port modeling technique,' in *Proc. Power Conversion and Intelligent Motion Conf.*, **2002**, pp. 387–392.
27. M. Hankaniemi, M. Karppanen, and T. Suntio, 'Dynamical characterization of voltage-mode controlled buck converter operating in CCM and DCM,' in *Proc. International Power Electronics and Motion Control Conf.*, **2006**, pp. 816–821.

4
Average and Small-Signal Modeling of Peak-Current-Mode Control

4.1
Introduction

Peak-current-mode (PCM) control was invented in 1970s [1, 2] and has been a popular method to control the switched-mode converter due to the dynamic features it provides, such as increased input-noise attenuation, effective first-order transfer functions, pulse-by-pulse current limiting, and easiness to parallel converters using a common reference current [3]. It has also been claimed that the PCM control would remove the effect of RHP zero [4], but this claim does not hold. Limited duty-ratio range at the basic switching frequency, high open-loop output impedance, and sensitivity to high-frequency noise in the current loop have been considered as its main disadvantages [3, 5]. The operation of the converter in the subharmonic mode has been considered earlier unstable [6–8] but proved to be stable by applying chaos theories [9, 10].

The dynamics associated with the PCM control has fascinated the engineers since its invention. The first dynamic models appeared in the late 1970s [6] but were rather inaccurate. The modeling method presented in [11, 12] has been considered to be the most accurate one. The basic idea behind the method is to address the existence of the mode limit to sampling effect, which causes a resonant pole pair at half the switching frequency, and consequently, high gain in the inductor-current loop. Applying the proposed method [11] would, however, produce models, which do not correspond to the measured frequency responses. The author of [11] has also clearly observed the same discrepancies and manipulated certain transfer functions from those resulting by applying the method in order to have better model accuracy.

The observed phenomena in the behavior of a PCM-controlled converter are caused by an infinite frequency-independent small-signal duty-ratio gain in the inductor-current loop both in continuous (CCM) [13–17] and discontinuous (DCM) [21] conduction modes. The correctness of the proposed models has been widely disputed [18, 19] based on different arguments, but the consistency and accuracy of the modeling have been, however, clearly proved in [17, 20].

Dynamic Profile of Switched-Mode Converter. Teuvo Suntio
© 2009 WILEY-VCH Verlag GmbH & Co. KGaA, Weinheim
ISBN: 978-3-527-40708-8

122 | *4 Average and Small-Signal Modeling of Peak-Current-Mode Control*

4.2
PCM-Control Principle

Under PCM control, the on-time (t_{on}) or duty ratio (d) is generated comparing the inductor-current and the control current (i_{co}) (i.e., $i_{co} = u_{co}/R_s$ in Figure 4.1, where R_s is the equivalent inductor-current sensing resistor). In order to extend the duty ratio (d) beyond the mode limit, the control current has to be compensated using an artificial ramp M_c. In practice, the compensation ramp is added to the sensed inductor-current signal, but the analysis would be more convenient considering the compensation ramp to be subtracted from the control current as depicted in Figure 4.1. The duty ratio (d) is established when the inductor current reaches the compensated control current (Figure 4.1b).

The PCM control is a method to change the internal dynamics of a converter by applying feedback from the inductor current as depicted in Figure 4.1: as

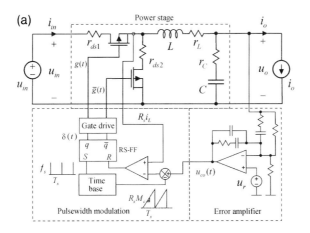

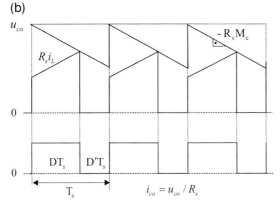

Figure 4.1 PCM-control principles: (a) circuit schematics with a buck converter and (b) duty-ratio generation.

a result of the inductor-current feedback, the dynamics associated with the on-time (t_{on}) or duty ratio (d) would be changed from the dynamics under VM control. Therefore, it may be obvious that the analytical models can be derived from the models or the state-space representation of the corresponding VMC converter [22] by replacing the perturbed duty ratio ($\hat{d}$) with the expression describing its dependence on the new constellation. This dependence is known as duty-ratio constraints [8]. The duty-ratio constraints are typically given [8] by

$$\hat{d} = F_m \left(\hat{i}_{co} - q_c \hat{i}_L - q_o \hat{u}_o - q_i \hat{u}_{in} \right) \quad (4.1)$$

where F_m is the duty-ratio gain and q_k is the feedforward or feedback gain from the defined variable. If the duty-ratio constraints are applied to the corresponding VMC state space, the perturbed output voltage ($\hat{u}_o$) has to be substituted with its relation to the state and input variables [17] yielding the second-order converters

$$\hat{d} = F_m^{sp} \left(\hat{i}_{co} - q_c^{sp} \hat{i}_L - q_o^{sp} \hat{u}_c - q_i^{sp} \hat{u}_{in} - q_{io}^{sp} \hat{i}_o \right) \quad (4.2)$$

where the superscript "sp" means state space. As a consequence, the fundamental issue in the PCM modeling is to find the proper definition for the duty-ratio constraints.

The time-averaged inductor current $\langle i_L \rangle$ contributes to the dynamics of the converter as discussed in Chapter 3. Therefore, the desired feedback signal would be also $\langle i_L \rangle$, but the real duty-ratio generation takes place as illustrated in Figure 4.2. From Figure 4.2, we can compute that at $t = (k+d)T_s$ holds

$$i_{co} - m_c d T_s = \langle i_L \rangle + \Delta i_L \quad (4.3)$$

where Δi_L is the difference between the peak inductor current and its averaged value at $t = (k+d)T_s$. Equation (4.3) is termed as comparator equation due to its physical realization as shown in Figure 4.1a. This means that the duty-ratio constraints can be determined if Δi_L can be found. It may be obvious that Δi_L would be affected by the operation mode (i.e., CCM or DCM).

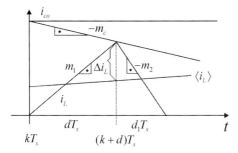

Figure 4.2 Duty-ratio generation based on the inductor-current up slope.

4.3
Modeling in CCM

In CCM, the averaged inductor current lies exactly in the middle of the ripple band. It may be approximated by means of a first-order function of time within a switching cycle. Its derivative can be approximated by means of the average slope of the instantaneous inductor current as derived in Chapter 3. Therefore, we may express $\langle i_L \rangle$ by

$$\langle i_L \rangle = (dm_1 - d'm_2)t + \frac{dd'T_s}{2}(m_1 + m_2) + i_L(kT_s) \tag{4.4}$$

where $i_L(kT_s)$ is the value of the averaged inductor current at the beginning of the cycle (Figure 4.2). (More detailed derivation of (4.4) can be found from [17]). The on-time instantaneous inductor current i_{L-on} (Figure 4.2) can be approximated by

$$i_{L-on} = m_1 t + i_L(kT_s) \tag{4.5}$$

As a consequence, Δi_L can be found by

$$\Delta i_L = i_{L-on}((k+d)T_s) - \langle i_L((k+d)T_s)_s \rangle \tag{4.6}$$

yielding

$$\Delta i_L = \frac{dd'T_s}{2}(m_1 + m_2) \tag{4.7}$$

and the comparator equation as

$$i_{co} - M_c dT_s = \langle i_L \rangle + \frac{dd'T_s}{2}(m_1 + m_2) \tag{4.8}$$

when the artificial compensation is assumed to be constant. The duty-ratio constraints can be developed from (4.8) by substituting the up and down slopes with the topology-dependent values and linearizing it at the defined operating point (i.e., developing the partial derivatives as instructed in Chapter 2). The same Δi_L as (4.7) is implicitly found also in [13–15] based on different methods.

If several inductor currents constitute the feedback signal, the overall $\Delta i_{L\Sigma}$ can be defined to be

$$\Delta i_{L\Sigma} = \frac{dd'T_s}{2} \cdot \sum_{i=1}^{n}(m_{1i} + m_{2i}) \tag{4.9}$$

According to (4.9), the PCMC modeling of the higher order converters can be easily made as demonstrated in Chapter 10. Essential is to remember that the up and down slopes are the local averages within the defined part of the switching cycle as instructed in Chapter 3.

Transformer isolation is often used for safety reasons and/or for scaling the input voltage to obtain more optimal duty-ratio range [8]. The inductor-current feedback is commonly taken from the transformer primary current containing the reflected inductor current (i'_L) and the transformer magnetizing current (i_{L_M}) as depicted in Figure 4.3 in the case of an active-reset *forward converter*. Similar conditions also apply to the other transformer-isolated converters, but the shape of magnetizing current may vary and, consequently, its effect on the duty-ratio constraints may be different [17]. It may be obvious that the magnetizing current would have similar effect as the artificial compensation has. As a consequence, the comparator equation may be given by

$$i_{co} - m_c dT_s - \frac{kdT_s u'_{in}}{L'_M} = \langle i_L \rangle + \Delta i_L \tag{4.10}$$

where u'_{in} and L'_M are the corresponding values given at the secondary side, and k is the coefficient taking into account the shape of the magnetizing current (i.e., *normal forward converter*: $k = 1$, *active-reset forward, full and half-bridge, push-pull converters*: $k = 1/2$). This means that the duty-ratio gain (F_m) and the input-voltage feedforward gain (q_i) are to be changed compared to the corresponding basic converter.

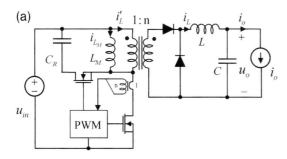

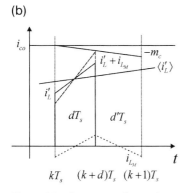

Figure 4.3 Active-reset forward converter: (a) schematics and (b) transformer on-time (dT_s) current.

4.3.1
Duty-Ratio Constraints for Buck, Boost, and Buck–Boost Converters

We develop the duty-ratio constraints for the basic converters corresponding to (4.1) and (4.2) utilizing the material given in Chapter 3 (Section 3.3). The corresponding converters are defined in Figures 3.6a (for buck), 3.8a (for boost), and 3.10a (for buck–boost).

4.3.1.1 Buck Converter
According to (3.11) and (3.12), $m_1 + m_2 = \frac{\langle u_{in} \rangle + U_D + (r_d - r_{ds1})\langle i_L \rangle}{L}$. Therefore, the comparator equation (4.8) becomes

$$i_{co} - M_c d T_s = \langle i_L \rangle + \frac{dd'T_s}{2L}(\langle u_{in} \rangle + U_D + (r_d - r_{ds1})\langle i_L \rangle) \tag{4.11}$$

The developing of the partial derivatives yields the duty-ratio constraints as

$$\hat{d} = \frac{1}{T_s \left(M_c + \dfrac{(D' - D)(U_{in} + U_D + (r_d - r_{ds1})I_L)}{2L} \right)}$$

$$\times \left(\hat{i}_{co} - \left(1 + \frac{DD'T_s}{2L}(r_d - r_{ds1})\right) \cdot \hat{i}_L - \frac{DD'T_s}{2L} \cdot \hat{u}_{in} \right) \tag{4.12}$$

from which we find that

$$F_m = \frac{1}{T_s \left(M_c + \dfrac{(D' - D)(U_{in} + U_D + (r_d - r_{ds1})I_L)}{2L} \right)}$$

$$q_c = 1 + \frac{DD'T_s}{2L}(r_d - r_{ds1})$$

$$q_o = 0$$

$$q_i = \frac{DD'T_s}{2L} \tag{4.13}$$

The duty-ratio constraints coefficients in (4.13) would define equally the coefficients in (4.1) and (4.2) (i.e., $F_m^{sp} = F_m$, $q_c^{sp} = q_c$, $q_o^{sp} = q_o$, $q_i^{sp} = q_i$, and $q_{io}^{sp} = 0$).

4.3.1.2 Boost Converter
According to (3.16) and (3.17), $m_1 + m_2 = \frac{\langle u_o \rangle + U_D + (r_d - r_{ds1})\langle i_L \rangle}{L}$ or $m_1 + m_2 = \frac{\langle u_C \rangle + U_D + (r_d + r_C - r_{ds1})\langle i_L \rangle - r_C \langle \hat{i}_o \rangle}{L}$, when the output voltage is substituted with (3.14). Therefore, the comparator equation can be given either as

$$i_{co} - M_c dT_s = \langle i_L \rangle + \frac{dd'T_s}{2L}(\langle u_o \rangle + U_D + (r_d - r_{ds1})\langle i_L \rangle) \tag{4.14}$$

or

$$i_{co} - M_c d T_s = \langle i_L \rangle + \frac{dd' T_s}{2L}(\langle u_c \rangle + U_D + (r_d + r_C - r_{ds1})\langle i_L \rangle - r_C \langle i_o \rangle) \tag{4.15}$$

Linearizing (4.14) yields

$$\hat{d} = \frac{1}{T_s \left(M_c + \dfrac{(D' - D)(U_o + U_D + (r_d - r_{ds1})I_L)}{2L} \right)}$$

$$\times \left(\hat{i}_{co} - \left(1 + \frac{DD'T_s}{2L}(r_d - r_{ds1})\right) \cdot \hat{i}_L - \frac{DD'T_s}{2L} \cdot \hat{u}_o \right) \tag{4.16}$$

from which the duty-ratio-constraints coefficients corresponding to (4.1) can be found to be

$$F_m = \frac{1}{T_s \left(M_c + \dfrac{(D' - D)(U_o + U_D + (r_d - r_{ds1})I_L)}{2L} \right)}$$

$$q_c = 1 + \frac{DD'T_s}{2L}(r_d - r_{ds1})$$

$$q_o = \frac{DD'T_s}{2L} \tag{4.17}$$

$$q_i = 0$$

Linearizing (4.15) yields

$$\hat{d} = \frac{1}{T_s \left(M_c + \dfrac{(D' - D)(U_o + U_D + (r_d + r_C - r_{ds1})I_L - r_C I_o)}{2L} \right)}$$

$$\times \left(\hat{i}_{co} - \left(1 + \frac{DD'T_s}{2L}(r_d + r_C - r_{ds1})\right) \cdot \hat{i}_L - \frac{DD'T_s}{2L} \cdot \hat{u}_C + \frac{DD'T_s}{2L} r_C \cdot \hat{i}_o \right) \tag{4.18}$$

from which the duty-ratio-constraints coefficients corresponding to (4.2) can be found to be

$$F_m^{sp} = \frac{1}{T_s \left(M_c + \dfrac{(D' - D)(U_o + U_D + (r_d + r_C - r_{ds1})I_L - r_C I_o)}{2L} \right)}$$

$$q_c^{sp} = 1 + \frac{DD'T_s}{2L}(r_d + r_C - r_{ds1})$$

$$q_o^{sp} = \frac{DD'T_s}{2L} \tag{4.19}$$

$$q_i^{sp} = 0$$

$$q_{io}^{sp} = -\frac{DD'T_s}{2L} r_C$$

4.3.1.3 Buck–Boost

According to (3.21) and (3.23), $m_1 + m_2 = \frac{\langle u_{in} \rangle + \langle u_o \rangle + U_D + (r_d - r_{ds1})\langle i_L \rangle}{L}$ or $m_1 + m_2 = \frac{\langle u_{in} \rangle + \langle u_c \rangle + U_D + (r_d + r_C - r_{ds1})\langle i_L \rangle - r_C \langle \hat{i}_o \rangle}{L}$, when the output voltage is substituted with (3.19). Therefore, the comparator equation (4.8) becomes

$$i_{co} - M_c d T_s = \langle i_L \rangle + \frac{dd' T_s}{2L}(\langle u_{in} \rangle + \langle u_o \rangle + U_D + (r_d - r_{ds1})\langle i_L \rangle) \quad (4.20)$$

or

$$i_{co} - M_c d T_s = \langle i_L \rangle + \frac{dd' T_s}{2L}(\langle u_{in} \rangle + \langle u_c \rangle + U_D + (r_d + r_C - r_{ds1})\langle i_L \rangle - r_C \langle i_o \rangle) \quad (4.21)$$

Linearizing (4.20) yields

$$\hat{d} = \frac{1}{T_s \left(M_c + \frac{(D' - D)(U_{in} + U_o + U_D + (r_d - r_{ds1})I_L)}{2L} \right)}$$
$$\times \left(\hat{i}_{co} - \left(1 + \frac{DD' T_s}{2L}(r_d - r_{ds1}) \right) \cdot \hat{i}_L - \frac{DD' T_s}{2L} \cdot \hat{u}_o - \frac{DD' T_s}{2L} \cdot \hat{u}_{in} \right)$$
$$(4.22)$$

from which the duty-ratio-constraints coefficients corresponding to (4.1) can be found to be

$$F_m = \frac{1}{T_s \left(M_c + \frac{(D' - D)(U_{in} + U_o + U_D + (r_d - r_{ds1})I_L)}{2L} \right)}$$
$$q_c = 1 + \frac{DD' T_s}{2L}(r_d - r_{ds1})$$
$$q_o = \frac{DD' T_s}{2L} \quad (4.23)$$
$$q_i = \frac{DD' T_s}{2L}$$

Linearizing (4.21) yields

$$\hat{d} = \frac{1}{T_s \left(M_c + \frac{(D' - D)(U_{in} + U_o + U_D + (r_d + r_C - r_{ds1})I_L - r_C I_o)}{2L} \right)}$$
$$\times \left(\hat{i}_{co} - \left(1 + \frac{DD' T_s}{2L}(r_d + r_C - r_{ds1}) \right) \cdot \hat{i}_L \right.$$
$$\left. - \frac{DD' T_s}{2L} \cdot \hat{u}_C - \frac{DD' T_s}{2L} \cdot \hat{u}_{in} + \frac{DD' T_s}{2L} r_C \cdot \hat{i}_o \right) \quad (4.24)$$

from which the duty-ratio-constraints coefficients corresponding to (4.2) can be found to be

$$F_m^{sp} = \frac{1}{T_s\left(M_c + \frac{(D'-D)(U_{in} + U_o + U_D + (r_d + r_C - r_{ds1})I_L - r_C I_o)}{2L}\right)}$$

$$q_c^{sp} = 1 + \frac{DD'T_s}{2L}(r_d + r_C - r_{ds1})$$

$$q_o^{sp} = \frac{DD'T_s}{2L} \tag{4.25}$$

$$q_i^{sp} = \frac{DD'T_s}{2L}$$

$$q_{io}^{sp} = -\frac{DD'T_s}{2L}r_C$$

According to the derived duty-ratio constraints, it may be obvious that the duty-ratio gain (F_m) would become infinite without compensation (i.e., $M_c = 0$) at $D = 0.5$, which is also the observed mode limit [8]. More detailed discussions on the subject will be given in Section 4.3.3.

4.3.1.4 General CCM Transfer Functions

The dynamic description of the basic converters can be given as a function of the corresponding VMC-converter transfer functions, the duty-ratio-constraints coefficients corresponding to (4.1), and the coefficients A and B based on the block diagrams shown in Figure 4.4, where Z_C is the impedance of the output capacitor, G_{cL} the control-to-inductor-current transfer function, and the coefficients A and B as follows: Buck: $A = 1$, $B = 0$; Boost and Buck–boost: $A = D'$, $B = I_L$. The other transfer functions are irrelevant and used only to detect the original VMC transfer functions.

From Figure 4.4, we may compute that the general transfer functions defining the internal dynamics of the PCM-controlled buck, boost, and buck–boost converters in CCM can be given as follows:

$$G_{io-o} = \frac{\left(1 + \frac{BF_m q_c}{A}\right)G_{io-o}^v - F_m q_i G_{co}^v}{1 + L_c(s) + L_v(s)}$$

$$Z_{o-o} = \frac{\left(1 + \frac{BF_m q_c}{A}\right)Z_{o-o}^v + \frac{F_m q_c}{A}G_{co}^v}{1 + L_c(s) + L_v(s)}$$

$$G_{co} = \frac{F_m G_{co}^v}{1 + L_c(s) + L_v(s)}$$

$$Y_{in-o} = Y_{in-o}^v - \frac{F_m\left(\left(q_o + \frac{q_c}{AZ_C}\right)G_{io-o}^v + q_i\right)G_{ci}^v}{1 + L_c(s) + L_v(s)}$$

(a)

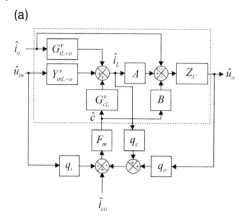

Figure 4.4 Block diagrams for PCM control in CCM for the basic converters: (a) output dynamics and (b) input dynamics.

(b)

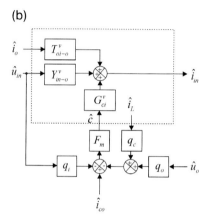

$$T_{\text{oi}-o} = T_{\text{oi}-o}^{\text{v}} + \frac{F_m\left(\left(q_o + \dfrac{q_c}{AZ_C}\right)Z_{o-o}^{\text{v}} - \dfrac{q_c}{A}\right)G_{\text{ci}}^{\text{v}}}{1 + L_c(s) + L_v(s)}$$

$$G_{\text{ci}} = \frac{F_m G_{\text{ci}}^{\text{v}}}{1 + L_c(s) + L_v(s)} \tag{4.26}$$

where the superscript "v" denotes the VMC transfer functions defined in Chapter 3. The internal inductor-current-loop gain $L_c(s)$, and the internal output-voltage-loop gain $L_v(s)$ are defined by

$$L_c(s) = F_m q_c G_{cL}^{\text{v}}$$
$$L_v(s) = F_m q_o G_{\text{co}}^{\text{v}} \tag{4.27}$$

where G_{cL}^{v} is the control-to-inductor-current transfer function.

The general PCMC transfer functions defined in (4.26) are actually quite useful, because several conclusions can be easily made based on them:

1. If the VMC control-to-output transfer function (G_{co}) contains, for instance, RHP zero, the same zero also exists in the PCMC converter with the corresponding control-bandwidth limitation.
2. If $q_i = 0$, then the input-to-output transfer function (G_{io-o}) cannot be made to be zero by means of the artificial compensation.
3. If $G_{io-o} = 0$, then $Y_{in-o} = Y_{in-o}^v - \frac{G_{io-o}^v G_{ci}^v}{G_{co}^v} = Y_{in-\infty}$, and $Y_{in-c} = Y_{in-\infty}$.

4.3.2
Specific Transfer Functions for the Basic Converters

The dynamic description of the PCMC converter can be derived from the corresponding VMC state-space representation given in Chapter 3 by substituting the perturbed duty ratio $\left(\hat{d}\right)$ with the corresponding state-space duty-ratio constraints (4.2) given in Subsection 4.3.1.

From Figure 4.2, we may expect that the PCM control would provide extra damping into the process similar to the operation in DCM and make the roots of the characteristic equation real, because the control current (i_{co}) would prevent the inductor current to move as freely as under the direct-duty-ratio control, where such a limiting does not exist. In small-signal sense, the increase in damping would be caused by a lossless resistor connected in series with the equivalent inductor, which can be generally given by $F_m^{sp} U_E q_c^{sp}$ and is typically much larger than the other restive losses. When the converter approaches the mode limit, the duty-ratio gain F_m approaches infinity. As a consequence, the inductor would be effectively disconnected, that is, the time-averaged inductor current would loose its status as a state variable. The PCMC transfer functions are given below, and their dynamical implications are discussed and compared to the corresponding VMC transfer functions in Section 4.5.

4.3.2.1 Buck Converter

The schematic of the diode-switched buck converter is given in Chapter 3 in Figure 3.6a. Substituting the perturbed duty ratio $\left(\hat{d}\right)$ with the derived duty-ratio constraints

$$\hat{d} = F_m^{sp}\left(\hat{i}_{co} - q_c^{sp}\hat{i}_L - q_i^{sp}\hat{u}_{in}\right)$$

$$F_m^{sp} = \frac{1}{T_s\left(M_c + \frac{(D' - D)U_E}{2L}\right)}$$

$$q_c^{sp} = 1 + \frac{DD'T_s}{2L}(r_d - r_{ds1})$$

$$q_i^{sp} = \frac{DD'T_s}{2L}$$

(4.28)

yields the PCMC small-signal state space as

$$\frac{d\hat{i}_L}{dt} = -\frac{\left(r_E + r_C + F_m^{sp} U_E q_c^{sp}\right)}{L} \cdot \hat{i}_L - \frac{1}{L} \cdot \hat{u}_C$$

$$+ \frac{D - F_m^{sp} U_E q_i^{sp}}{L} \cdot \hat{u}_{in} + \frac{r_C}{L} \cdot \hat{i}_o + \frac{F_m^{sp} U_E}{L} \cdot \hat{i}_{co}$$

$$\frac{d\hat{u}_C}{dt} = \frac{\hat{i}_L}{C} - \frac{\hat{i}_o}{C} \tag{4.29}$$

$$\hat{i}_{in} = \left(D - F_m^{sp} I_L q_c^{sp}\right) \cdot \hat{i}_L + F_m^{sp} I_L \cdot \hat{i}_{co}$$

$$\hat{u}_o = \hat{u}_C + r_C C \frac{d\hat{u}_C}{dt}$$

$$U_E = U_{in} + U_D + (r_d - r_{ds1}) I_L$$

$$r_E = r_L + D r_{ds1} + D' r_d$$

Applying *Laplace* transformation and matrix algebra to (4.29), the input dynamics of the PCMC buck converter can be solved to be

$$Y_{in-o} = \frac{\dfrac{\left(D - F_m^{sp} U_E q_i^{sp}\right)\left(D - F_m^{sp} I_L q_c^{sp}\right) s}{L}}{\Delta} - F_m^{sp} q_i^{sp} I_L$$

$$T_{oi-o} = \frac{\dfrac{\left(D - F_m^{sp} I_L q_c^{sp}\right)(1 + s \cdot r_C C)}{LC}}{\Delta} \tag{4.30}$$

$$G_{ci} = \frac{\dfrac{F_m^{sp} U_E \left(D - F_m^{sp} I_L q_c^{sp}\right) s}{L}}{\Delta} + F_m^{sp} I_L$$

and the output dynamics as

$$G_{io-o} = \frac{\dfrac{\left(D - F_m^{sp} U_E q_i^{sp}\right)(1 + s \cdot r_C C)}{LC}}{\Delta}$$

$$Z_{o-o} = \frac{\dfrac{\left(r_E + F_m^{sp} U_E q_c^{sp} + sL\right)(1 + s \cdot r_C C)}{LC}}{\Delta} \tag{4.31}$$

$$G_{co} = \frac{\dfrac{F_m^{sp} U_E (1 + s \cdot r_C C)}{LC}}{\Delta}$$

where the denominator (Δ) is defined by

$$\Delta = s^2 + s \cdot \frac{r_E + F_m^{sp} U_E q_c^{sp} + r_C}{L} + \frac{1}{LC} \tag{4.32}$$

4.3.2.2 Boost Converter

The schematic of the diode-switching boost converter is given in Chapter 3 in Figure 3.8a. Substituting the perturbed duty ratio $\left(\hat{d}\right)$ in the VMC small-signal state space with the derived duty-ratio constraints

$$\hat{d} = F_m^{\text{sp}} \left(\hat{i}_{\text{co}} - q_c^{\text{sp}} \hat{i}_L - q_o^{\text{sp}} \hat{u}_C - q_{\text{io}}^{\text{sp}} \hat{i}_o \right)$$

$$F_m^{\text{sp}} = \frac{1}{T_s \left(M_c + \dfrac{(D' - D) U_E}{2L} \right)}$$

$$q_c^{\text{sp}} = 1 + \frac{DD'T_s}{2L}(r_d + r_C - r_{\text{ds1}}) \tag{4.33}$$

$$q_o^{\text{sp}} = \frac{DD'T_s}{2L}$$

$$q_{\text{io}}^{\text{sp}} = -\frac{DD'T_s}{2L} r_C$$

yields the PCMC small-signal state space as

$$\frac{d\hat{i}_L}{dt} = -\frac{r_E + D'^2 r_C + F_m^{\text{sp}} U_E q_c^{\text{sp}}}{L} \cdot \hat{i}_L - \frac{D' + F_m^{\text{sp}} U_E q_o^{\text{sp}}}{L} \cdot \hat{u}_C + \frac{1}{L} \cdot \hat{u}_{\text{in}}$$

$$+ \frac{D' r_C - F_m^{\text{sp}} U_E q_{\text{io}}^{\text{sp}}}{L} \cdot \hat{i}_o + \frac{F_m^{\text{sp}} U_E}{L} \cdot \hat{i}_{\text{co}}$$

$$\frac{d\hat{u}_C}{dt} = \frac{D' + F_m^{\text{sp}} I_L q_c^{\text{sp}}}{C} \cdot \hat{i}_L + \frac{F_m^{\text{sp}} I_L q_o^{\text{sp}}}{C} \cdot \hat{u}_C - \frac{1 - F_m^{\text{sp}} I_L q_{\text{io}}^{\text{sp}}}{C} \cdot \hat{i}_o - \frac{F_m^{\text{sp}} I_L}{C} \cdot \hat{i}_{\text{co}}$$

$$\hat{i}_{\text{in}} = \hat{i}_L \tag{4.34}$$

$$\hat{u}_o = \hat{u}_C + r_C C \frac{d\hat{u}_C}{dt}$$

$$U_E = U_o + U_D + (r_d + r_C - r_{\text{ds1}}) I_L - r_C I_o$$

$$r_E = r_L + D r_{\text{ds1}} + D'(r_d + D r_C)$$

Applying *Laplace* transformation and matrix algebra to (4.34), the input dynamics of the PCMC boost converter can be solved to be as

$$Y_{\text{in-o}} = \frac{\dfrac{1}{L}\left(s - \dfrac{F_m^{\text{sp}} I_L q_o^{\text{sp}}}{C}\right)}{\Delta}$$

$$T_{\text{oi-o}} = \frac{\dfrac{D' r_C - F_m^{\text{sp}} U_E q_{\text{io}}^{\text{sp}}}{L}\left(s + \dfrac{(D' + F_m^{\text{sp}} U_E q_o^{\text{sp}}) L}{(D' r_C - F_m^{\text{sp}} U_E q_{\text{io}}^{\text{sp}}) C}\right)}{\Delta} \tag{4.35}$$

$$G_{\text{ci}} = \frac{\dfrac{F_m^{\text{sp}} U_E}{L}\left(s + \dfrac{D' I_L}{C U_E}\right)}{\Delta}$$

and the output dynamics as

$$G_{io-o} = \frac{\dfrac{(D' + F_m^{sp} I_L q_c^{sp})}{LC}(1 + s \cdot r_C C)}{\Delta}$$

$$Z_{o-o} = \frac{\dfrac{(1 - F_m^{sp} I_L q_{io}^{sp})}{LC}\left(\dfrac{q_{io}^{sp}\left(D' U_E - I_L(r_E + D'^2 r_C)\right) - q_c^{sp} I_L D' r_C}{1 - F_m^{sp} I_L q_{io}^{sp}} + s \cdot L\right)(1 + s \cdot r_C C)}{\Delta}$$

$$G_{co} = \frac{\dfrac{F_m^{sp}\left(D' U_E - I_L(r_E + D'^2 r_C)\right)}{LC}\left(1 - s \cdot \dfrac{L I_L}{D' U_E - I_L(r_E + D'^2 r_C)}\right)(1 + s \cdot r_C C)}{\Delta}$$

(4.36)

where the denominator (Δ) is defined by

$$\Delta = s^2 + s \cdot \left(\frac{r_E + D'^2 r_C + F_m^{sp} U_E q_c^{sp}}{L} - \frac{F_m^{sp} I_L q_o^{sp}}{C}\right)$$
$$+ \frac{D'^2 + D' F_m^{sp}\left(U_E q_o^{sp} + I_L q_c^{sp}\right) - F_m^{sp} I_L q_o^{sp}(r_E + D'^2 r_C)}{LC}$$

(4.37)

4.3.2.3 Buck–Boost Converter

The schematic of the diode-switched buck–boost converter is given in Chapter 3 in Figure 3.10a. Substituting the perturbed duty ratio $\left(\hat{d}\right)$ with the defined duty-ratio constraints

$$\hat{d} = F_m\text{sp}\left(\hat{i}_{co} - q_c^{sp}\hat{i}_L - q_o^{sp}\hat{u}_C - q_{io}^{sp}\hat{i}_o\right)$$

$$F_m^{sp} = \frac{1}{T_s\left(M_c + \dfrac{(D' - D) U_E}{2L}\right)}$$

$$q_c^{sp} = 1 + \frac{DD'T_s}{2L}(r_d + r_C - r_{ds1})$$

$$q_o^{sp} = \frac{DD'T_s}{2L}$$

$$q_i^{sp} = \frac{DD'T_s}{2L}$$

$$q_{io}^{sp} = -\frac{DD'T_s}{2L} r_C$$

(4.38)

yields the PCMC small-signal state space as

$$\frac{d\hat{i}_L}{dt} = -\frac{r_E + D'^2 r_C + F_m^{sp} U_E q_c^{sp}}{L} \cdot \hat{i}_L - \frac{D' + F_m^{sp} U_E q_o^{sp}}{L} \cdot \hat{u}_C$$
$$+ \frac{D - F_m^{sp} U_E q_i^{sp}}{L} \cdot \hat{u}_{in} + \frac{D' r_C - F_m^{sp} U_E q_{io}^{sp}}{L} \cdot \hat{i}_o + \frac{F_m^{sp} U_E}{L} \cdot \hat{i}_{co}$$

$$\frac{d\hat{u}_C}{dt} = \frac{D' + F_m^{sp} I_L q_c^{sp}}{C} \cdot \hat{i}_L + \frac{F_m^{sp} I_L q_o^{sp}}{C} \cdot \hat{u}_C + \frac{F_m^{sp} I_L q_i^{sp}}{C} \cdot \hat{u}_{in}$$
$$- \frac{1 - F_m^{sp} I_L q_{io}^{sp}}{C} \cdot \hat{i}_o - \frac{F_m^{sp} I_L}{C} \cdot \hat{i}_{co} \qquad (4.39)$$

$$\hat{i}_{in} = (D - F_m^{sp} I_L q_c^{sp}) \cdot \hat{i}_L - F_m^{sp} I_L q_o^{sp} \cdot \hat{u}_C - F_m^{sp} I_L q_i^{sp} \cdot \hat{u}_{in}$$
$$- F_m^{sp} I_L q_{io}^{sp} \cdot \hat{i}_o + F_m^{sp} I_L \cdot \hat{i}_{co}$$

$$\hat{u}_o = \hat{u}_C + r_C C \frac{d\hat{u}_C}{dt}$$

$$U_E = U_{in} + U_o + U_D + (r_d + r_C - r_{ds1})I_L - r_C I_o$$

$$r_E = r_L + D r_{ds1} + D'(r_d + D r_C)$$

Applying *Laplace* transformation and matrix algebra to (4.39), the input dynamics of the PCMC buck–boost converter can be solved, but the explicit transfer functions are very complex and long and, therefore, not given here. The transfer functions describing the output dynamics are, however, manageable and defined as

$$G_{io-o} = \frac{\dfrac{F_m^{sp} I_L q_i^{sp}}{C}\left(s + \dfrac{D\left(D' + F_m^{sp} I_L q_c^{sp}\right) + F_m^{sp} q_i^{sp}\left((r_E + D'^2 r_C)I_L - D' U_E\right)}{L F_m^{sp} I_L q_i^{sp}}\right)(1 + s \cdot r_C C)}{\Delta}$$

$$Z_{o-o} = \frac{\dfrac{(1 - F_m^{sp} I_L q_{io}^{sp})}{LC}\left(\dfrac{r_E + F_m^{sp} U_E q_c^{sp} + F_m^{sp}\left(q_{io}^{sp}(D' U_E - I_L(r_E + D'^2 r_C)) - q_c^{sp} I_L D' r_C\right)}{1 - F_m^{sp} I_L q_{io}^{sp}} + s \cdot L\right)(1 + s \cdot r_C C)}{\Delta}$$

$$G_{co} = \frac{\dfrac{F_m^{sp}(D' U_E - I_L(r_E + D'^2 r_C))\left(1 - s \cdot \dfrac{L I_L}{D' U_E - I_L(r_E + D'^2 r_C)}\right)(1 + s \cdot r_C C)}{LC}}{\Delta}$$

$$(4.40)$$

where the denominator (Δ) is defined by

$$\Delta = s^2 + s \cdot \left(\frac{r_E + D'^2 r_C + F_m^{sp} U_E q_c^{sp}}{L} - \frac{F_m^{sp} I_L q_o^{sp}}{C} \right)$$

$$+ \frac{D'^2 + D' F_m \left(U_E q_o^{sp} + I_L q_c^{sp} \right) - F_m I_L q_o^{sp} \left(r_E + D'^2 r_C \right)}{LC} \quad (4.41)$$

It may be obvious that the symbolic forms of Z_{o-o}, G_{co}, and the characteristic equation (Δ) are same for the boost and buck–boost converters.

4.3.3
Origin and Consequences of Mode Limit in CCM

We may compute according to the averaged comparator Eq. (4.8) at steady state that the difference between the control current (I_{co}) and the average inductor current (I_L) in terms of the duty ratio (D) can be given by

$$I_{co} - I_L = -\frac{(M_1 + M_2) T_s}{2} \cdot D^2 + \left(M_c + \frac{M_1 + M_2}{2} \right) T_s \cdot D \quad (4.42)$$

which has a minima according to

$$|I_{co} - I_L|_{\min} = \frac{M_c T_s}{2} \left(1 + \frac{M_c}{M_1 + M_2} \right) + \frac{(M_1 + M_2) T_s}{8} \quad (4.43)$$

at the duty ratio (D)

$$D = \frac{1}{2} + \frac{M_c}{M_1 + M_2} \quad (4.44)$$

The duty ratio (4.44), where the minima takes place, is the same value, where the small-signal duty-ratio gain would become infinite, because F_m can be given generally as

$$F_m = \frac{1}{T_s \left(M_c + \frac{(D' - D)(M_1 + M_2)}{2} \right)} \quad (4.45)$$

according to (4.8). This means also that the 100% of the duty-ratio range requires the compensation to be designed as

$$M_c = \frac{M_1 + M_2}{2} \quad (4.46)$$

The parabola shape of (4.42) dictates that the difference would decrease along the increase in the duty ratio until the duty ratio defined by (4.44) is reached. After that, the difference should start increasing again, but it is physically impossible within a single cycle, because the increasing duty ratio requires

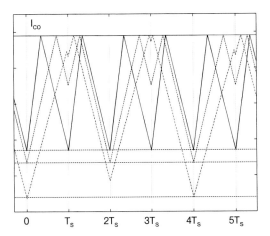

Figure 4.5 Simulated inductor-current waveforms normal (solid line) and subharmonic modes (second-harmonic: dashed line, fourth-harmonic: dash-dot line) based on an uncompensated buck converter.

the difference to keep on decreasing as well. As a consequence, the converter is forced to enter into the second harmonic mode of operation, where the difference naturally increases and a stable operating point would be found as shown in Figure 4.5. The possible stable subharmonic operation modes may exist at $f_s/2^n$, where f_s is the switching frequency and $n = 1, 2, 3, \ldots$, until the converter enters into a chaotic operation mode [17].

Equation (4.42) may be developed also in terms of D as

$$D^2 - \left(1 + \frac{2M_c}{M_1 + M_2}\right) \cdot D + \frac{2(I_{co} - I_L)}{T_s(M_1 + M_2)} = 0 \quad (4.47)$$

having roots at

$$D_{1,2} = \frac{1}{2} + \frac{M_c}{M_1 + M_2} \pm \sqrt{\left(\frac{1}{2} + \frac{M_c}{M_1 + M_2}\right)^2 - \frac{2(I_{co} - I_L)}{T_s(M_1 + M_2)}} \quad (4.48)$$

If the difference $(I_{co} - I_L)$ in (4.48) is substituted with $|I_{co} - I_L|_{\min}$ (4.43), the quadratic equation in (4.47) would have a double root coinciding with (4.44), which means that the real-valued solution exists only up to the duty ratio equaling (4.44). Therefore, it may be obvious that the average comparator Eq. (4.8) and the small-signal duty-ratio gain (F_m) would predict correctly the location and existence of the mode limit as also discussed in [20].

It is observed that the inductor current up (M_1) and down (M_2) slopes maintain certain relation in the subharmonic mode, which is clearly visible in Figure 4.6 as well. The formula defining mathematically the relation can be derived as follows: the small-signal inductor-current loop (see Figure 4.4) has infinite duty-ratio gain (i.e., $F_m = \infty$) at the mode limit. As a consequence, the perturbation in the inductor current would follow exactly the perturbation in the control current, which is zero at open loop. This means that the derivative of the time-averaged inductor current ($\langle i_L \rangle$) has to be zero, i.e.,

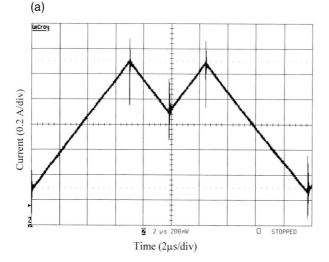

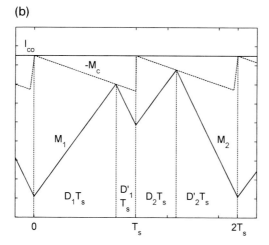

Figure 4.6 Inductor-current waveforms in second-harmonic mode: (a) measured inductor current with zero compensation and (b) simulated inductor current with certain amount of compensation.

$d_{\max}m_1 - d'_{\max}m_2 = 0$, where $d_{\max}$ equals (4.44). If we substitute $d_{\max}$ with (4.44), we get

$$M_2 = M_1 + 2M_c \tag{4.49}$$

which shows that the absolute values of the up and down slopes are same in an uncompensated (i.e., $M_c = 0$) converter as also shown in Figure 4.6a.

From Figure 4.6b, we may compute that

$$(M_1 + M_c)D_1 = M_2 D_2' + M_c D_2$$
$$(M_1 + M_c)D_2 = M_2 D_1' + M_c D_1 \quad (4.50)$$

which yields the average duty ratio (D_{av}) as

$$D_{av} = \frac{D_1 + D_2}{2} = \frac{M_2}{M_1 + M_2} \quad (4.51)$$

If we denote $M = U_o/U_{in}$ and consider the ideal converters, then we will get by applying (4.51) that $M = D_{av}$ for a buck converter, $M = 1/D_{av}'$ for a boost converter, and $M = D_{av}/D_{av}'$ for a buck–boost converter, which are similar to the ideal modulos (i.e., $M(D)$) defined in [8]. The average duty ratio (4.51) can be shown to be equal to the maximum duty ratio defined in (4.44) by substituting M_2 in (4.51) with (4.49) and applying some mathematical manipulation. This means in practice that the output dynamics of the converter entered into a subharmonic mode would be limited by D_{av}.

4.4
Modeling in DCM

The modeling of PCM control in DCM is the similar process as in CCM. The same comparator equation (4.3) applies, and the main task is to find a proper definition for Δi_L, which can be solved from the inductor current waveforms shown in Figure 4.7 according to its definition at $t = (k + d)T_s$. This procedure yields

$$\Delta i_L = m_1 d T_s - \frac{m_1 d(d + d_1) T_s}{2}, \quad (4.52)$$

where the first term corresponds to the peak inductor current and the last term to the average inductor current [16, 21].

The unknown duty ratio (d_1) can be solved from the equality $dm_1 = d_1 m_2$, which yields

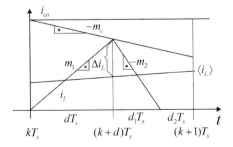

Figure 4.7 Inductor-current waveforms in DCM.

$$d_1 = \frac{m_1}{m_2} \cdot d \qquad (4.53)$$

Substituting d_1 in (4.52) with (4.53) yields

$$\Delta i_L = m_1 dT_s - \frac{m_1(m_1 + m_2)d^2 T_s}{2m_2} \qquad (4.54)$$

and consequently the comparator equation (4.8) becomes as

$$i_{co} - M_c dT_s = \langle i_L \rangle + m_1 dT_s - \frac{m_1(m_1 + m_2)d^2 T_s}{2m_2} \qquad (4.55)$$

The coefficients of the duty-ratio constraints of (4.1) or (4.2) can be solved from (4.55) by substituting the up and down slopes with their topology-based values introduced in Chapter 3 (Section 3.5), and developing the proper partial derivatives. It may be obvious that the small-signal duty-ratio gain (F_m) can be generally given by

$$F_m = \frac{1}{T_s \left(M_c + \dfrac{M_1(M_2 - (M_1 + M_2)D)}{M_2} \right)} \qquad (4.56)$$

It is obvious that the duty-ratio gain (F_m) may become infinite also in DCM similarly to CCM, when the duty ratio is

$$D = \frac{M_2}{M_1 + M_2} + \frac{M_2 M_c}{M_1(M_1 + M_2)} \qquad (4.57)$$

The value $M_2/(M_1 + M_2)$ actually defines the mode limit between the DCM and CCM operation and is symbolically same as the similar maximum value in CCM (4.51). The simulated inductor-current waveforms shown in Figure 4.8 based on an uncompensated DCM buck converter having $K = 0.45$ (i.e., $L = 9\,\mu H$, $R_{eq} = 4\,\Omega$, $f_s = 100\,kHz$) prove that the subharmonic operation would also take place in DCM [21]. The possible subharmonic frequencies are all the even and odd harmonics of the switching frequencies.

4.4.1
Duty-Ratio Constraints for Basic Converters

We develop the duty-ratio constraints for the basic converters corresponding to (4.1) by utilizing the material provided in Chapter 3 (Section 3.5) and omitting the effect of circuit parasitic elements.

4.4.1.1 Buck Converter

The inductor-current up slope is $m_1 = (\langle u_{in} \rangle - \langle u_o \rangle)/L$ and down slope $m_2 = \langle u_o \rangle/L$. As a consequence, the averaged comparator equation can be

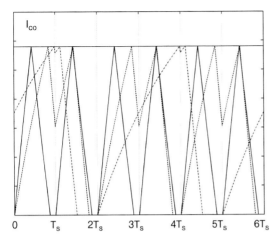

Figure 4.8 Simulated inductor-current waveforms in subharmonic mode in a DCM buck converter (basic switching frequency: solid line, second-harmonic: dashed line, third-harmonic: dash-dot line).

given by

$$i_{co} - M_c dT_s = \langle i_L \rangle + \frac{(\langle u_{in} \rangle - \langle u_o \rangle)dT_s}{L} - \frac{(\langle u_{in} \rangle - \langle u_o \rangle)\langle u_{in} \rangle d^2 T_s}{2L\langle u_o \rangle} \tag{4.58}$$

and the corresponding small-signal duty-ratio constraints by

$$\hat{d} = \frac{1}{T_s\left(M_c + \dfrac{U_{in}(1-M)(M-D)}{LM}\right)} \left(\hat{i}_{co} - \hat{i}_L - \frac{DT_s}{L}\left(\frac{D}{2M^2} - 1\right)\cdot \hat{u}_o \right.$$
$$\left. - \frac{DT_s}{L}\left(1 - \frac{(2-M)D}{2M}\right)\cdot \hat{u}_{in}\right) \tag{4.59}$$

where $M = U_o/U_{in}$.

According to (4.59), the duty-ratio-constraints coefficients are:

$$F_m = \frac{1}{T_s\left(M_c + \dfrac{U_{in}(1-M)(M-D)}{LM}\right)}$$

$$q_c = 1$$

$$q_o = \frac{DT_s}{L}\left(\frac{D}{2M^2} - 1\right)$$

$$q_i = \frac{DT_s}{L}\left(1 - \frac{(2-M)D}{2M}\right)$$

(4.60)

4.4.1.2 Boost Converter

The inductor-current up slope is $m_1 = \langle u_{in}\rangle/L$ and down slope $m_2 = (\langle u_o\rangle - \langle u_{in}\rangle)/L$. As a consequence, the averaged comparator equation can be given by

$$i_{co} - M_c d T_s = \langle i_L\rangle + \frac{\langle u_{in}\rangle d T_s}{L} - \frac{\langle u_{in}\rangle \langle u_o\rangle d^2 T_s}{2L(\langle u_o\rangle - \langle u_{in}\rangle)} \tag{4.61}$$

and the corresponding small-signal duty-ratio constraints by

$$\hat{d} = \frac{1}{T_s\left(M_c + \dfrac{U_{in}(D'M-1)}{L(M-1)}\right)}$$

$$\times \left(\hat{i}_{co} - \hat{i}_L - \frac{D^2 T_s}{2L(M-1)^2}\cdot \hat{u}_o - \frac{DT_s}{L}\left(1 - \frac{DM^2}{2(M-1)}\right)\cdot \hat{u}_{in}\right) \tag{4.62}$$

where $M = U_o/U_{in}$.

According to (4.62), the duty-ratio-constraints coefficients are:

$$\begin{aligned}
F_m &= \frac{1}{T_s\left(M_c + \dfrac{U_{in}(D'M-1)}{L(M-1)}\right)} \\
q_c &= 1 \\
q_o &= \left(\frac{D^2 T_s}{2L(M-1)^2}\right) \\
q_i &= \frac{DT_s}{L}\left(1 - \frac{DM^2}{2(M-1)}\right)
\end{aligned} \tag{4.63}$$

4.4.1.3 Buck–boost Converter

The inductor-current up slope is $m_1 = \langle u_{in}\rangle/L$ and down slope $m_2 = \langle u_o\rangle/L$. As a consequence, the averaged comparator equation can be given by

$$i_{co} - M_c dT_s = \langle i_L\rangle + \frac{\langle u_{in}\rangle dT_s}{L} - \frac{\langle u_{in}\rangle(\langle u_{in}\rangle + \langle u_o\rangle)d^2 T_s}{2L\langle u_o\rangle} \tag{4.64}$$

and the corresponding small-signal duty-ratio constraints by

$$\hat{d} = \frac{1}{T_s\left(M_c + \dfrac{U_{in}(D'M - D)}{LM}\right)}$$

$$\times \left(\hat{i}_{co} - \hat{i}_L - \frac{D^2 T_s}{2LM^2}\cdot \hat{u}_o - \frac{DT_s}{L}\left(1 - \frac{D(2+M)}{2M}\right)\cdot \hat{u}_{in}\right) \tag{4.65}$$

where $M = U_o/U_{in}$.

According to (4.65), the duty-ratio-constraints coefficients are:

$$F_m = \frac{1}{T_s\left(M_c + \dfrac{U_{in}(D'M - D)}{LM}\right)}$$

$$q_c = 1$$

$$q_o = \frac{D^2 T_s}{2LM^2}$$

$$q_i = \frac{DT_s}{L}\left(1 - \frac{D(2+M)}{2M}\right)$$

(4.66)

4.4.2
Small-Signal PCMC State Spaces for the Basic Converters

The state spaces are derived by substituting the perturbed duty ratio $\left(\hat{d}\right)$ in the corresponding VMC state spaces (Chapter 3, Section 3.5) with

$$\hat{d} = F_m \left(\hat{i}_{co} - \hat{i}_L - q_o \cdot \hat{u}_C - q_i \cdot \hat{u}_{in}\right) \qquad (4.67)$$

where the coefficients are derived in Section 4.4.1 for the basic converters. The transfer functions describing the converter dynamics can be solved naturally from the presented state spaces, but the solving is left for the reader. The dynamics of the PCMC converter in DCM is highly damped due to the upper limit provided by the control current and the bottom limit provided by the zero level. Therefore, the transfer functions would have mainly a first-order nature [21].

Buck:

$$\frac{d\hat{i}_L}{dt} = -\left(\frac{R_{eq}}{L}\sqrt{\frac{K}{1-M}} + \frac{2F_m U_{in}}{L}\right) \cdot \hat{i}_L - \left(\frac{1}{L(1-M)}\sqrt{\frac{K}{1-M}} + \frac{2F_m q_o}{L}\right) \cdot \hat{u}_C$$

$$+ \left(\frac{(2-M)M}{L(1-M)}\sqrt{\frac{K}{1-M}} - \frac{2F_m U_{in} q_i}{L}\right) \cdot \hat{u}_{in} + \frac{2F_m U_{in}}{L} \cdot \hat{i}_{co}$$

$$\frac{d\hat{u}_C}{dt} = \frac{\hat{i}_L}{C} - \frac{\hat{i}_o}{C}$$

$$\hat{i}_{in} = -\frac{2F_m U_o}{R_{eq}}\sqrt{\frac{1-M}{K}} \cdot \hat{i}_L - \left(\frac{M^2}{R_{eq}(1-M)} + \frac{2F_m U_o q_o}{R_{eq}}\sqrt{\frac{1-M}{K}}\right) \cdot \hat{u}_C$$

$$+ \left(\frac{M^2}{R_{eq}(1-M)} - \frac{2F_m U_o}{R_{eq}}\sqrt{\frac{1-M}{K}}\right) \cdot \hat{u}_{in} + \frac{2F_m U_o}{R_{eq}}\sqrt{\frac{1-M}{K}} \cdot \hat{i}_{co}$$

$$\hat{u}_o = \hat{u}_C + r_C C \frac{d\hat{u}_C}{dt} \qquad (4.68)$$

Boost:

$$\frac{d\hat{i}_L}{dt} = -\left(\frac{R_{eq}}{L}\sqrt{\frac{K(M-1)}{M}} + \frac{2F_m U_o}{L}\right)\cdot \hat{i}_L - \left(\frac{1}{L}\sqrt{\frac{KM}{M-1}} + \frac{2F_m U_o q_o}{L}\right)\cdot \hat{u}_C$$

$$+ \left(\frac{M^2}{L}\sqrt{\frac{KM}{M-1}} - \frac{2F_m U_o q_i}{L}\right)\cdot \hat{u}_{in} + \frac{2F_m U_o}{L}\cdot \hat{i}_{co}$$

$$\frac{d\hat{u}_C}{dt} = \left(1 + \frac{2F_m U_o}{R_{eq}}\sqrt{\frac{M-1}{KM}}\right)\cdot \frac{\hat{i}_L}{C} + \frac{2F_m U_o q_o}{R_{eq} C}\sqrt{\frac{M-1}{KM}}\cdot \hat{u}_C - \frac{1}{R_{eq} C}$$

$$\times \left(M(M-1) - 2F_m U_o q_o \sqrt{\frac{M-1}{KM}}\right)\cdot \hat{u}_{in} - \frac{\hat{i}_o}{C} - \frac{2F_m U_o}{R_{eq} C}\sqrt{\frac{M-1}{KM}}\cdot \hat{i}_{co}$$

$$\hat{i}_{in} = \hat{i}_L$$

$$\hat{u}_o = \hat{u}_C + r_C C \frac{d\hat{u}_C}{dt} \tag{4.69}$$

Buck–Boost:

$$\frac{d\hat{i}_L}{dt} = -\left(\frac{R_{eq}\sqrt{K}}{L} + \frac{2F_m(U_{in}+U_o)}{L}\right)\cdot \hat{i}_L - \left(\frac{\sqrt{K}}{L} + \frac{2F_m(U_{in}+U_o)q_o}{L}\right)\cdot \hat{u}_C$$

$$+ \left(\frac{M(M+2)\sqrt{K}}{L} - \frac{2F_m(U_{in}+U_o)q_i}{L}\right)\cdot \hat{u}_{in} + \frac{2F_m(U_{in}+U_o)}{L}\cdot \hat{i}_{co}$$

$$\frac{d\hat{u}_C}{dt} = \left(1 + \frac{2F_m U_o}{R_{eq}\sqrt{K}}\right)\cdot \frac{\hat{i}_L}{C} + \frac{2F_m U_o q_o}{R_{eq} C\sqrt{K}}\cdot \hat{u}_C - \frac{1}{R_{eq} C}\left(M^2 - \frac{2F_m U_o q_i}{\sqrt{K}}\right)\cdot \hat{u}_{in}$$

$$- \frac{\hat{i}_o}{C} - \frac{2F_m U_o}{R_{eq} C\sqrt{K}}\cdot \hat{i}_{co}$$

$$\hat{i}_{in} = -\frac{2F_m U_o}{R_{eq}\sqrt{K}}\cdot \hat{i}_L - \frac{2F_m U_o}{R_{eq}\sqrt{K}}\cdot \hat{u}_C + \left(\frac{M^2}{R_{eq}} - \frac{2F_m U_o q_i}{R_{eq}\sqrt{K}}\right)\cdot \hat{u}_{in} + \frac{2F_m U_o}{R_{eq}\sqrt{K}}\cdot \hat{i}_{co}$$

$$\hat{u}_o = \hat{u}_C + r_C C \frac{d\hat{u}_C}{dt} \tag{4.70}$$

4.4.3
Origin and Consequences of Mode Limit in DCM

The averaged comparator equation (4.55) may be presented also at steady state, when the compensation is set to zero, by

$$I_{co} - I_L = M_1 T_s \cdot D - \frac{M_1(M_1+M_2)T_s}{2M_2}\cdot D^2 \tag{4.71}$$

Equation (4.71) is a similar quadratic equation as in CCM (4.42), which dictates that the difference ($I_{co} - I_L$) shall decrease along the increase in the duty ratio until the minimum value

$$|I_{co} - I_L|_{min} = \frac{M_1 M_2 T_s}{2(M_1 + M_2)} \quad (4.72)$$

is reached. The corresponding maximum duty ratio (D_{max}) is

$$D_{max} = \frac{M_2}{M_1 + M_2} \quad (4.73)$$

which is known to define the mode limit between the DCM and CCM operation [8]. The averaged comparator equation can also be developed in terms of duty ratio as

$$D^2 - \frac{2M_2}{M_1 + M_2} \cdot D + \frac{2M_2(I_{co} - I_L)}{T_s M_1 (M_1 + M_2)} = 0 \quad (4.74)$$

It may be obvious that (4.74) has a real-valued solution only up to the duty ratio corresponding to (4.73).

The steady-state comparator equation (4.55) can also be developed in terms of the output-to-input relation (M), when taking into account that $D = M\sqrt{\frac{KM}{1-M}}$ for a buck converter, $D = \sqrt{KM(M-1)}$ for a boost converter, and $D = M\sqrt{K}$ for a buck–boost converter. This procedure yields at resistive load for a buck converter

$$M^3 - M^2 + K\left(\frac{I_{co} R}{U_{in}}\right)^2 = 0 \quad (4.75)$$

for a boost converter

$$M^2 - M - \frac{K}{U_{in}}\left(\frac{I_{co} R}{2}\right)^2 = 0 \quad (4.76)$$

and for a buck–boost converter

$$M = \frac{\sqrt{K} \cdot I_{co}}{2 U_{in}} \quad (4.77)$$

Equation (4.75) can be developed further yielding

$$\left(M - \frac{2}{3}\right)^2 \left(M - \left(\frac{3 I_{co} R}{4 U_{in}}\right)^2\right) = 0 \quad (4.78)$$

which shows that there exists a double root at $M = \frac{2}{3}$, which means that there are no real-valued solutions for M after $M = \frac{2}{3}$ at open loop as also discussed in [8]. It can be proved theoretically that the mode limit at $M = \frac{2}{3}$ exists only at open loop: according to the inductor-current waveforms (Figure 4.6, $M_{co} = 0$), we can compute substituting D with $M\sqrt{\frac{K}{1-M}}$ and D_1 with $\sqrt{K(1-M)}$ that a real-valued solution for the control current can always be found as

$$I_{co} = \frac{2U_o}{R}\sqrt{\frac{1-\frac{U_o}{U_{in}}}{K}} \tag{4.79}$$

when the feedback loop is connected. This proves that the corresponding mode limit does not exist under feedback control. According to (4.76) and (4.77), any special open-loop mode limit does not exist in boost and buck–boost converters, because the roots of the corresponding equations are always real. This also means that the artificial compensation is not necessary in DCM, when the converters are designed to operate in such a way that their duty ratio is always less than D_{max} defined in (4.73).

4.5 Dynamic Review

The buck and boost converters operating in CCM will be analyzed in detail. The dynamical features of the PCM control in a buck converter are compared experimentally to the corresponding VMC buck converter as analyzed in Chapter 3 (Section 3.6). The PCM-controlled DCM converters are not analyzed dynamically at all due to their limited importance in the field.

A PCM-controlled converter cannot operate at open loop at a constant-current load due to their current-output nature. This means that the internal dynamics are not directly measurable, but we have to measure first the load-affected transfer functions and then compute the internal transfer functions from the measured data by applying the load-affected transfer functions defined in Chapter 2 (Eq. (2.30)). The commercially available electronic loads do not usually provide the ideal features they are supposed to provide. Therefore, it is highly recommended to use a pure resistor as a load. The output impedance can be, however, measured directly, which would also make the computations more deterministic.

As an example, the measured and computed control-to-output transfer functions of a PCMC buck converter at the duty ratio of 0.3 and 0.4 are shown in Figure 4.9, where the artificial compensation (M_c) is set to zero. In practice, the measured control-to-output transfer functions also contain the effect of the PWM modulator and the equivalent inductor-current-sensing resistor, which are removed computationally from the responses shown in Figure 4.9. At low frequencies, the load resistor totally hides the real behavior of the control-to-output transfer function but not anymore at the higher frequencies.

The effect of the increasing duty ratio on the dynamics of the PCMC buck converter is clearly visible in the internal control-to-output transfer function: the DC gain increases, the low-frequency pole moves toward origin, and the

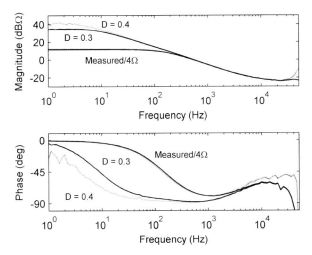

Figure 4.9 Measured control-to-output transfer function versus the internal control-to-output transfer function at the duty ratio of 0.3 and 0.4, where the load resistor has been 4 Ω.

high-frequency pole toward infinity. If the conclusions are drawn based on the load-affected transfer functions, then it is obvious that the low-frequency gain corresponds to the magnitude of the load resistor (i.e., ≈ 12 dBΩ) and the low-frequency pole is totally controlled by the load resistor (i.e., $f_{\text{low}} = 1/2\pi RC$). The moving of the high-frequency pole to the higher frequencies is visible but could be interpreted as a consequence of some other phenomenon as for example the sampling effect [11].

4.5.1
Buck Converter

The dynamic profile of the PCM-controlled buck converter as shown in Figure 4.10 is analyzed in CCM and compared to the dynamics associated with the VMC converter reviewed earlier in Chapter 3 (Section 3.6.1). The PCM control is implemented using a commercial PCM modulator UC3842 where the PWM modulator contributes an extra gain of 1/3. The on-time inductor current is measured using a current transformer having the turns ratio of 1/100 as shown in Figure 4.10. The equivalent inductor-current sensing resistor (R_s) equals 75 mΩ. The used artificial compensation (M_c) equals $1.45 \cdot U_o/2L$.

The analysis would be done at the input voltages of 20 and 50 V, but the practical measurements have been carried out at the input voltage of 20 V. The value of the duty ratio (D) at the input voltage of 20 V is 0.53 and 0.21 at 50 V as defined in Chapter 3 (Section 3.6.1). As a consequence, the duty-ratio gain (F_m^{sp}), the inductor-current-feedback gain (q_c^{sp}), and the input-voltage-feedforward gain (q_i^{sp}) (4.28) are at 20 V as

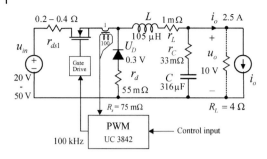

Figure 4.10 Experimental PCM-controlled buck converter.

$$F_m^{sp} = 1.576728$$
$$q_c^{sp} = 0.99709 \quad (4.80)$$
$$q_i^{sp} = 1.186 \times 10^{-2}$$

and at 50 V as

$$F_m^{sp} = 0.484776$$
$$q_c^{sp} = 0.99806 \quad (4.81)$$
$$q_i^{sp} = 7.9 \times 10^{-3}$$

4.5.1.1 Control-to-Output Transfer Function

The control-to-output transfer function (G_{co}) is symbolically presented in (4.82). The increased damping due to the lossless resistor ($F_m^{sp} U_E q_c^{sp}$) means that the roots of the characteristic equation (i.e., the denominator of (4.82)) or the poles are real and well separated.

$$\frac{F_m^{sp} U_E (1 + s \cdot r_C C)}{LC} \Big/ \left(s^2 + s \cdot \frac{r_E + F_m^{sp} U_E q_c^{sp} + r_C}{L} + \frac{1}{LC} \right) \quad (4.82)$$

As a consequence, the low-frequency pole can be approximated to be as

$$f_{p-\text{low}} \approx \frac{1}{2\pi} \cdot \frac{1}{F_m^{sp} U_E q_c^{sp} C} \quad (4.83)$$

and the high-frequency pole as

$$f_{p-\text{high}} \approx \frac{1}{2\pi} \cdot \frac{F_m^{sp} U_E q_c^{sp}}{L} \quad (4.84)$$

The low-frequency gain of the transfer function equals $F_m^{sp} U_E$, respectively.

4.5 Dynamic Review

The corresponding resistive-load-affected transfer function can be given as

$$G_{co}^{R} = \frac{\dfrac{R}{R+r_C} \cdot \dfrac{F_m^{sp} U_E (1 + s \cdot r_C C)}{LC}}{s^2 + s \cdot \dfrac{(R(r_E + F_m^{sp} U_E q_c^{sp} + r_C) + r_C(r_E + F_m^{sp} U_E q_c^{sp}))C + L}{LC(R+r_C)} + \dfrac{R + r_E + F_m^{sp} U_E q_c^{sp}}{LC(R+r_C)}} \quad (4.85)$$

from which the low- and high-frequency poles and the low-frequency gain can be approximated, when $F_m^{sp} U_E \gg r_E$, to be as

$$\begin{aligned}
f_{p-\text{low}} &\approx \frac{1}{2\pi} \cdot \left(\frac{1}{F_m^{sp} U_E q_c^{sp} C} + \frac{1}{RC} \right) \\
f_{p-\text{high}} &\approx \frac{1}{2\pi} \cdot \frac{F_m^{sp} U_E q_c^{sp}}{L} \\
|G_{co}^R|_{\text{low}} &\approx \frac{R F_m^{sp} U_E}{R + F_m^{sp} U_E q_c^{sp}}
\end{aligned} \quad (4.86)$$

Typically, $F_m^{sp} U_E \gg R$. Therefore, $f_{p-\text{low}}$ approaches to $1/2\pi RC$, and the low-frequency gain to R as demonstrated in Figure 4.9. As a consequence, the internal dynamics is not fully observable.

The measured control-to-output transfer functions at $D = 0.3$ and 0.4 (Figure 4.8) are repeated in Figure 4.11, where the dots and squares are the corresponding predictions using (4.82). The modeling accuracy is obvious.

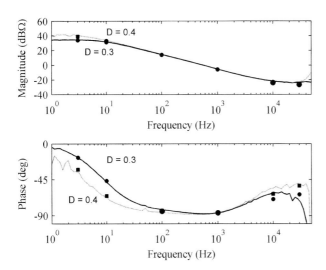

Figure 4.11 The measured internal control-to-output transfer functions at $D = 0.3$ and 0.4, where the dots and squares denote the predicted responses.

In order to compare the dynamic changes imposed by the PCM control to the VMC converter, we give the control-to-output transfer functions of both of the converters at 20 V as

$$G_{co}^{VMC} = \frac{s \cdot 6.2 \times 10^3 + 5.93 \times 10^8}{s^2 + s \cdot 2.08 \times 10^3 + 3 \times 10^7}$$

$$G_{co}^{PCMC} = \frac{s \cdot 9.756 \times 10^3 + 9.356 \times 10^8}{s^2 + s \cdot 2.9687 \times 10^5 + 3 \times 10^7}$$

(4.87)

and at 50 V as

$$G_{co}^{VMC} = \frac{s \cdot 1.6 \times 10^4 + 1.5 \times 10^9}{s^2 + s \cdot 1.33 \times 10^3 + 3 \times 10^7}$$

$$G_{co}^{PCMC} = \frac{s \cdot 7.57 \times 10^3 + 7.26 \times 10^8}{s^2 + s \cdot 2.303 \times 10^5 + 3 \times 10^7}$$

(4.88)

The damping factor (ζ) (Chapter 2, Section 2.6.2) varies from 0.19 to 0.12, and from 27.1 to 21.0, for the VMC and PCMC converters, respectively, where the first value corresponds to 20 V and the second value to 50 V. The main reason for the increased damping is the lossless series resistor ($F_m^{sp} U_E q_c^{sp}$), which is equal to 30.95 Ω at 20 V and 24.04 Ω at 50 V. As a consequence, the poles of the PCMC converter are real and well separated. The corresponding frequency responses are shown in Figure 4.12: typical to the PCMC transfer function is the first-order nature and the high independence from the changes in the input voltage. This means that the crossover frequency of the voltage-loop gain does not change when the input voltage changes as is the case with the VMC transfer function. The high-frequency phase lag of the PCMC transfer

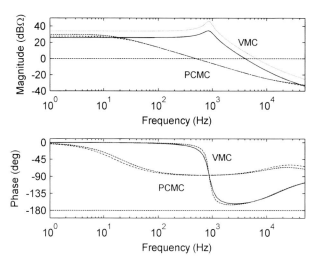

Figure 4.12 The predicted control-to-output transfer functions of a VMC and PCMC buck converters at the input voltages of 20 and 50 V.

function is highest at the high-line condition and, therefore, the controller should be designed also at the high line to ensure robust stability.

4.5.1.2 Output Impedance

The open-loop output impedance (Z_{o-o}) of the PCMC converter is presented symbolically in (4.89). Its low-frequency value equals $r_E + F_m^{sp} U_E q_c^{sp}$, where r_E is the low-frequency output impedance of the corresponding VMC converter. Its high-frequency value equals r_C. It may be obvious that the high-frequency pole (4.84) and the inductor-based zero would effectively cancel each other, when $F_m^{sp} U_E q_c^{sp} \gg r_E$.

$$\frac{\frac{(r_E + F_m^{sp} U_E q_c^{sp} + sL)(1 + s \cdot r_C C)}{LC}}{s^2 + s \cdot \frac{r_E + F_m^{sp} U_E q_c^{sp} + r_C}{L} + \frac{1}{LC}} \quad (4.89)$$

The measured internal output impedances corresponding to the control-to-output transfer functions of Figure 4.9 are shown in Figure 4.13, where the dots and the squares denote the predicted responses. It is obvious that any resonant-like behavior does not exist at half the switching frequency, as the method used in [11] would predict. The output impedance has clearly first-order nature as discussed above. The model-based prediction (i.e., the dots ($D = 0.3$) and squares ($D = 0.4$)) would also produce quite accurate results. Along the increase in the duty ratio, the low-frequency gain increases and the low-frequency pole moves closer to the origin.

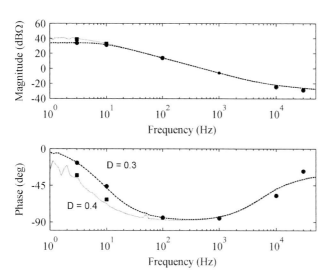

Figure 4.13 Directly measured internal output impedance at $D = 0.3$ and 0.4. The dots and squares denote the predicted responses.

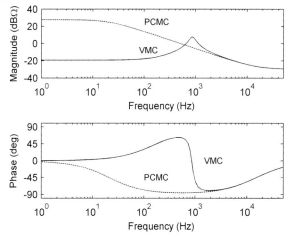

Figure 4.14 The predicted internal output impedances of VMC and PCMC buck converter at the input voltage of 50 V.

The VMC and PCMC output impedances are given numerically in (4.90) and the corresponding frequency responses in Figure 4.14 only at the input voltage of 50 V due to minimal changes in their behavior with respect to the input voltage.

$$Z_{o-o}^{\text{VMC}} = \frac{s^2 \cdot 3.3 \times 10^{-2} + s \cdot 3.2 \times 10^3 + 3.23 \times 10^6}{s^2 + s \cdot 1.33 \times 10^3 + 3 \times 10^7}$$

$$Z_{o-o}^{\text{PCMC}} = \frac{s^2 \cdot 3.3 \times 10^{-2} + s \cdot 1.075 \times 10^4 + 7.28 \times 10^8}{s^2 + s \cdot 2.303 \times 10^5 + 3 \times 10^7}$$

(4.90)

The load impedance will affect the converter dynamics as if the open-loop output impedance would behave as a weighting gain as discussed in Chapters 2 and 3 (i.e., $G_{co}^L = G_{co}/(1 + Z_{o-o}/Z_L)$). Therefore, we can predict the load sensitivity of the converter according to the shape of the open-loop output impedance (i.e., both the magnitude and phase), [23]: the VMC converter is sensitive to capacitive load especially at the frequencies close to the resonant frequency. The PCMC converter cannot become unstable due to capacitive load, although its voltage-loop crossover frequency may be reduced easily. We will discuss more in detail the dynamical load effects in Chapter 8.

4.5.1.3 Input-to-Output Transfer Function

The open-loop input-to-output transfer function (G_{io-o}) can be given symbolically by

$$G_{io-o} = \frac{\dfrac{(D - F_m^{\text{sp}} U_E q_i^{\text{sp}})(1 + s \cdot r_C C)}{LC}}{s^2 + s \cdot \dfrac{r_E + F_m^{\text{sp}} U_E q_c^{\text{sp}} + r_C}{L} + \dfrac{1}{LC}}$$

(4.91)

It may be obvious that the input noise attenuation may be theoretically very high if $D - F_m^{sp} U_E q_i^{sp} = 0$. This means that the artificial compensation (M_c) had to be set to $DU_E/2L$. According to (4.29), U_E is dependent on the operating point and, therefore, the ideal compensation cannot be maintained. This compensation value is known as an optimal compensation and is typically expressed as $U_o/2L$[8]. Physically, it means that the average inductor current is maintained constant at open loop and, therefore, the output voltage does not change when the input voltage changes. As we discussed earlier, we had to use light overcompensation (i.e., $M_c \approx 1.45 \cdot U_o/2L$) in order to ensure signal integrity in the inductor-current loop and, therefore, the attenuation would also be finite.

The numerical values of the input-to-output transfer functions of the VMC and PCMC converters are given at 20 V by

$$G_{io-o}^{VMC} = \frac{s \cdot 167 + 1.6 \times 10^7}{s^2 + s \cdot 2.08 \times 10^3 + 3 \times 10^7}$$
$$G_{io-o}^{PCMC} = \frac{s \cdot 50.84 + 4.876 \times 10^6}{s^2 + s \cdot 2.9687 \times 10^5 + 3 \times 10^7}$$
(4.92)

and at 50 V by

$$G_{io-o}^{VMC} = \frac{s \cdot 66 + 6.3 \times 10^6}{s^2 + s \cdot 1.33 \times 10^3 + 3 \times 10^7}$$
$$G_{io-o}^{PCMC} = \frac{s \cdot 6.19 + 5.94 \times 10^5}{s^2 + s \cdot 2.303 \times 10^5 + 3 \times 10^7}$$
(4.93)

The corresponding frequency responses are shown in Figure 4.15a, where the upper curve within the group (i.e., VMC or PCMC) corresponds to 20 V and the lower curve to 50 V. It is obvious that the PCM control provides much higher attenuation at all frequencies than the VM control. The frequency responses measured (solid lines) and predicted (dashed line) at 20 V are shown in Figure 4.15b. The reason for the deviation between the measured and predicted responses of the PCMC converter is the assumed level of artificial compensation, which deviates from the real compensation.

The input-to-output transfer function determines the level of the source interactions and the propagation of the load interactions through the converter [24–26]. Therefore, it may be obvious that the interactions taken place at the frequencies close to the resonant frequency would affect easily the VMC converter but not the PCMC converter. The interactions are treated more in detail in Chapter 8.

4.5.1.4 Input Admittance

The open-loop input admittance (Y_{in-o}) is symbolically given in (4.94). If the converter were optimally compensated (i.e., $M_c = DU_E/2L$), then $Y_{in-o} = -F_m^{sp} q_i^{sp} I_L = -DI_L/U_E$. In practice, such a condition is not possible as

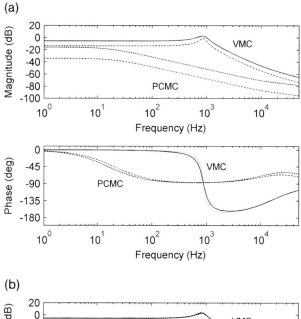

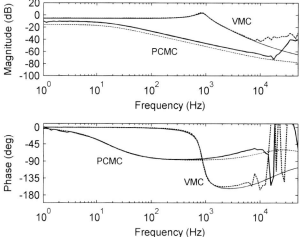

Figure 4.15 The input-to-output transfer functions of the VMC and PCMC converters: (a) the predicted transfer functions at the input voltages of 20 and 50 V (the upper curves within the group correspond to 20 V) and (b) the measured and predicted transfer functions at 20 V.

discussed earlier.

$$\frac{\left(D - F_m^{\mathrm{sp}} U_E q_i^{\mathrm{sp}}\right)\left(D - F_m^{\mathrm{sp}} I_L q_c^{\mathrm{sp}}\right) s}{L} - F_m^{\mathrm{sp}} q_i^{\mathrm{sp}} I_L \quad (4.94)$$

The corresponding numerical values of the input admittances of the VMC and PCMC converters are given at 20 V by

$$Y_{\text{in}-o}^{\text{VMC}} = \frac{s \cdot 2.86 \times 10^3}{s^2 + s \cdot 2.08 \times 10^3 + 3 \times 10^7}$$

$$Y_{\text{in}-o}^{\text{PCMC}} = -\frac{s^2 \cdot 4.676 \times 10^{-2} + s \cdot 1.91 \times 10^4 + 1.4 \times 10^6}{s^2 + s \cdot 2.9687 \times 10^5 + 3 \times 10^7}$$

(4.95)

and at 50 V by

$$Y_{\text{in}-o}^{\text{VMC}} = \frac{s \cdot 4.2 \times 10^2}{s^2 + s \cdot 1.33 \times 10^3 + 3 \times 10^7}$$

$$Y_{\text{in}-o}^{\text{PCMC}} = -\frac{s^2 \cdot 9.58 \times 10^{-3} + s \cdot 4.08 \times 10^3 + 2.872 \times 10^5}{s^2 + s \cdot 2.303 \times 10^5 + 3 \times 10^7}$$

(4.96)

The corresponding frequency responses are shown in Figure 4.16a, where the upper curves within the group (i.e., VMC or PCMC) correspond to 20 V and the bottom curves to 50 V. The frequency responses measured (dashed lines) and predicted (solid lines) are shown in Figure 4.16b. The high-frequency deviation between the measured and predicted responses is due to the circuit parasitic elements in the practical converter.

The open-loop input admittance or its inverse impedance is one of the decisive factors describing the effect of the source interactions in a converter [24, 26]. The lack of resonant peaking in the PCMC converter implies improved insensitivity to the source interactions compared to the VMC converter. More detailed treatment of the subject is presented in Chapter 8.

4.5.1.5 Ideal Input Admittance

The ideal admittance ($Y_{\text{in}-\infty}$) (i.e., $Y_{\text{in}-o} - \frac{G_{\text{io}-o} G_{\text{ci}}}{G_{\text{co}}}$) is same for the VMC and PCMC converters and can be given by

$$-\frac{DI_L}{U_E}$$

(4.97)

The corresponding measured-data-based frequency responses are shown in Chapter 3 (Section 3.6.1, Figure 3.23). The open-loop input admittance of the PCMC buck converter is quite close to the ideal input admittance, even if the artificial compensation is not ideal as shown in Figure 4.17 because of a rather high attenuation provided by the open-loop input-to-output transfer function.

4.5.1.6 Short-Circuit Input Admittance

The short-circuit input admittance ($Y_{\text{in}-\text{sc}}$) (i.e., $Y_{\text{in}-o} + \frac{G_{\text{io}-o} T_{\text{oi}-o}}{Z_{o-o}}$) can be symbolically given by

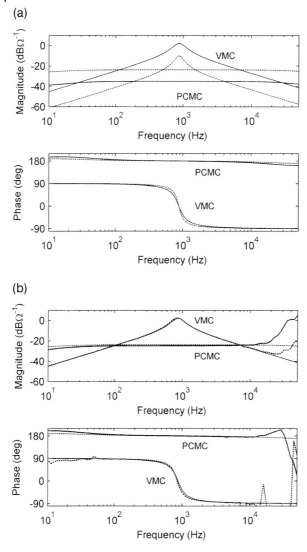

Figure 4.16 The input admittances of the VMC and PCMC converters: (a) the predicted responses at the input voltages of 20 V (the upper curves within the group) and 50 V (the bottom curves within the group) and (b) the measured and predicted responses at 20 V.

$$\frac{\left(D - F_m^{sp} U_E q_i^{sp}\right)\left(D - F_m^{sp} I_L q_c^{sp}\right)}{r_E + F_m^{sp} U_E q_c^{sp} + sL} - F_m^{sp} q_i^{sp} I_L \tag{4.98}$$

$Y_{\text{in}-sc}$ equals $Y_{\text{in}-\infty}$ for an optimally compensated converter due to $G_{\text{io}-o} = 0$. The corresponding numerical values of the VMC and PCMC converters are

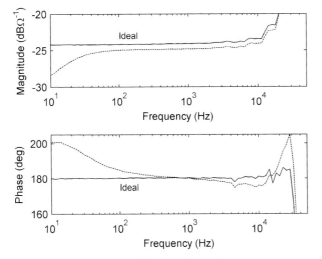

Figure 4.17 The ideal (solid line) and open-loop input admittance (dashed line) of the PCMC converter at the input voltage of 20 V.

given at 20 V by

$$Y_{in-sc}^{VMC} = \frac{1.505}{1 + s \cdot 5.65 \times 10^{-4}}$$

$$Y_{in-sc}^{PCMC} = -\frac{s \cdot 1.58 \times 10^{-7} + 6.44 \times 10^{-2}}{1 + s \cdot 3.37 \times 10^{-6}}$$

(4.99)

and at 50 V by

$$Y_{in-sc}^{VMC} = \frac{0.409}{1 + s \cdot 9.77 \times 10^{-4}}$$

$$Y_{in-sc}^{PCMC} = -\frac{s \cdot 4.16 \times 10^{-8} + 1.039 \times 10^{-2}}{1 + s \cdot 4.34 \times 10^{-6}}$$

(4.100)

The corresponding predicted frequency responses are given in Figure 4.18, where the upper curves within the group correspond to 20 V and the bottom curves to 50 V.

4.5.2
Boost Converter

The dynamic profile of the PCM-controlled boost converter shown in Figure 4.19 is analyzed in CCM and compared to the dynamics associated with the corresponding VMC converter reviewed earlier in Chapter 3 (Section 3.6.2). The PCM control is assumed to be implemented using the same commercial

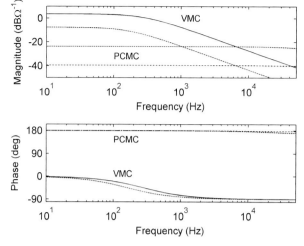

Figure 4.18 The predicted frequency responses of the short-circuit input admittance of the VMC and PCMC converters at the input voltages of 20 V (the upper curves within the group) and at 50 V (the bottom curves within the group).

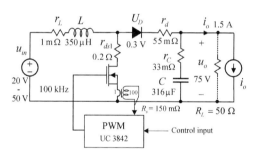

Figure 4.19 PCM-controlled boost converter.

PWM modulator as in the case of the buck converter. The on-time inductor current is measured using the equivalent sensing resistor of 150 mΩ as shown in Figure 4.19. The artificial compensation (M_c) is set to $U_o/2L$, which would provide a 100% duty-ratio range for all the input-voltage values. The required compensation value can be computed by setting the denominator of the duty-ratio gain (F_m) to zero.

The analysis would be done at the input voltages of 20 and 50 V, where the duty ratio (D) equals 0.748 and 0.339, and the inductor current (I_L) to 5.95 A and 2.27 A, respectively, as defined in Chapter 3 (Section 3.6.2). As a consequence, the duty-ratio gain (F_m^{sp}), the inductor-current-feedback gain (q_c^{sp}), the output-voltage-feedback gain (q_o^{sp}), and the output-current-feedforward gain (q_{io}^{sp}) (4.33) are at 20 V as

$$F_m^{sp} = 1.8411$$

$$q_c^{sp} = 0.9997$$

$$q_o^{sp} = 2.694 \times 10^{-3}$$
$$q_{io}^{sp} = -8.889 \times 10^{-5} \tag{4.101}$$

and at 50 V as

$$F_m^{sp} = 0.70644$$
$$q_c^{sp} = 0.99964$$
$$q_o^{sp} = 3.1992 \times 10^{-3} \tag{4.102}$$
$$q_{io}^{sp} = -1.056 \times 10^{-4}$$

4.5.2.1 Control-to-Output Transfer Function

The control-to-output transfer function (G_{co}) is symbolically presented in (4.103). The increased damping due to the lossless resistor ($F_m^{sp} U_E q_c$) means that the roots of the characteristic equation (i.e., the denominator of (4.103)) or the poles are real and well separated. The RHP zero present in the transfer function is the same as the RHP zero of the corresponding VMC converter.

$$\frac{\dfrac{F_m^{sp}(D'U_E - I_L(r_E + D'^2 r_C))}{LC}\left(1 - s \cdot \dfrac{LI_L}{D'U_E - I_L(r_E + D'^2 r_C)}\right)(1 + s \cdot r_C C)}{\Delta}$$

$$\Delta = s^2 + s \cdot \left(\frac{r_E + D'^2 r_C + F_m^{sp} U_E q_c^{sp}}{L} - \frac{F_m^{sp} I_L q_o^{sp}}{C}\right)$$

$$+ \frac{D'^2 + D' F_m^{sp}(U_E q_o^{sp} + I_L q_c^{sp}) - F_m^{sp} I_L q_o^{sp}(r_E + D'^2 r_C)}{LC} \tag{4.103}$$

Because of the well-separated roots, the low-frequency pole can be approximated by

$$f_{p-\text{low}} \approx \frac{D'^2 + D' F_m^{sp}(U_E q_o^{sp} + I_L q_c^{sp})}{F_m U_E q_c^{sp} C} \tag{4.104}$$

and the high-frequency pole by

$$f_{p-\text{high}} \approx \frac{F_m^{sp} U_E q_c^{sp}}{L} \tag{4.105}$$

The low-frequency gain can be approximated by

$$|G_{co}|_{\text{low}} \approx \frac{U_E}{U_E q_o^{sp} + I_L q_c^{sp}} \tag{4.106}$$

According to (4.106), we may conclude that the increase in the duty-ratio gain (F_m), when approaching the mode limit, is not clearly visible in the low-frequency gain similarly to the gain of the PCMC buck converter (4.86). It is, however, obvious that the low-frequency pole (4.104) would move toward the origin and the high-frequency pole (4.105) toward the infinity, when the duty-ratio gain increases.

In order to compare the dynamics imposed by the PCM control in the VMC boost converter, we give numerically the control-to-output transfer functions of both of the converters at 20 V as

$$G_{co}^{VMC} = \frac{-s^2 \cdot 0.2 - s \cdot 1.72 \times 10^2 + 1.61 \times 10^8}{s^2 + s \cdot 494 + 5.75 \times 10^5}$$
$$G_{co}^{PCMC} = \frac{-s^2 \cdot 0.36 - s \cdot 3.158 \times 10^4 + 2.96 \times 10^8}{s^2 + s \cdot 3.926 \times 10^5 + 2.633 \times 10^7}$$

(4.107)

and at 50 V as

$$G_{co}^{VMC} = \frac{-s^2 \cdot 7.49 \times 10^{-2} - s \cdot 2.54 \times 10^3 + 4.45 \times 10^8}{s^2 + s \cdot 363 + 3.95 \times 10^6}$$
$$G_{co}^{PCMC} = \frac{-s^2 \cdot 5.29 \times 10^{-2} - s \cdot 1.794 \times 10^3 + 3.146 \times 10^8}{s^2 + s \cdot 1.517 \times 10^5 + 1.453 \times 10^7}$$

(4.108)

The corresponding frequency responses of the PCMC control-to-output transfer functions are shown in Figure 4.20, where the dots and squares are the switching-model-based simulated responses. The match between the predictions and simulated responses is obvious.

The comparison between the VMC and PCMC transfer functions are shown in Figure 4.21: the high-frequency magnitude behavior of the PCMC transfer function at 20 V implies that the voltage-loop crossover frequency would be

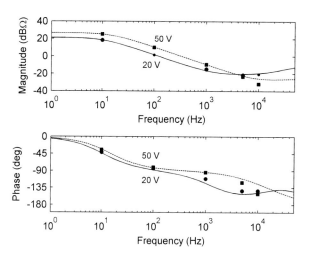

Figure 4.20 Predicted and switching-model-based simulated (dots and squares) frequency responses of the control-to-output transfer function of the PCMC boost converter at the input voltages of 20 V (solid line) and 50 V (dashed line).

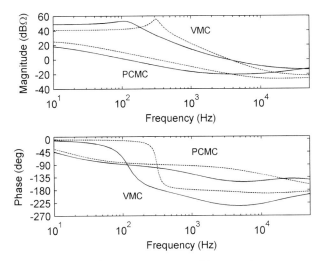

Figure 4.21 The frequency responses of the control-to-output transfer functions of the VMC and the PCMC boost converters at the input voltages of 20 V (solid lines) and 50 V (dashed lines).

shorter in the PCMC converter than in the VMC converter in order to maintain sufficient gain margin. If considering the situation from the viewpoint of the phase, the conclusion might be different but the magnitude behavior would be most decisive. The stability of the converter would also be quite sensitive to the size of the ESR (r_C) of the output capacitor: an increase in the ESR is directly reflected in an increase in the high-frequency gain.

4.5.2.2 Output Impedance

The open-loop output impedance (Z_{o-o}) of the PCMC boost converter is presented symbolically in (4.109). Its low-frequency value can be approximated to equal $F_m^{sp} U_E q_c^{sp}/(D'^2 + D' F_m^{sp}(U_E q_o^{sp} + I_L q_c^{sp}))$, and the high-frequency value to r_C. It may be obvious that the high-frequency zero and pole would cancel each other effectively and, therefore, the response would resemble a first-order response.

$$\frac{\frac{1}{LC}\left(r_E + F_m^{sp} U_E q_c^{sp} + F_m^{sp}\begin{pmatrix} q_{io}^{sp}(D' U_E - I_L(r_E + D'^2 r_C)) \\ - q_c^{sp} I_L D' r_C \end{pmatrix}\right) + s \cdot L}{\Delta}(1 + s \cdot r_C C)$$

$$\Delta = s^2 + s \cdot \left(\frac{r_E + D'^2 r_C + F_m^{sp} U_E q_c^{sp}}{L} - \frac{F_m^{sp} I_L q_o^{sp}}{C}\right)$$

$$+ \frac{D'^2 + D' F_m^{sp}\left(U_E q_o^{sp} + I_L q_c^{sp}\right) - F_m^{sp} I_L q_o^{sp}\left(r_E + D'^2 r_C\right)}{LC} \tag{4.109}$$

4 Average and Small-Signal Modeling of Peak-Current-Mode Control

In order to compare the dynamic changes imposed by the PCM control in the VMC boost converter, we give numerically the output impedances of both of the converters at 20 V as

$$Z_{o-o}^{\text{VMC}} = \frac{s^2 \cdot 3.3 \times 10^{-2} + s \cdot 3.18 \times 10^3 + 1.54 \times 10^6}{s^2 + s \cdot 494 + 5.75 \times 10^5}$$
$$Z_{o-o}^{\text{PCMC}} = \frac{s^2 \cdot 3.3 \times 10^{-2} + s \cdot 1.612 \times 10^4 + 1.24 \times 10^9}{s^2 + s \cdot 3.926 \times 10^5 + 2.633 \times 10^7}$$
(4.110)

and at 50 V as

$$Z_{o-o}^{\text{VMC}} = \frac{s^2 \cdot 3.3 \times 10^{-2} + s \cdot 3.18 \times 10^3 + 1.02 \times 10^6}{s^2 + s \cdot 363 + 3.95 \times 10^6}$$
$$Z_{o-o}^{\text{PCMC}} = \frac{s^2 \cdot 3.3 \times 10^{-2} + s \cdot 8.17 \times 10^3 + 4.8 \times 10^8}{s^2 + s \cdot 1.517 \times 10^5 + 1.453 \times 10^7}$$
(4.111)

The frequency responses of the output impedance at the input voltages of 20 V (solid line) and 50 V (dashed line) are shown in Figure 4.22, where the switching-model-based responses are denoted by dots and squares. The match with the analytical predictions and the switch-model-based responses is obvious as well as the first-order nature.

The comparison between the VMC and PCMC output impedances is shown in Figure 4.23. The effect of PCM control is essentially similar to the corresponding effect in the buck converter: the low-frequency gain increases and the output impedance changes to have more capacitive nature compared

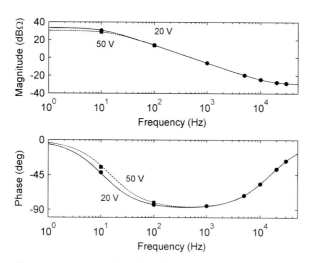

Figure 4.22 Predicted and switching-model-based (dots and squares) frequency responses of the output impedance of the PCMC boost converter at the input voltages of 20 V (solid line) and 50 V (dashed line).

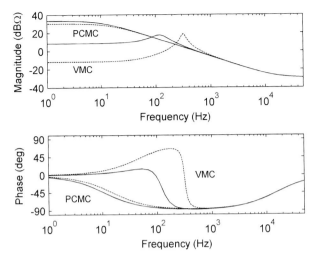

Figure 4.23 The frequency responses of the output impedances of the VMC and the PCMC boost converters at the input voltages of 20 V (solid lines) and 50 V (dashed lines).

to the corresponding VMC converter. The capacitive nature of the output impedance means that the instability sensitivity to the capacitive load naturally reduces.

4.5.2.3 Input-to-Output Transfer Function

The input-to-output transfer function (G_{io-o}) can be symbolically given by

$$\frac{\frac{(D' + F_m^{sp} I_L q_c^{sp})}{LC}(1 + s \cdot r_C C)}{\Delta}$$

$$\Delta = s^2 + s \cdot \left(\frac{r_E + D'^2 r_C + F_m^{sp} U_E q_c^{sp}}{L} - \frac{F_m^{sp} I_L q_o^{sp}}{C} \right)$$

$$+ \frac{D'^2 + D' F_m^{sp} (U_E q_o^{sp} + I_L q_c^{sp}) - F_m^{sp} I_L q_o^{sp}(r_E + D'^2 r_C)}{LC} \quad (4.112)$$

In order to compare the dynamic changes imposed by the PCM control into the VMC converter, we give numerically the input-to-output transfer functions of both the converters at 20 V as

$$G_{io-o}^{VMC} = \frac{s \cdot 17.79 + 1.7 \times 10^6}{s^2 + s \cdot 494 + 5.75 \times 10^5}$$

$$G_{io-o}^{PCMC} = \frac{s \cdot 1.056 \times 10^3 + 1.013 \times 10^8}{s^2 + s \cdot 3.926 \times 10^5 + 2.633 \times 10^7} \quad (4.113)$$

and at 50 V as

$$G_{\text{io-o}}^{\text{VMC}} = \frac{s \cdot 21.14 + 2.03 \times 10^6}{s^2 + s \cdot 363 + 3.95 \times 10^6}$$
$$G_{\text{io-o}}^{\text{PCMC}} = \frac{s \cdot 2.134 \times 10^2 + 2.047 \times 10^7}{s^2 + s \cdot 1.517 \times 10^5 + 1.453 \times 10^7}$$

(4.114)

The corresponding frequency responses are given in Figure 4.24 at the input voltages of 20 and 50 V. It is obvious that low-frequency input-noise attenuation does not exist, but it does at the high input voltage in the VMC converter. This means that the boost converter is sensitive to the source interactions and the load interactions would also easily make the source interactions more severe.

4.5.2.4 Input Admittance
The open-loop input admittance ($Y_{\text{in-o}}$) can be symbolically given by

$$\frac{\frac{1}{L}\left(s - \frac{F_m^{\text{sp}} I_L q_o^{\text{sp}}}{C}\right)}{\Delta}$$

$$\Delta = s^2 + s \cdot \left(\frac{r_E + D'^2 r_C + F_m^{\text{sp}} U_E q_c^{\text{sp}}}{L} - \frac{F_m^{\text{sp}} I_L q_o^{\text{sp}}}{C}\right)$$

$$+ \frac{D'^2 + D' F_m^{\text{sp}} \left(U_E q_o^{\text{sp}} + I_L q_c^{\text{sp}}\right) - F_m^{\text{sp}} I_L q_o^{\text{sp}} \left(r_E + D'^2 r_C\right)}{LC}$$

(4.115)

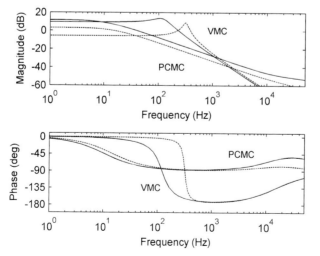

Figure 4.24 The frequency responses of the input-to-output transfer functions of the VMC and the PCMC boost converters at the input voltages of 20 V (solid lines) and 50 V (dashed lines).

The VMC and PCMC input admittances are given numerically for comparison at 20 V as

$$Y_{in-o}^{VMC} = \frac{s \cdot 2.86 \times 10^3}{s^2 + s \cdot 494 + 5.75 \times 10^5}$$
$$Y_{in-o}^{PCMC} = \frac{s \cdot 2.86 \times 10^3 - 2.668 \times 10^5}{s^2 + s \cdot 3.926 \times 10^5 + 2.633 \times 10^7}$$
(4.116)

and at 50 V as

$$Y_{in-o}^{VMC} = \frac{s \cdot 2.86 \times 10^3}{s^2 + s \cdot 363 + 3.95 \times 10^6}$$
$$Y_{in-o}^{PCMC} = \frac{s \cdot 2.86 \times 10^3 - 4.64 \times 10^4}{s^2 + s \cdot 1.517 \times 10^5 + 1.453 \times 10^7}$$
(4.117)

The corresponding frequency responses are shown in Figure 4.25. The lack of resonant peaking in the PCMC converter would make the converter less sensitive to the source interactions than the corresponding VMC converter. The effect is similar as in the buck converter.

4.5.2.5 Ideal Input Admittance

The ideal input admittance ($Y_{in-\infty}$) can be given by

$$-\frac{I_L}{D'U_E - I_L\left(r_E + D'^2 r_C\right)} \cdot \frac{1}{1 - s \cdot \dfrac{LI_L}{D'U_E - I_L\left(r_E + D'^2 r_C\right)}} \quad (4.118)$$

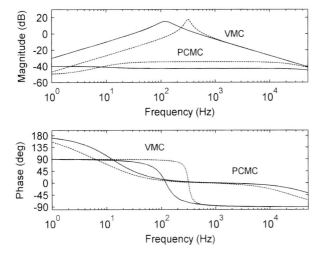

Figure 4.25 The frequency responses of the input admittances of the VMC and the PCMC boost converters at the input voltages of 20 V (solid lines) and 50 V (dashed lines).

4.5.2.6 Short-Circuit Input Admittance

The short-circuit admittance ($Y_{\text{in-sc}}$) can be symbolically given by

$$\frac{1 - F_m^{\text{sp}} I_L q_{\text{io}}^{\text{sp}}}{r_E + F_m^{\text{sp}} U_E q_c^{\text{sp}} + F_m^{\text{sp}} \left(q_{\text{io}}^{\text{sp}} \left(D' U_E - I_L (r_E + D'^2 r_C) \right) - q_c^{\text{sp}} I_L D' r_C \right) + s \cdot L \left(1 - F_m^{\text{sp}} I_L q_{\text{io}}^{\text{sp}} \right)} \tag{4.119}$$

The VMC and PCMC short-circuit admittances are given for comparison at 20 V as

$$Y_{\text{in-sc}}^{\text{VMC}} = \frac{5.86}{1 + s \cdot 2.05 \times 10^{-3}}$$
$$Y_{\text{in-sc}}^{\text{PCMC}} = \frac{7.276 \times 10^{-3}}{1 + s \cdot 2.547 \times 10^{-6}} \tag{4.120}$$

and at 50 V as

$$Y_{\text{in-sc}}^{\text{VMC}} = \frac{8.9}{1 + s \cdot 3.11 \times 10^{-3}}$$
$$Y_{\text{in-sc}}^{\text{PCMC}} = \frac{1.887 \times 10^{-2}}{1 + s \cdot 6.604 \times 10^{-6}} \tag{4.121}$$

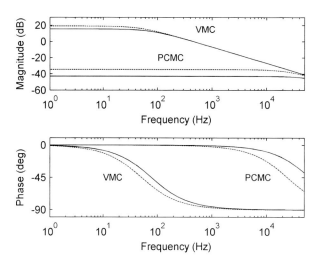

Figure 4.26 The frequency responses of the short-circuit input admittances of the VMC and the PCMC boost converters at the input voltages of 20 V (solid lines) and 50 V (dashed lines).

as well as the corresponding frequency responses in Figure 4.26. The much lower value of the short-circuit input admittance in the PCMC converter compared to the VMC converter implies reduced sensitivity to the source interactions.

References

1. C.W. Deisch, 'Simple switching control method changes power converter into a current source,' in *Proc. IEEE Power Electronics Specialists Conf.*, **1978**, pp. 300–306.
2. A. Capel, G. Ferrante, D. O'Sullivan, and A. Wienberg, 'Application of the injected current model for the dynamic analysis of switching regulators with the new concept of LC^3 modulator,' in *Proc. IEEE Power Electronics Specialists Conf.*, **1978**, pp. 135–147.
3. R. Redl and N.O. Sokal, 'What a design engineer should know about current mode control,' in *Proc. IEEE Power Electronics Design Conference*, **1985**, pp. 18–33.
4. D.M. Mitchell, 'Tricks of the trade: understanding the right-half-plane zero in small-signal DC–DC converter models,' *IEEE Power Electronics Society Newsletter*, January **2001**, pp. 5–6.
5. G.K. Schoneman and D.M. Mitchell, 'Output impedance considerations for switching regulators with current injected control,' *IEEE Trans. Power Electron.*, vol. 4, no. 1, **1989**, pp. 25–35.
6. S.-H. Hsu, A. Brown, L. Rensink, and R.D. Middlebrook, 'Modeling and analysis of switching DC-to-DC converters in constant-frequency current programmed mode,' in *Proc. IEEE Power Electronics Specialists Conf.*, **1979**, pp. 248–301.
7. A.S. Kislovski, 'Unified model for open-loop instability,' in *Proc. IEEE Applied Power Electronics Conf.*, **1991**, pp. 459–465.
8. R.W. Erickson and D. Maksimovic, *Fundamentals of Power Electronics*, Kluwer, Norwell, MA, USA, **2001**, 2nd Edition.
9. S. Banerjee and G.C. Verghese, *Nonlinear Phenomena in Power Electronics*, IEEE Press, NJ, USA, **2001**.
10. C.K. Tse, *Complex Behaviour of Switching Power Converters*, CRC Press, Boca Raton, FL, USA, **2004**.
11. R.B. Ridley, 'A new continuous-time model for current-mode control,' *IEEE Trans. Power Electron.*, vol. 6, no. 2, **1991**, pp. 271–280.
12. R.B. Ridley, 'A new continuous-time model for current-mode control with constant frequency, constant on-time, and constant off-time, in CCM and DCM,' in *Proc. IEEE Power Electronics Specialists Conf.*, **1990**, pp. 382–389.
13. C.P. Schultz, 'A unified model of constant frequency switching regulators using multiloop feedback control,' in *Proc. Power Conversion and Intelligent Motion Conf.*, **1993**, pp. 319–329.
14. F.D. Tan and R.D. Middlebrook, 'A unified model for current-programmed converters,' *IEEE Trans Power Electron.*, vol. 10, no. 4, July **1995**, pp. 397–408.
15. J. Sun and R.M. Bass, 'A new approach to average modeling of PWM converters with current-mode control,' in *Proc. IEEE Industrial Electronics Annual Conf.*, **1997**, pp. 599–604.
16. T. Suntio, 'Unified derivation and analysis of duty-ratio constraints for peak-current-mode control in continuous and discontinuous modes,' in *Proc. IEEE Industrial Electronics Annual Conf.*, **2002**, pp. 1398–1403.
17. T. Suntio, M. Hankaniemi, and T. Roinila 'Dynamical modelling of peak-current-mode controlled converter in continuous conduction mode,' *J. Simul. Modelling Pract.*

18. M.K. Kazimierczuk, 'Transfer function of current modulator in PWM converters with current-mode control,' *IEEE Trans. Circuits Syst. I, Fundam. Theory Appl.*, vol. 47, no. 9, **2000**, pp. 1407–1412.
19. B. Johansson, 'A comparison and improvement of two-continuous time models for current-mode control,' in *Proc. IEEE International Telecommunications Energy Conf.*, **2002**, pp. 552–559.
20. J. Sun and B. Choi, 'Average modeling and switching instability prediction for peak-current control,' in *Proc. IEEE Power Electronics Specialists Conf.*, **2005**, pp. 2764–2770.
21. T. Suntio, 'Analysis and modelling of peak-current-mode controlled buck converter in DICM,' *IEEE Trans. Ind. Electron.*, vol. 48, no. 1, **2001**, pp. 127–135.
22. T. Suntio, 'Unified average and small-signal modeling of direct-on-time control,' *IEEE Trans. Ind. Electron.*, vol. 53, no. 1, **2006**, pp. 287–295.
23. M. Hankaniemi, M. Karppanen, and T. Suntio, 'Load imposed instability and performance degradation in a regulated converter,' *IEE Proc. Electr. Power Appl.*, vol. 153, no. 6, **2006**, pp. 781–786.
24. T. Suntio, M. Hankaniemi, and M. Karppanen, 'Analysing dynamics of regulated converters,' *IEE Proc. Elect. Power Appl.*, vol. 153, no. 6, **2006**, pp. 905–910.
25. T. Suntio, K. Kostov, T. Tepsa, and J. Kyyrä, 'Using input invariance as a method to facilitate system design in DPS applications,' *J. Circuits Syst. Comput.*, vol. 13, no. 4, **2004**, pp. 707–723.
26. M. Hankaniemi, M. Karppanen, T. Suntio, A. Altowati, and K. Zenger, 'Source-reflected load interactions in a regulated converter,' in *Proc. IEEE Industrial Electronics Annual Conf.*, **2006**, pp. 2893–2898.

5
Average and Small-Signal Modeling of Average-Current-Mode Control

5.1
Introduction

Average-current-mode (ACM) control was introduced in the early 1990s [1] to alleviate the disadvantages related to the peak-current-mode (PCM) control such as the limited-duty-ratio range and high-frequency noise sensitivity [2, 3]. ACM control is typically used in power-factor-correction applications [4] and in the converters interfacing the solar panels [5, 6]. The duty-ratio generation is basically identical to the method utilized in the VMC control, but the control signal contains both the output-voltage-loop control signal and the averaged inductor current. The internal dynamics of the converter would naturally change by application of ACM control but may resemble either the dynamics of VM or PCM control depending on the level of inductor-current ripple left in the duty-ratio generation. It may be obvious that the dynamical modeling is quite similar to the PCM modeling introduced in Chapter 4. Therefore, it is also natural that the proposed modeling methods in [7, 8, 11, 13, and, 14] would utilize the modeling of PCM control [9, 12, 15]. We will introduce the ACM modeling only in the continuous conduction mode. The DCM models can be naturally derived by applying the proposed method and using the results developed for the PCM control in DCM (Chapter 4, Section 4.4).

5.2
ACM-Control Principle

Under ACM control, the duty ratio (d) is generated comparing the output signal (u_{ca}) of the current-loop amplifier and the constant ramp signal ($R_s M_c$) provided by the PWM modulator as shown in Figure 5.1, where R_s is the inductor-current equivalent sensing resistor, and M_c the slope of the PWM ramp in current domain. The duty ratio is established when the output signal (u_{ca}) of the current-loop amplifier reaches the PWM ramp signal. The output signal of the current-loop amplifier (u_{ca}) can be given by

$$u_{ca} = u_{co} + G_{ca}(u_{co} - R_s i_L) \qquad (5.1)$$

Dynamic Profile of Switched-Mode Converter. Teuvo Suntio
© 2009 WILEY-VCH Verlag GmbH & Co. KGaA, Weinheim
ISBN: 978-3-527-40708-8

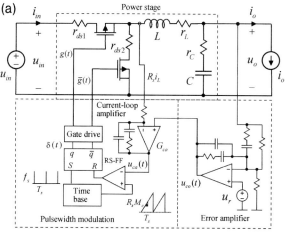

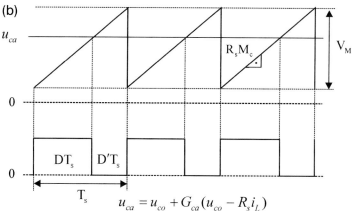

$$u_{ca} = u_{co} + G_{ca}(u_{co} - R_s i_L)$$

Figure 5.1 ACM-control principles: (a) circuit schematics with a buck converter and (b) duty-ratio generation.

where u_{co} is the output signal of the voltage-loop amplifier and G_{ca} the transfer function of the current-loop amplifier. The instantaneous inductor current (i_L) can be presented as a sum of the time-averaged inductor current $\langle i_L \rangle$ and the triangular-shaped switching ripple ($i_{L\text{-ripple}}$) ripple. Therefore, the current-amplifier output signal can be given, according to [13], by

$$u_{ca} = u_{co} + G_{ca}(u_{co} - R_s \langle i_L \rangle) - G_{ca} i_{L\text{-ripple}} \tag{5.2}$$

The current-loop amplifier is typically a PI-type controller with an extra high-frequency pole as depicted in Figure 5.1a. It is obvious that the inductor-current ripple may affect the converter dynamics, if the extra pole does not remove the ripple-current term in (5.2): the basic dynamics of the converter is defined assuming that the duty ratio is generated based on the averaged behavior of (5.2), i.e., $\langle u_{ca}\rangle = \langle u_{co}\rangle + G_{ca}(\langle u_{co}\rangle - \langle i_L\rangle)$. If the ripple term is not zero, the dynamics would be changed. The dynamical models presented in [10] are based on the assumption of the zero-ripple conditions.

In practice, the proper dynamical models exist only either with the full- [13] or with zero-ripple conditions [10]. The full-ripple effect would take place when the current-loop amplifier is implemented as a perfect PI controller without the extra high-frequency pole.

5.3
Modeling with Full-Ripple-Current Feedback

ACM control is also a method to change the dynamics imposed by the direct-duty-ratio or VM control similar to PCM control. Therefore, the fundamental issue is to find the proper duty-ratio constraints.

The inductor-current loop (Figure 5.1a) is provided with an average filter or PI controller, which is shown in Figure 5.2a in current domain, i.e., the voltage signals are replaced with the corresponding current signals to facilitate the development of the duty-ratio constraints. Its transfer function (G_{ca}) can be given by

$$G_{ca} = \frac{1 + s \cdot R_f C_f}{s \cdot R_{in}(C_f + C_p)\left(1 + s \cdot R_f \frac{C_f C_p}{C_f + C_p}\right)} \qquad (5.3)$$

and the corresponding frequency response is presented conceptually in Figure 5.2b, where the effect of the location of the extra high-frequency pole (f_p) is explicitly shown. When doing the modeling, we assume that f_p is much higher than the switching frequency (f_s). This means that the amplifier gain at the switching frequency is $K_f \approx R_f/R_{in}$ (sec Figure 5.2b), and consequently, the inductor-current switching-frequency ripple would be weighted by K_f. Therefore, the amplifier output signal (5.2) can be presented in current domain by

$$i_{ca} = (1 + G_{ca})i_{co} - G_{ca}\langle i_L\rangle - K_f i_{L\text{-ripple}} \qquad (5.4)$$

The corresponding duty-ratio generation process, when taking into account (5.4), is shown in Figure 5.3, where $\langle i_{ca}\rangle = (1 + G_{ca})\langle i_{co}\rangle - G_{ca}\langle i_L\rangle$, m_1 and m_2 are the corresponding inductor-current up and down slopes as well as K_f is the current-loop-amplifier gain at the switching frequency as defined above.

According to the above-presented duty-ratio-generation process, it may be obvious that Δi_L is the same value as defined in the conjunction with the PCM modeling, that is, $\Delta i_L = dd'T_s/2(m_1 + m_2)$. This means that the duty-ratio constraints can be given by

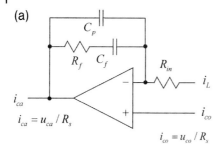

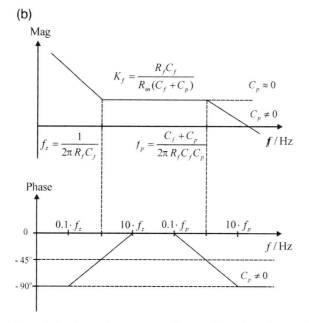

Figure 5.2 Current-loop PI controller: (a) physical amplifier and (b) its frequency response.

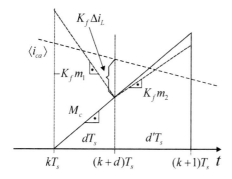

Figure 5.3 The duty-ratio generation process in current domain.

5.3 Modeling with Full-Ripple-Current Feedback

$$\hat{d} = \cfrac{1}{T_s \left(M_c + \cfrac{K_f (D' - D) U_E}{2L} \right)}$$

$$\times \left((1 + G_{ca}) \cdot \hat{i}_{co} - G_{ca} \cdot \hat{i}_L - K_f q_o \cdot \hat{u}_o - K_f q_i \cdot \hat{u}_{in} \right) \quad (5.5)$$

by applying the duty-ratio constraints developed for the PCM control, where U_E, q_o, and q_i are defined in Chapter 4 (Section 4.3.1) for the buck, boost, and buck–boost converters.

According to (5.5), it is obvious that the current-loop-amplifier gain at the switching frequency would have a fundamental effect on the ripple effects. It may also be obvious that the similar mode limit observed in the PCM control does not easily exist, because the converter is automatically compensated by means of M_c. Reference [11] provides excessive data on determining the value of F_m for an ACM-controlled buck converter, and coming up to the conclusion that $F_m = 1/R_s T_s M_c = 1/V_M$, where V_M is the peak-to-peak voltage of the PWM ramp. The tests were carried out at the operating point close to $D = 0.5$, and in addition by using very low value of K_f. According to (5.5), the conclusion is evident, because $M_c = V_M/R_s T_s$.

The ACM transfer functions cannot be easily obtained anymore by substituting the perturbed duty ratio ($\hat{d}$) in the corresponding VMC small-signal state space by means of (5.5) because of the current-loop-amplifier transfer function but the block-diagram technique introduced in conjunction with the PCM modeling has to be applied.

Figure 5.4 shows the required block diagrams, where the coefficients A and B for the basic converters are as follows: buck $A = 1$, $B = 0$, boost and buck–boost $A = D'$, $B = I_L$, and the output-voltage-feedback gain (q_o) and the input-voltage-feedforward gain (q_i) as defined above based on the corresponding PCM gains.

From Figure 5.4a, we may compute the transfer functions defining the internal dynamics of the ACM-controlled converter as a function of the VMC transfer functions and the duty-ratio-constraints coefficients to be as follows:

$$G_{io-o} = \cfrac{\left(1 + \cfrac{B F_m G_{ca}}{A}\right) G^v_{io-o} - F_m q_i G^v_{co}}{1 + L_c(s) + L_v(s)}$$

$$Z_{o-o} = \cfrac{\left(1 + \cfrac{B F_m G_{ca}}{A}\right) Z^v_{o-o} + \cfrac{F_m G_{ca}}{A} G^v_{co}}{1 + L_c(s) + L_v(s)}$$

$$G_{co} = \cfrac{(1 + G_{ca}) F_m G^v_{co}}{1 + L_c(s) + L_v(s)}$$

$$Y_{in-o} = Y^v_{in-o} - \cfrac{F_m \left(\left(q_o + \cfrac{G_{ca}}{A Z_C}\right) G^v_{io-o} + q_i \right) G^v_{ci}}{1 + L_c(s) + L_v(s)}$$

$$T_{oi-o} = T_{oi-o}^v + \frac{F_m((q_o + \frac{G_{ca}}{AZ_C})Z_{o-o}^v - \frac{G_{ca}}{A})G_{ci}^v}{1 + L_c(s) + L_v(s)}$$

$$G_{ci} = \frac{(1 + G_{ca})F_m G_{ci}^v}{1 + L_c(s) + L_v(s)} \quad (5.6)$$

where the inductor-current-loop gain ($L_c(s)$), and the output-voltage-loop gain ($L_v(s)$) are as follows:

$$\begin{aligned} L_c(s) &= G_{ca} F_m G_{cL}^v \\ L_v(s) &= F_m q_o G_{co}^v \end{aligned} \quad (5.7)$$

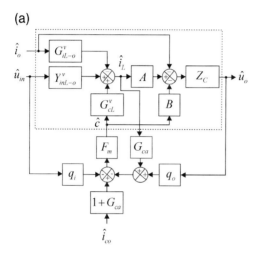

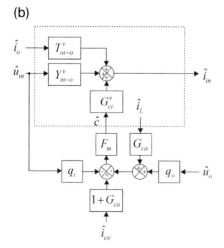

Figure 5.4 Block diagrams for ACM control in CCM: (a) output dynamics and (b) input dynamics.

According to (5.6), we may conclude the following:

1. The control-to-output transfer function (G_{co}) incorporates the same RHP zero as in the corresponding VMC converter if any.
2. The buck converter may have high input-noise attenuation similar to the corresponding PCM converter depending on K_f.
3. If K_f is small, then the transfer functions will have features resembling the VM control.
4. If K_f is high, then the transfer functions will have features resembling the PCM control.

5.4
Dynamic Review

The dynamic effects imposed by the ACM control are analyzed by using the buck converter as shown in Figure 5.5. The peak-to-peak voltage swing (V_M) of the PWM modulator is set to 3 V, and consequently, $M_c = 40$ A/10 µs (i.e., $V_M/R_s T_s$). The proposed dynamical models (5.6) are verified by means of simulated frequency responses. The dynamics of the ACM-controlled converter is compared to the corresponding VM- and PCM-controlled converters. The internal transfer functions cannot be obtained directly by simulation because of the current-output nature of the converter at open loop, but they have to be computed similarly to the PCM transfer functions from the measured data.

The current-loop amplifier is designed to have a zero (i.e., $1/R_f C_f$, $R_f = 10$ kΩ, $C_f = 47$ nF) at half the resonant frequency (i.e., 339 Hz). Its high-frequency gain (R_f/R_{in}) is varied from 1 to 10 by changing R_{in} accordingly as shown in Figure 5.6. The high-frequency pole is placed to infinity (i.e., $C_p = 0$).

The corresponding inductor-current-loop gains (i.e., $L_c(s)$ (5.7)) are shown in Figure 5.7 at the input voltages of 20 and 50 V. It may be obvious that K_f

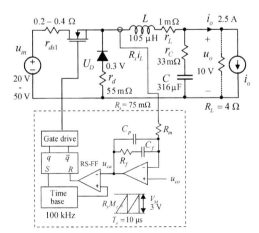

Figure 5.5 ACM-controlled buck converter.

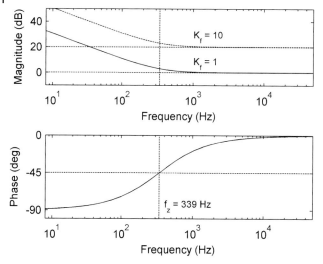

Figure 5.6 Current-loop-amplifier frequency responses.

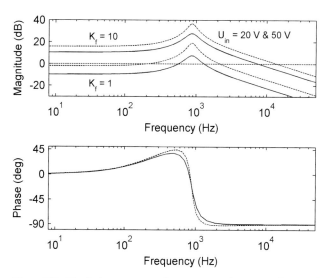

Figure 5.7 The inductor-current-loop gains at the input voltages of 20 and 50 V with varying current-amplifier high-frequency gain (K_f).

should be higher than unity for a proper low-frequency gain but not much higher than 10 to ensure proper crossover frequency. It is defined in [1] that the maximum gain should not exceed the value given by $K_f = M_c/M_2$, where M_2 is the down slope of the inductor current yielding 42. The loop phase margin is always at least 90°.

5.4.1
Control-to-Output Transfer Function

The control-to-output transfer function (G_{co}) can be presented symbolically by

$$\frac{(1+G_{ca})F_m G_{co}^v}{1+G_{ca}F_m G_{cL}^v} \tag{5.8}$$

because the output-voltage feedback gain (q_o) is zero for a buck converter, where G_{co}^v and G_{cL}^v are the VMC control-to-output and control-to-inductor current transfer functions, respectively. In this case, the control-to-output transfer function will be given in voltage domain by G_{co}/R_s, where $R_s = 75$ mΩ as shown in Figure 5.5, because M_c is dependent of R_s. The predicted (solid and dashed lines) and simulated (dots and squares) internal transfer functions are shown in Figure 5.8 at the input voltages of 20 and 50 V. It is obvious that the predictions and simulated responses have a good match.

Figure 5.9 shows the load-affected (dash-dot lines) and internal (solid and dashed lines) transfer functions at the input voltage of 20 V. The differences are observable at the frequencies lower than the resonant frequency. If making conclusions based on the load-affected transfer functions (i.e., $Z_L = 4\,\Omega$), the low-frequency phase of $-90°$ implying tendency to low-frequency instability in the case of traditional PI-type controllers would not be noticed.

Figure 5.10 shows the control-to-output transfer function of the ACM, VM, and PCM-controlled converters in voltage domain at the input voltage of 20 V. It may be obvious that the high value of K_f induces PCM-like features and the low value of K_f induces VM-like features.

5.4.2
Output Impedance

The output impedance (Z_{o-o}) can be given symbolically by

$$\frac{Z_{o-o}^v + F_m G_{ca} G_{co}^v}{1 + F_m G_{ca} G_{cL}^v} \tag{5.9}$$

The predicted and simulated (dots, squares, and triangles) frequency responses at the input voltages of 20 V (solid lines) and 50 V (dashed lines) are shown in Figure 5.11. The simulated responses match perfectly the corresponding predicted responses. The ACM control clearly increases the low-frequency impedance similarly to the PCM control.

The comparison between the corresponding VM, PCM, and ACM (solid lines) output impedances at the output voltage of 20 V are shown in Figure 5.12 verifying the PCM features of the ACM output impedance.

The load-affected ($Z_L = 4\,\Omega$) (dashed and dash-dot lines) and the internal (solid lines) output impedances are shown in Figure 5.13. The differences are

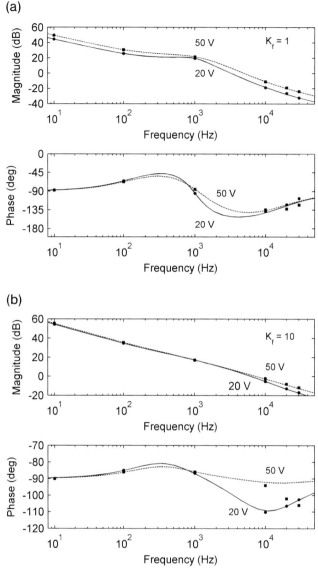

Figure 5.8 The frequency responses of the internal control-to-output transfer functions at the input voltages of 20 and 50 V: (a) $K_f = 1$ and (b) $K_f = 10$.

observable up to the frequencies lower than the resonant frequency of the output filter. The differences between the load-affected and the internal output impedances are significant and could lead to totally erroneous analytical conclusions on the load interactions if based on the load-affected transfer functions.

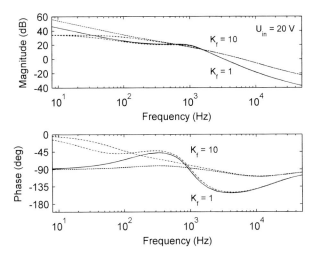

Figure 5.9 Load-affected (dash-dot lines) and internal (solid and dashed lines) transfer functions at the input voltage of 20 V.

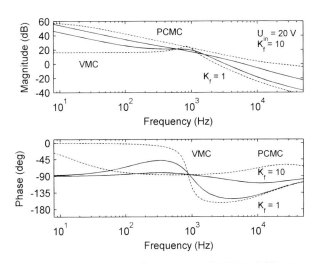

Figure 5.10 Comparison of VM, PCM, and ACM (solid lines) transfer functions at the input voltage of 20 V.

5.4.3
Input-to-Output Transfer Function

The input-to-output transfer function ($G_{\text{io}-o}$) can be given symbolically by

$$\frac{G^{\text{v}}_{\text{io}-o} - F_m q_i G^{\text{v}}_{\text{co}}}{1 + F_m G_{\text{ca}} G^{\text{v}}_{cL}} \tag{5.10}$$

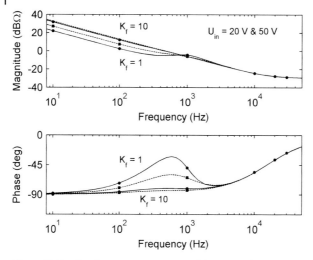

Figure 5.11 The frequency responses of the internal output impedances at the input voltages of 20 V (solid lines) and 50 V (dashed lines), when K_f varies from 1 to 10.

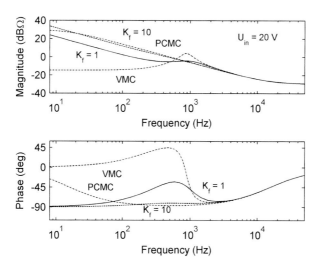

Figure 5.12 The comparison of VM, PCM, and ACM (solid lines) output impedances at the input voltage of 20 V, when K_f is varying from 1 to 10.

The predicted and simulated (dots, squares, and triangles) internal input-to-output transfer functions at the input voltages of 20 V (solid lines) and 50 V (dashed lines) are shown in Figure 5.14. The predicted and simulated responses have very good match validating the accuracy of the proposed modeling method. It may be obvious that the increase in the ripple feedback also increases the input-noise attenuation.

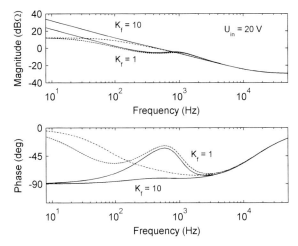

Figure 5.13 Load-affected (dashed lines) and internal (solid lines) output impedances at the output voltage of 20 V, when K_f varies from 1 to 10.

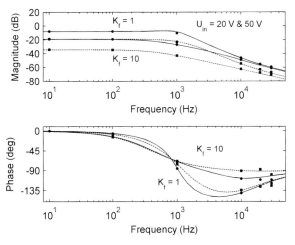

Figure 5.14 The predicted and simulated (dots and squares) frequency responses of the internal input-to-output transfer function at the input voltage of 20 V (solid lines) and 50 V (dashed lines), when K_f varies from 1 to 10.

The comparison of the corresponding VM, PCM, and ACM (solid lines) transfer functions at the input voltage of 20 V is shown in Figure 5.15, which clearly indicates how the low value of K_f makes the converter resemble VM converter and the higher value of K_f makes the converter resemble PCM converter. It may be obvious that the PCM control has the highest high-frequency attenuation, which would

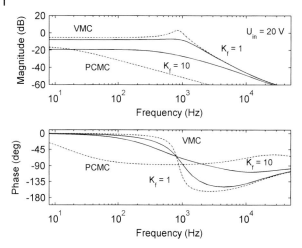

Figure 5.15 The comparison of the VM, PCM, and ACM (solid lines) input-to-output transfer functions at the input voltage of 20 V, when K_f varies from 1 to 10.

make it more insensitive to the source interactions than the other control modes.

The load-affected ($Z_L = 4\,\Omega$) (dashed and dash-dot lines) and the internal (solid lines) input-to-output transfer functions are shown in Figure 5.16. The differences are observable up to the resonant frequency of the output filter. The low-frequency input-noise attenuation of the load-affected transfer functions is much higher than the real internal transfer functions due to the effect of

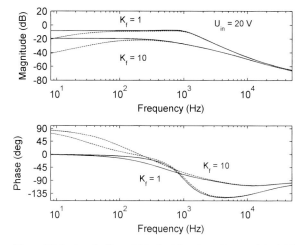

Figure 5.16 Load-affected (dashed lines) and internal (solid lines) input-to-output transfer functions at the output voltage of 20 V, when K_f varies from 1 to 10.

$R_L/(R_L + Z_{o-o})$, which may lead to erroneous conclusions on the possible source and reflected-load interactions.

5.4.4
Input Admittance

The input admittance (Y_{in}) can be symbolically given by

$$Y_{in-o}^v = \frac{F_m(\frac{G_{ca}}{Z_C} \cdot G_{io-o}^v + q_i)G_{ci}^v}{1 + F_m G_{ca} G_{cL}^v} \qquad (5.11)$$

The predicted internal input admittance (solid line $K_f = 1$, dashed line $K_f = 10$) and the corresponding VM and PCM input admittances (dash-dot lines) are shown in Figure 5.17. Depending on K_f, the ACM input admittance may have feature resembling VM converter or PCM converter. The source interactions may, however, be worse than that in the corresponding VM or PCM converter.

5.5
Effect of Current-Loop High-Frequency Pole

The ripple component of the inductor current is a triangle-shaped signal, which can be presented by means of its Fourier components by

$$i_{L\text{-ripple}} = \sum_{n=1}^{\infty} \frac{\Delta i_{L\text{-pp}}}{2\pi^2 n^2 DD'}(-2\sin^2 n\pi D \cdot \cos n\omega_o t$$
$$+ \sin n2\pi D \cdot \sin n\omega_o t) \qquad (5.12)$$

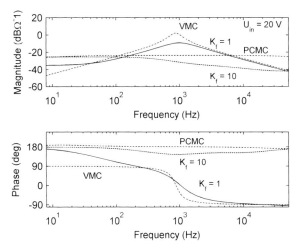

Figure 5.17 The comparison of the VM, PCM, and ACM (solid and dashed lines) input admittances at the input voltage of 20 V, when K_f varies from 1 to 10.

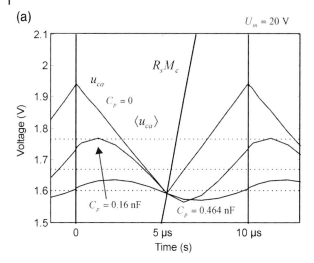

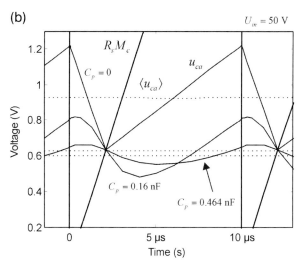

Figure 5.18 The simulated duty-ratio generation process, where the current-amplifier high-frequency pole (f_p) is placed at infinity, switching frequency, and 1/4 of the switching frequency. (a) $U_{in} = 20$ V and (b) $U_{in} = 50$ V.

where $\Delta i_{L\text{-pp}}$ is the peak-to-peak value of the ripple current. The inverse relation to the square of n indicates that the first harmonic components have a very strong effect and, therefore, the high-frequency pole (f_p) would shape easily the ripple component and, consequently, affect the dynamics of the ACM-controlled converter more than would be expected. We do not give analytical models capable to perfectly predict the converter dynamics. The boundaries for the ripple effects are naturally the full-ripple

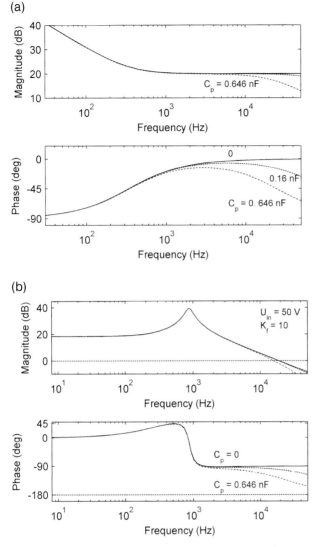

Figure 5.19 Effect of current-amplifier high-frequency pole on (a) amplifier transfer function and (b) inductor-current loop gain.

effects [13] introduced in this chapter and zero-ripple effects introduced in [10] setting $K_f = 0$.

The simulated duty-ratio generation process (5.4) at the input voltages of 20 and 50 V are shown in Figure 5.18, where $C_p = 0$ corresponds to $f_p = \infty$, $C_p = 160$ pF to $f_p = f_s$, and $C_p = 646$ pF to $f_p = 0.25 \cdot f_s$. The dashed horizontal lines correspond to the averaged current-amplifier output signal ($\langle u_{ca} \rangle$). The control voltage (u_{co}) has been kept constant. The PWM ramp signal ($R_s M_c$)

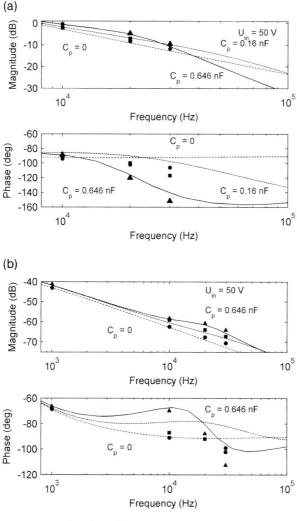

Figure 5.20 The effect of high-frequency pole (f_p) in the inductor-current loop on (a) control-to-output transfer function and (b) input-to-output transfer function, where the dash-dot lines correspond to $f_p = \infty$, dashed lines to $f_p = f_s$, and solid lines to $f_p = 0.25 \cdot f_s$.

has a voltage swing of 3 V. The strong attenuating effect of the high-frequency pole is obvious as discussed above.

The corresponding current-amplifier transfer function (G_{ca}) and the inductor-current loop gain ($L_c(s)$) are shown in Figure 5.19, when $K_f = 10$. It may be obvious that the crossover frequency and phase margin of the inductor-current loop would reduce along the decrease in the high-frequency pole (f_p), which are 15.8 kHz and 57.3° (Figure 5.19b) for $f_p = 0.25 \cdot f_s$, respectively.

The effect of the location of the high-frequency pole on the control-to-output transfer function is shown in Figure 5.20a and on the input-to-output transfer function in Figure 5.20b. The simulated frequency responses are marked with dots ($C_p = 0$), squares ($C_p = 160$ pF), and triangles ($C_p = 646$ pF). The predicted responses marked with dash-dot line ($C_p = 0$) are computed by using $K_f = 10$. The other predicted responses ($C_p = 160$ pF and $C_p = 646$ pF) are computed by using $K_f = 0$. According to the simulated responses, the high-frequency pole affects mainly the high-frequency behavior of the corresponding transfer functions. The observed peaking at 15.8 kHz is the consequence of the phase margin (i.e., <60°) in the inductor-current loop as discussed earlier. It may be obvious that the use of $K_f = 0$, that is, omitting the ripple effect, would produce quite accurate predictions.

References

1. L.H. Dixon, 'Average current mode control of switching power supplies,' in *Proc. Power Supply Design Seminar SEM 700*, Unitrode Corporation, **1990**, pp. 5.1–5.14.
2. C.W. Deisch, 'Simple switching control method changes power converter into a current source,' in *Proc. IEEE Power Electronics Specialists Conf.*, **1978**, pp. 300–306.
3. R. Redl and N.O. Sokal, 'What a design engineer should know about current mode control,' in *Proc. IEEE Power Electronics Design Conference*, **1985**, pp. 18–33.
4. R. Redl, 'Power-factor correction in single-phase switching-mode power supplies – An overview,' *Int. J. Electron.*, vol. 77, no. 5, **1994**, pp. 555–582.
5. F. Tonicello, 'The control problem of maximum point power tracking in power systems,' in *Proc. Seventh European Space Power Conference*, **2005**, 7 pp.
6. J.H. Lee, H.S. Bae, S.H. Park, and B.H. Cho, 'Constant resistance control of solar array regulator using average current mode control,' in *Proc. IEEE Applied Power Electronics Conf.*, **2006**, pp. 1544–1548.
7. W. Tang, F.C. Lee, and R.B. Ridley, 'Small-signal modeling of average current-mode control,' *IEEE Trans. Power Electron.*, vol. 8, no. 2, **1993**, pp. 112–119.
8. Y.-S. Jung and M.-J. Youn, 'Sampling effect in continuous-time small-signal modeling of average-current mode control,' *IEE Proc. Electr. Power Appl.*, vol. 149, no. 4, **2002**, pp. 311–316.
9. R.B. Ridley, 'A new continuous-time model for current-mode control,' *IEEE Trans. Power Electron.*, vol. 6, no. 2, **1991**, pp. 271–280.
10. J. Sun and R.M. Bass, 'Modeling and practical design issues for average current control,' in *Proc. IEEE Applied Power Electronics Specialists Conf.*, **1999**, pp. 980–986.
11. P. Cooke, 'Modeling average current mode control,' in *Proc. IEEE Applied Power Electronics Specialists Conf.*, **2000**, pp. 256–262.
12. R.W. Erickson and D. Maksimovic, *Fundamentals of Power Electronics*, Kluwer, Norwell, MA, USA, **2001**, 2nd Edition.
13. T. Suntio, J. Lempinen, I. Gadoura, and K. Zenger, 'Dynamic effects of inductor current ripple in average current mode control,' in *Proc. IEEE Power Electronics Specialists Conf.*, **2001**, pp. 1259–1264.
14. T. Suntio, M. Rahkala, I. Gadoura, and K. Zenger, 'Dynamic effects of

inductor current ripple in peak-current and average-current mode control,' in *Proc. IEEE Industrial Electronics Society Annual Conf.*, **2001**, pp. 1072–1077.

15. T. Suntio, 'Unified derivation and analysis of duty-ratio constraints for peak-current-mode control in continuous and discontinuous modes,' in *Proc. IEEE Industrial Electronics Society Annual Conf.*, **2002**, pp. 1398–1403.

6
Average and Small-Signal Modeling of Self-Oscillation Control

6.1
Introduction

Self-oscillation control is typically used in the applications requiring low-EMI or low-cost solutions such as power-factor correction [1] as well as mobile phone charging and notebook battery charging [2–4]. Boost and flyback converters [1, 4] are the most common converter topologies applied under the self-oscillation control. The self-oscillation control is a form of peak-current-mode control, where the cycle is initiated, when the inductor current reaches the zero level and terminated, when the inductor current reaches the control current. A determined delay is usually applied before switching on the next cycle to prevent the switching frequency to increasing indefinitely. The resulting switching frequency is naturally variable. The operation mode is commonly known as boundary conduction mode (BCM), critical conduction or transition mode. A simple self-oscillating flyback converter used in the mobile phone applications is shown in Figure 6.1 [3] illustrating the implementation of the control principle. A detailed operational analysis of the self-oscillating flyback converter can be found from [4].

It has been usually assumed that the dynamics associated with the self-oscillation control is similar to the fixed-frequency VM control [5, 6]. The PCM control and variable switching frequency would, however, change the dynamics requiring special modeling methods to capture it [7–10]. The method presented in this chapter is based on [8–11]. We present the dynamic review only for buck and buck–boost or flyback converters.

6.2
Self-Oscillation Modeling

The self-oscillation control is a derivative of PCM control in which the resulting inductor-current waveforms are as shown in Figure. 6.2, when a determined switching delay (T_D) is applied. Similarly to PCM control, we have to find the proper duty-ratio constraints under variable switching frequency. However, the

Dynamic Profile of Switched-Mode Converter. Teuvo Suntio
© 2009 WILEY-VCH Verlag GmbH & Co. KGaA, Weinheim
ISBN: 978-3-527-40708-8

6 Average and Small-Signal Modeling of Self-Oscillation Control

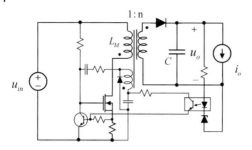

Figure 6.1 A simple self-oscillating flyback converter.

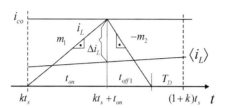

Figure 6.2 Inductor-current waveforms under self-oscillation control.

small-signal direct-on-time-controlled state space has to be solved before we can proceed with the PCM modeling. The direct-on-time control is not a physical control mode but a fictive mode serving only the purposes of modeling [11].

6.2.1
Averaged Direct-on-Time Model

According to Chapter 3, the derivative of the average inductor current $\langle i_L \rangle$ can be given by

$$\frac{d\langle i_L \rangle}{dt} = \frac{t_{\text{on}}}{t_s} \cdot m_1 - \frac{t_{\text{off1}}}{t_s} \cdot m_2 \tag{6.1}$$

From the inductor-current waveforms in Figure 6.2, we may solve t_{off1} by means of $\langle i_L \rangle$ yielding

$$t_{\text{off1}} = \frac{2\langle i_L \rangle}{m_2 \left(1 - \dfrac{T_D}{t_s}\right)} \tag{6.2}$$

Substituting t_{off1} in (6.1) with (6.2) yields

$$\frac{d\langle i_L \rangle}{dt} = \frac{t_{\text{on}}}{t_s} \cdot m_1 - \frac{2}{t_s - T_D} \cdot \langle i_L \rangle \tag{6.3}$$

The derivative for the average capacitor voltage ($\langle u_C \rangle$) can be generally given by

$$\frac{d\langle u_C \rangle}{dt} = \frac{q_1 \langle i_L \rangle}{C} - \frac{\langle i_o \rangle}{C} \tag{6.4}$$

where the coefficient q_1 is as follows: buck $q_1 = 1$, boost and buck–boost $q_1 = 1 - t_{\text{on}}/(t_s - T_D)$.

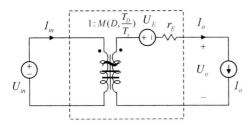

Figure 6.3 The steady-state equivalent circuit for the self-oscillating converters.

The average output voltage ($\langle u_o \rangle$) can be given by

$$\langle u_o \rangle = \langle u_C \rangle + r_C C \frac{d\langle u_C \rangle}{dt} \tag{6.5}$$

and the average input current ($\langle i_{in} \rangle$) by

$$\langle i_{in} \rangle = q_2 \langle i_L \rangle \tag{6.6}$$

where the coefficient q_2 is as follows: buck and buck–boost $q_2 = t_{on}/(t_s - T_D)$, boost $q_2 = 1$.

As a summary, the general direct-on-time averaged state space can be given by

$$\begin{aligned} \frac{d\langle i_L \rangle}{dt} &= \frac{t_{on}}{t_s} \cdot m_1 - \frac{2}{t_s - T_D} \cdot \langle i_L \rangle \\ \frac{d\langle u_C \rangle}{dt} &= \frac{q_1 \langle i_L \rangle}{C} - \frac{\langle i_o \rangle}{C} \\ \langle i_{in} \rangle &= q_2 \langle i_L \rangle \\ \langle u_o \rangle &= \langle u_C \rangle + r_C C \frac{d\langle u_C \rangle}{dt} \end{aligned} \tag{6.7}$$

where q_1 and q_2 are as defined above.

The steady-sate operating point of the converter can be solved from (6.7) setting the derivatives to zero and substituting the inductor current up slope (m_1) with its topology-based values given in Chapter 3 (Section 3.3), which can be given in the form of an equivalent circuit shown in Figure 6.3.

The coefficients of the equivalent circuit and the other operating-point parameters can be defined for the basic converters as follows:

Buck:

$$M\left(D, \frac{T_D}{T_s}\right) = \frac{D}{1 - \frac{T_D}{T_s}}$$

$$U_E = \frac{D' - \frac{T_D}{T_s}}{1 - \frac{T_D}{T_s}} \cdot U_D \text{ or } 0$$

$$r_E = r_L + \frac{D}{1 - \frac{T_D}{T_s}} \cdot r_{ds1} + \frac{D' - \frac{T_D}{T_s}}{1 - \frac{T_D}{T_s}} \cdot (r_d \text{ or } r_{ds2})$$

$$I_L = I_o$$

$$I_{in} = \frac{D}{1 - \frac{T_D}{T_s}} \cdot I_o$$

Boost:

$$M\left(D, \frac{T_D}{T_s}\right) = \frac{1 - \frac{T_D}{T_s}}{D' - \frac{T_D}{T_s}}$$

$$U_E = U_D \text{ or } 0$$

$$r_E = \frac{\left(1 - \frac{T_D}{T_s}\right)^2}{\left(D' - \frac{T_D}{T_s}\right)^2} \cdot r_L + \frac{D\left(1 - \frac{T_D}{T_s}\right)}{\left(D' - \frac{T_D}{T_s}\right)^2} \cdot r_{ds1}$$

$$+ \frac{1 - \frac{T_D}{T_s}}{D' - \frac{T_D}{T_s}} \cdot (r_d \text{ or } r_{ds2}) + \frac{D}{D' - \frac{T_D}{T_s}} \cdot r_C$$

$$I_L = \frac{1 - \frac{T_D}{T_s}}{D' - \frac{T_D}{T_s}} \cdot I_o$$

$$I_{in} = \frac{1 - \frac{T_D}{T_s}}{D' - \frac{T_D}{T_s}} \cdot I_o$$

Buck–Boost:

$$M\left(D, \frac{T_D}{T_s}\right) = -\frac{D}{D' - \frac{T_D}{T_s}}$$

$$U_E = U_D \text{ or } 0$$

$$r_E = \frac{\left(1 - \dfrac{T_D}{T_s}\right)^2}{\left(D' - \dfrac{T_D}{T_s}\right)^2} \cdot r_L + \frac{D\left(1 - \dfrac{T_D}{T_s}\right)}{\left(D' - \dfrac{T_D}{T_s}\right)^2} \cdot r_{ds1} + \frac{1 - \dfrac{T_D}{T_s}}{D' - \dfrac{T_D}{T_s}}$$

$$\cdot (r_d \text{ or } r_{ds2}) + \frac{D}{D' - \dfrac{T_D}{T_s}} \cdot r_C$$

$$I_L = \frac{1 - \dfrac{T_D}{T_s}}{D' - \dfrac{T_D}{T_s}} \cdot I_o$$

$$I_{in} = \frac{D}{D' - \dfrac{T_D}{T_s}} \cdot I_o$$

The steady-state cycle time (T_s) can be solved from the inductor current waveforms in Figure 6.2 yielding

$$T_s^2 - \left(2T_D + \frac{2I_L(M_1 + M_2)}{M_1 M_2}\right) \cdot T_s + T_D^2 = 0 \tag{6.8}$$

and substituting the topology-based up and down slopes with the values defined in Chapter 3 (Section 3.3).

6.2.2
Small-Signal Direct-on-Time Model

The small-signal state space can be derived from the averaged state space (6.7) by developing the proper partial derivatives as instructed in Chapter 2. Under self-oscillation control or generally in the variable-frequency-operation mode, the cycle time (t_s) is also variable. Therefore, we also have to develop proper dynamic constraints (i.e., cycle-time constraints) for the varying cycle time in order to introduce its effect on the dynamics of the converter.

The cycle-time constraints may be obtained from the definition of the cycle time $t_s = t_{on} + t_{off1} + T_D$ by substituting t_{off1} with (6.2) yielding

$$t_s = t_{on} + \frac{2\langle i_L \rangle}{m_2\left(1 - \dfrac{T_D}{t_s}\right)} + T_D \tag{6.9}$$

and by developing the proper partial derivatives, and formulating the results as

$$\hat{t}_s = F_m^c(\hat{t}_{on} + q_c^c \cdot \hat{i}_L + q_o^c \cdot \hat{u}_o + q_i^c \cdot \hat{u}_{in}) \tag{6.10}$$

6 Average and Small-Signal Modeling of Self-Oscillation Control

The final small-signal state space results when the perturbed cycle time ($\hat{t}_s$) is substituted with (6.10).

It has been observed that the circuit parasitic elements and the delay (T_D) do not contribute much onto the converter dynamics [10]. Therefore, we may omit their effect and develop the constraints coefficients in a most convenient form. The parasitic elements and the delay (T_D) would, however, significantly affect the steady-state operating point, and should not be omitted when defining the steady-state parameters.

The coefficients of the cycle-time constraints (6.10) can be given for the basic converters as shown in Table 6.1.

We do not give explicitly the small-signal state spaces but would proceed directly to the required PCM modeling.

6.2.3
Small-Signal PCM Models

In order to obtain the PCM small-signal state space, we have to develop the on-time constraints relating the perturbed on-time ($\hat{t}_{on}$) to the control current ($\hat{i}_{co}$), the other circuit variables and elements similarly to the fixed-frequency PCM modeling treated in Chapter 4. From Figure 6.2, the comparator equation defining the length of the on-time at $t = kt_s + t_{on}$ can be given by

$$i_{co} = \langle i_L \rangle + \Delta i_L \tag{6.11}$$

The difference (Δi_L) between the peak inductor current and its average value can be computed to be

$$\Delta i_L = \frac{t_{on} m_1}{2} \left(1 + \frac{T_D}{t_s}\right) \tag{6.12}$$

Therefore, the comparator equation (6.11) becomes

$$i_{co} = \langle i_L \rangle + \frac{t_{on} m_1}{2} \left(1 + \frac{T_D}{t_s}\right) \tag{6.13}$$

Table 6.1 Cycle-time-constraints coefficients for buck, boost, and buck–boost converters.

Converter	F_m^0	q_c^0	q_o^0	q_i^0
Buck	$\dfrac{2L}{U_m - U_o}$	1	$-\dfrac{DT_s}{2L}$	$\dfrac{DT_s}{2L}$
Boost	$\dfrac{2L}{U_m}$	1	0	$\dfrac{DT_s}{2L}$
Buck–boost	$\dfrac{2L}{U_m}$	1	0	$\dfrac{DT_s}{2L}$

Table 6.2 On-time-constraints coefficients for buck, boost, and buck–boost converters.

Converter	F_m^c	q_c^c	q_o^c	q_i^c
Buck	1	$\dfrac{2L}{U_0}$	$-\dfrac{D(U_m - U_0)T_s}{U_0^2}$	0
Boost	1	$\dfrac{2L}{U_0 - U_m}$	$-\dfrac{DU_m T_s}{(U_0 - U_m)^2}$	$\dfrac{DU_m T_s}{(U_0 - U_m)^2}$
Buck–boost	1	$\dfrac{2L}{U_0}$	$-\dfrac{DU_m T_s}{U_0^2}$	0

from which the on-time constraints can be solved by developing the proper partial derivatives and applying the cycle-time constraints (6.10) yielding

$$\hat{t}_{on} = F_m^o \left(\hat{i}_{co} - q_c^o \cdot \hat{i}_L - q_o^o \cdot \hat{u}_o - q_i^o \cdot \hat{u}_{in} \right) \tag{6.14}$$

The coefficients of the on-time constraints (6.14) can be given for the basic converters as shown in Table 6.2.

The small-signal state spaces and the corresponding sets of the transfer functions corresponding to the self-oscillation control can be given for the basic converters as follows:

Buck:

$$\begin{aligned}
\frac{d\hat{i}_L}{dt} &= -\frac{4}{T_s} \cdot \hat{i}_L + \frac{2}{T_s} \cdot \hat{i}_{co} \\
\frac{d\hat{u}_C}{dt} &= \frac{1}{C} \cdot \hat{i}_L - \frac{1}{C} \cdot \hat{i}_o \\
\hat{i}_{in} &= D(2D-1) \cdot \hat{i}_L + \frac{DD'T_s}{2L} \cdot \hat{u}_C - \frac{D^2 D' T_s}{2L} \cdot \hat{u}_{in} + DD' \cdot \hat{i}_{co} \\
\hat{u}_o &= \hat{u}_C + r_C C \frac{d\hat{u}_C}{dt}
\end{aligned} \tag{6.15}$$

$$\begin{bmatrix} Y_{in-o} & T_{oi-o} \\ G_{io-o} & -Z_{o-o} \end{bmatrix} = \begin{bmatrix} -\dfrac{D^2 D' T_s}{2L} \left(= -\dfrac{DI_L}{U_{in}} \right) & -\dfrac{DD' T_s}{s \cdot 2LC} \left(= -\dfrac{DI_L}{s \cdot U_o C} \right) \\ 0 & -\dfrac{1 + s \cdot r_C C}{sC} \end{bmatrix} \tag{6.16}$$

$$\begin{bmatrix} G_{ci} \\ G_{co} \end{bmatrix} = \begin{bmatrix} \dfrac{DD'\left(s^2 + s \cdot \dfrac{2}{D'T_s} + \dfrac{1}{LC}\right)}{s\left(s + \dfrac{4}{T_s}\right)} \\ \dfrac{2(1 + s \cdot r_C C)}{s \cdot T_s C \left(s + \dfrac{4}{T_s}\right)} \end{bmatrix} \tag{6.17}$$

where we have applied the identity $T_s = 2LI_L U_{in}/(U_{in} - U_o)U_o$.

In boundary mode, the average inductor current stays constant at half the peak inductor current, which is defined by the control current. Therefore, it is natural that the input-to-output transfer function (G_{io-o}) of a buck converter is zero, because $I_L = I_o$, and the input admittance (Y_{in-o}) equals the ideal input admittance ($Y_{in-\infty} \approx -DI_L/U_{in}$). The transfer functions are mainly of first order except the control-related transfer functions, which are the expected results due to the tightly bounded inductor current.

Boost:

$$\frac{d\hat{i}_L}{dt} = -\frac{4}{T_s} \cdot \hat{i}_L + \frac{2}{T_s} \cdot \hat{i}_{co}$$

$$\frac{d\hat{u}_C}{dt} = \frac{D'(2D+1)}{C} \cdot \hat{i}_L - \frac{DD'^2 T_s}{2LC} \cdot \hat{u}_C + \frac{DD'T_s}{2LC} \cdot \hat{u}_{in} - \frac{1}{C} \cdot \hat{i}_o - \frac{DD'}{C} \cdot \hat{i}_{co}$$

$$\hat{i}_{in} = \hat{i}_L$$

$$\hat{u}_o = \hat{u}_C + r_C C \frac{d\hat{u}_C}{dt} \tag{6.18}$$

$$\begin{bmatrix} Y_{in-o} & T_{oi-o} \\ G_{io-o} & -Z_{o-o} \end{bmatrix} = \begin{bmatrix} 0 & 0 \\ \dfrac{DD'T_s(1 + s \cdot r_C C)}{2LC\left(s + \dfrac{DD'^2 T_s}{2LC}\right)} & -\dfrac{1 + s \cdot r_C C}{C\left(s + \dfrac{DD'^2 T_s}{2LC}\right)} \end{bmatrix} \tag{6.19}$$

$$\begin{bmatrix} G_{ci} \\ G_{co} \end{bmatrix} = \begin{bmatrix} \dfrac{2}{T_s\left(s + \dfrac{4}{T_s}\right)} \\ \dfrac{2D'\left(1 - s \cdot \dfrac{DT_s}{2}\right)(1 + s \cdot r_C C)}{T_s C\left(s + \dfrac{DD'^2 T_s}{2LC}\right)\left(s + \dfrac{4}{T_s}\right)} \end{bmatrix} \tag{6.20}$$

where we have applied the identity $T_s = 2LI_L U_o/(U_o - U_{in})U_{in}$. As discussed earlier, the average inductor current stays constant at $I_{co}/2$. Therefore, it is natural that the input admittance (Y_{in-o}) and the output-to-input transfer function (T_{oi-o}) are zero as well as the control-to-input transfer function (G_{ci}) is of first order. The control-to-output transfer function (G_{co}) contains an RHP zero at $2/DT_s$, which is symbolically same as the RHP zero in the DCM VM control-to-output transfer function. It may be obvious that the RHP zero does not much affect the magnitude but does affect the phase behavior (i.e., the phase starts affecting from $\omega = 0.2/DT_s$, which corresponds to $f = (0.064/D) \cdot f_s$). It may be obvious that the maximum control bandwidth would be limited, and the stability of the converter would be sensitive to the variation in the output-capacitor ESR (r_C).

Buck–Boost:

$$\frac{d\hat{i}_L}{dt} = -\frac{4}{T_s} \cdot \hat{i}_L + \frac{2}{T_s} \cdot \hat{i}_{co}$$

$$\frac{d\hat{u}_C}{dt} = \frac{D'(2D+1)}{C} \cdot \hat{i}_L - \frac{DD'^2 T_s}{2LC} \cdot \hat{u}_C + \frac{D^2 D' T_s}{2LC} \cdot \hat{u}_{in} - \frac{1}{C} \cdot \hat{i}_o - \frac{DD'}{C} \cdot \hat{i}_{co}$$

$$\hat{i}_{in} = D(2D-1) \cdot \hat{i}_L + \frac{DD'^2 T_s}{2L} \cdot \hat{u}_C - \frac{D^2 D' T_s}{2L} \cdot \hat{u}_{in} + DD' \cdot \hat{i}_{co}$$

$$\hat{u}_o = \hat{u}_C + r_C C \frac{d\hat{u}_C}{dt} \tag{6.21}$$

$$\begin{bmatrix} Y_{in-o} & T_{oi-o} \\ G_{io-o} & -Z_{o-o} \end{bmatrix} = \begin{bmatrix} -\dfrac{D^2 D' T_s \cdot s}{2L\left(s + \dfrac{DD'^2 T_s}{2LC}\right)} & -\dfrac{DD'^2 T_s \cdot s}{2LC\left(s + \dfrac{DD'^2 T_s}{2LC}\right)} \\ \dfrac{D^2 D' T_s (1 + s \cdot r_C C)}{2LC\left(s + \dfrac{DD'^2 T_s}{2LC}\right)} & -\dfrac{1 + s \cdot r_C C}{C\left(s + \dfrac{DD'^2 T_s}{2LC}\right)} \end{bmatrix} \tag{6.22}$$

$$\begin{bmatrix} G_{ci} \\ G_{co} \end{bmatrix} = \begin{bmatrix} \dfrac{DD'\left(s^2 + s \cdot \dfrac{2}{D'T_s} + \dfrac{D'}{LC}\right)}{\left(s + \dfrac{DD'^2 T_s}{2LC}\right)\left(s + \dfrac{4}{T_s}\right)} \\ \dfrac{2D'\left(1 - s \cdot \dfrac{DT_s}{2}\right)(1 + s \cdot r_C C)}{T_s C\left(s + \dfrac{DD'^2 T_s}{2LC}\right)\left(s + \dfrac{4}{T_s}\right)} \end{bmatrix} \tag{6.23}$$

where we have applied the identity $T_s = 2LI_L(U_{in} + U_o)/U_{in} U_o$. The control-to-output transfer function (G_{co}) contains an RHP zero at $2/DT_s$, which is symbolically same as the RHP zero in the DCM VM control-to-output transfer function. It may be obvious that the RHP zero does not much affect the magnitude but does affect the phase behavior (i.e., the phase starts affecting from $\omega = 0.2/DT_s$, which corresponds to $f = (0.064/D) \cdot f_s$). The buck–boost converter does not have similar special features as the buck and boost converters, because both the input and output currents are discontinuous. It may be obvious that the maximum control bandwidth would be limited and the stability of the converter would be sensitive to the variation in the output-capacitor ESR (r_C) as discussed in [10]. The first-order nature of (6.22) is natural due to tightly bounded inductor current.

6.3
Dynamic Review

The dynamic effects imposed by the self-oscillation control are reviewed by using the same buck converter as in the other chapters and a flyback or buck–boost converter introduced in [8, 10]. The converter can operate at open loop only at resistive load similarly to the other current-mode-controlled converters. Therefore, the internal transfer functions have to be computed from the measured or simulated load-affected responses. The resistive load would effectively hide the dynamic low-frequency behavior.

6.3.1
Buck Converter

The buck converter shown in Figure 6.4 is subjected to the dynamic analyses. There are available commercial critical (CM) or transition (TM) mode PWM modulators as introduced in [1], which can be used to implement the self-oscillation control.

The required control input (I_{co}) in current domain can be approximated by means of

$$I_{co}^2 - 2I_o \cdot I_{co} - \frac{2I_o M_1 M_2}{M_1 + M_2} \cdot T_D = 0 \tag{6.24}$$

where M_1 and M_2 are the up and down slopes of the inductor current, respectively. Similarly the cycle time (T_s) can be solved by means of

$$T_s = T_D + \frac{M_1 M_2}{M_1 + M_2} \cdot I_{co} \tag{6.25}$$

and the duty ratio (D) by means of

$$D = \frac{I_{co}}{M_1 T_s} \tag{6.26}$$

Equations (6.24)–(6.26) are solved from the inductor current waveforms shown in Figure 6.2.

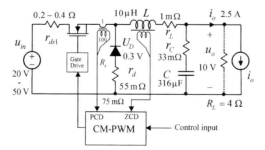

Figure 6.4 A buck converter under self-oscillation control.

The buck converter is analyzed at the input voltage of 20 and 50 V by varying the delay time (T_D) from 0 to 2 µs. The resulting switching frequencies and duty ratios are as follows:

20 V : 98 kHz, 0.53($T_D = 0$), 72 kHz, 0.454($T_D = 2$ µs)

50 V : 165 kHz, 0.21($T_D = 0$), 104 kHz, 0.17($T_D = 2$ µs)

6.3.1.1 Control-to-Output Transfer Function

The internal control-to-output transfer function (G_{co}) can be given symbolically by

$$\frac{2(1 + s \cdot r_C C)}{s \cdot T_s C \left(s + \dfrac{4}{T_s}\right)} \tag{6.27}$$

having a pole at the origin, which implies possibility of low-frequency instability, if conventional PI or PID controllers are used due to of phase lag $-180°$ at low frequencies. The frequency responses of the control-to-output transfer function at the input voltage of 20 V (solid lines) and 50 V (dashed lines) are shown in Figure 6.5, where the dots and squares denote the simulated responses at the input voltage of 20 V with $T_D = 0$ and $T_D = 2$ µs, respectively. The dash-dot lines of the phase response denote the effect of the delay (i.e., $T_D = 2$ µs corresponding to 20% of the cycle time). The simple model predicts quite well the control-to-output transfer function, but the phase has less phase lag than the model predicts at high frequency due to the modulation-process nonlinearities as discussed in [12].

The load-affected (dashed line) and internal (solid line) control-to-output transfer functions at the input voltage of 20V are shown in Figure 6.6. It may be obvious that the resistive load effectively hides the real behavior of the transfer function at the low frequencies and, therefore, the critical low-frequency phase behavior would not be visible as discussed earlier.

Figure 6.7 shows the comparison of the VM, PCM, and self-oscillation control-to-output transfer functions at the input voltage of 20 V. It may be obvious that the self-oscillation control imposes similar features than the PCM control except the low-frequency phase.

6.3.1.2 Output Impedance

The output impedance (Z_{o-o}) can be symbolically given by

$$\frac{1 + s \cdot r_C C}{sC} \tag{6.28}$$

which is the impedance of the output capacitor. It may be obvious that the result is as could be expected, because the average inductor current would stay

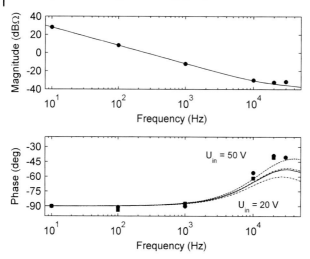

Figure 6.5 The predicted and simulated (dots and squares) frequency responses of the control-to-output transfer function at the input voltage of 20 V (solid lines) and 50 V (dashed lines). The dots correspond to $T_D = 0$ and the squares to $T_D = 2\ \mu s$, respectively.

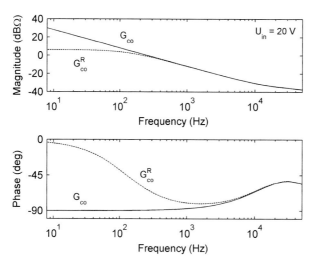

Figure 6.6 The resistive-load-affected and internal control-to-output transfer function at the input voltage of 20 V and $T_D = 0$.

constant at half the control current. This also means that the output impedance does not change along the operation point as is usual with the other control modes. The predicted (solid line) and simulated (dots) frequency responses of the output impedance are shown in Figure 6.8a confirming the validity of the

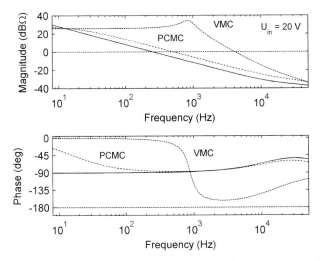

Figure 6.7 The comparison of the VM (dashed line), PCM (dash-dot line), and self-oscillation (solid line) control-to-output transfer functions at the input voltage of 20 V.

prediction. The corresponding VM and PCM output impedances compared to the self-oscillation output impedance at the input voltage of 20 V are shown in Figure 6.8b revealing that the PCM output impedance also resembles greatly the impedances of its output capacitor.

6.3.1.3 Input-to-Output Transfer Function

The input-to-output transfer function ($G_{\text{io}-o}$) is naturally zero, because the average inductor current and, consequently, the output current would stay constant regardless the changes in the input voltage.

6.3.1.4 Input Admittances

The open-loop input admittance ($Y_{\text{in}-o}$), the ideal input admittance ($Y_{\text{in}-\infty}$), and the short-circuit input admittance ($Y_{\text{in}-sc}$) are equal because $G_{\text{io}-o} = 0$. They can be symbolically given by

$$-\frac{DI_o}{\left(1 - \dfrac{T_D}{T_s}\right) U_{\text{in}}} \tag{6.29}$$

which equals the general ideal input admittance ($-I_{\text{in}}/U_{\text{in}}$) of the buck converter.

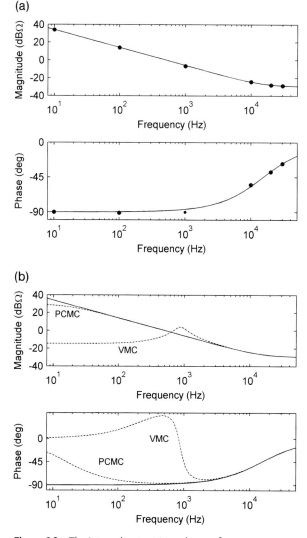

Figure 6.8 The internal output impedance of (a) self-oscillation-controlled buck converter (solid lines), and (b) compared to the output impedances of the corresponding VM and PCM buck converters (dashed lines). The dots denote the simulated responses.

6.3.2
Flyback Converter

The flyback or buck–boost converter shown in Figure 6.9 is subjected to the dynamic analyses. The experimental frequency responses are shown only for

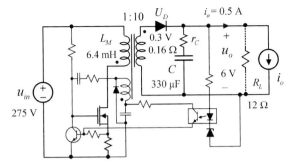

Figure 6.9 A flyback converter.

the output-voltage loop gain due to the difficulties to measure the open-loop responses reliably.

The required control current (I_{co}) can be approximated by means of

$$I_{co}^2 - \frac{2I_o(M_1 + M_2)}{M_1} \cdot I_{co} - \frac{2I_o M_2}{L} \cdot T_D = 0 \tag{6.30}$$

and the cycle time (T_s) and the duty ratio (D) according to (6.25) and (6.26), respectively.

The open-loop transfer functions are analyzed assuming that the input-side parameters of the flyback converter (Figure 6.9) are transformed into the secondary side: $U_{in} = 27.5$ V and $L = 64$ μH. The measured voltage loop gain ($L(s)$) is shown in Figure 6.10a by varying the delay time from 0 to 2 μs [10], and the corresponding predictions in Figure 6.10b. The switching frequency was approximately 64 kHz and the load of a 12 Ω resistor. It may be obvious that the delay does not much affect the dynamics, when its value is less than 12% of the cycle time. The extra low-frequency ripple in the measurements is due to the rectified line voltage having frequency of 50 Hz.

6.3.2.1 Control-to-Output Transfer Function

The control-to-output transfer function (G_{co}) can be symbolically given by

$$\frac{2D'\left(1 - s \cdot \frac{DT_s}{2}\right)(1 + s \cdot r_C C)}{T_s C \left(s + \frac{DD'^2 T_s}{2LC}\right)\left(s + \frac{4}{T_s}\right)} \tag{6.31}$$

where the low-frequency value $L/DD'T_s$ corresponds to $D'U_o/2DI_o$ or $U_{in}/2I_o$. This shows that the low-frequency gain would be highest at the high input voltage and low output current.

The converter is analyzed at the input voltage of 275 V, which corresponds to 27.5 V at the secondary side. The delay (T_D) is set to zero and 5 μs

(a)
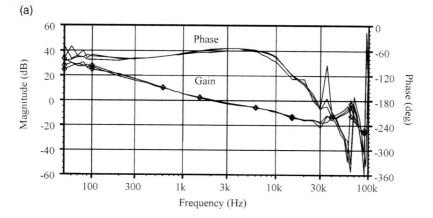

(b)
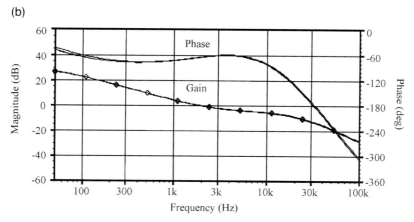

Figure 6.10 The frequency response of the voltage loop gain: (a) experimental measurement and (b) predictions.

corresponding to the switching frequencies of 63 and 40 kHz as well as the duty ratios of 0.19 and 0.15, respectively. The corresponding predicted (solid line) and simulated (dots: $T_D = 0$, squares: $T_D = 5$ μs) frequency responses are shown in Figure 6.11. The delay of 5 μs corresponds to about 20% of the cycle time. When the delay is small compared to the cycle time, the simple transfer function predicts the dynamics well. The tendency of the increasing delay is to lower the low-frequency gain and increase the phase lag as is visible in Figure 6.11 (the squares). The modulator nonlinearities would cause the high-frequency phase lag to be less than predicted as discussed in [12]. In the case of rather long delay, it is necessary to use the proposed modeling technique without simplifying the results in order to get more accurate predictions.

The flat high-frequency gain indicates that the control bandwidth cannot be designed very high in order to maintain sufficient gain margin. It may be also obvious that the high-frequency gain has very strong dependence on the ESR

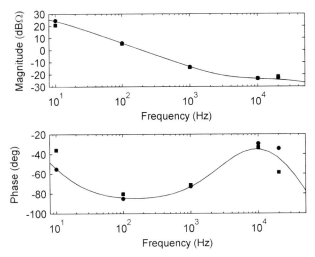

Figure 6.11 The predicted (solid line) and simulated (dots: $T_D = 0$, and squares: $T_D = 5$ μs) frequency responses of the control-to-output transfer function.

of the output capacitor, and the increase in it could reduce the gain margin to be unacceptable. As discussed above, the high input voltage and low output current would also increase the gain, which may lead to unacceptable low gain margin or even instability.

The load-affected transfer function compared to the internal transfer function is shown in Figure 6.12. In this case, the internal transfer function does not contain any surprising features as was the case with the buck converter.

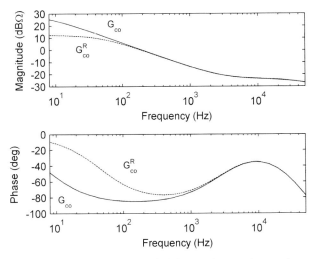

Figure 6.12 Load-affected (dashed line) and internal control-to-output transfer function.

6.3.2.2 Output Impedance

The output impedance (Z_{o-o}) can be symbolically given by

$$\frac{1 + s \cdot r_C C}{C\left(s + \dfrac{DD'^2 T_s}{2LC}\right)} \tag{6.32}$$

where the low-frequency value $2L/T_s DD'$ corresponds to U_o/DI_o [8]. This shows that the output impedance would increase, when the input voltage increases. The predicted (solid and dashed lines) and simulated (dots and squares) frequency responses of the output impedance are shown in Figure 6.13, where the solid lines and dots correspond to $T_D = 0$ and the dashed lines and squares to $T_D = 5\,\mu s$. It may be obvious that the proposed model would predict accurately the response at the low delay values but the longer delays would require the application of the proposed method without simplifications for obtaining more accurate predictions.

6.3.2.3 Input-to-Output Transfer Function

The input-to-output transfer function (G_{io-o}) can be symbolically given by

$$\frac{D^2 D' T_s (1 + s \cdot r_C C)}{2LC\left(s + \dfrac{DD'^2 T_s}{2LC}\right)} \tag{6.33}$$

The low-frequency value of the transfer function is D/D', which is the same as the value of the corresponding fixed-frequency converter. It means that

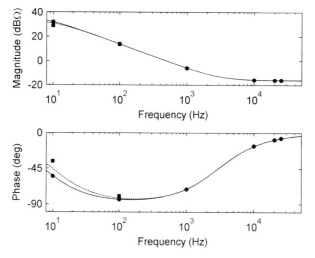

Figure 6.13 The predicted (solid line: $T_D = 0$, dashed line: $T_D = 5\,\mu s$) and simulated (dots: $T_D = 0$, and squares: $T_D = 5\,\mu s$) frequency responses of the output impedance.

the converter attenuates the input noise at the low frequencies, when the duty ratio is less than 0.5. The maximum designed duty ratio of the flyback converter (Figure 6.9) is typically 0.5 in order to keep the component stresses at a reasonable level.

The predicted (solid and dashed lines) and simulated (dots: $T_D = 0$, squares: $T_D = 5\,\mu s$) frequency responses of the input-to-output transfer function are shown in Figure 6.14. The high-frequency phase responses are missing due to the problems to extract them, when the switching-frequency ripple is much higher than the response signal. The proposed model accurately predicts the response, when the delay is small. The tendency of the delay is to increase the attenuation and decrease the phase at low frequencies.

The resistive-load-affected transfer function (dashed line) compared to the internal transfer function (solid line) is shown in Figure 6.15. It may be obvious that the load-affected transfer function would give misleading information on the attenuation potential of the converter. The extra attenuation is from the attenuating effect of the output impedance and the load resistor.

6.3.2.4 Input Admittance

The internal input admittance (Y_{in-o}) can be symbolically given by

$$-\frac{D^2 D' T_s \cdot s}{2L\left(s + \dfrac{DD'^2 T_s}{2LC}\right)} \tag{6.34}$$

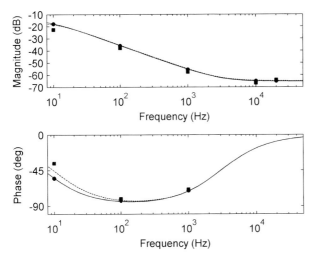

Figure 6.14 The predicted (solid line: $T_D = 0$, dashed line: $T_D = 5\,\mu s$) and simulated (dots: $T_D = 0$, and squares: $T_D = 5\,\mu s$) frequency responses of the input-to-output transfer function.

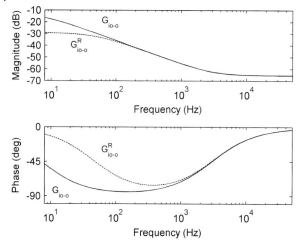

Figure 6.15 Load-affected (dashed line) and internal input-to-output transfer function.

The corresponding frequency responses are shown in Figure 6.16, where the solid and dashed lines correspond to the predicted responses ($Y_{\text{in}-o}$) (i.e., $T_D = 0$, and $T_D = 5\ \mu s$), the dash-dotted lines to the predicted resistive-load-affected transfer functions ($Y_{\text{in}-o}^R$), and the dots ($T_D = 0$) and squares ($T_D = 5\ \mu s$) the simulated frequency responses, respectively. The load affects the responses only at low frequencies. The simulated responses are the load-affected responses. It may be obvious that the effect of the delay is insignificant. The phase-behavior implies that the flyback converter can be unstable even at open loop when the input filter is connected, if the filter is not properly designed (i.e., no impedance overlap is allowed).

6.3.2.5 Ideal and Short-Circuit Admittances

The ideal input admittance ($Y_{\text{in}-\infty}$) can be symbolically given by

$$-\frac{D^2 T_s}{2L} \cdot \frac{1}{1 - s \cdot \frac{DT_s}{2}} \approx -\frac{DI_L}{U_{\text{in}}} \cdot \frac{1}{1 - s \cdot \frac{DT_s}{2}} \quad (6.35)$$

and the short-circuit admittance ($Y_{\text{in}-sc}$) by

$$-\frac{D^2 D' T_s}{2L} \approx -\frac{DD' I_L}{U_{\text{in}}} \quad (6.36)$$

The corresponding frequency responses are shown in Figure 6.17. The RHP zero of the control-to-output transfer function (G_{co}) forms the RHP pole in the ideal input admittance (6.35). The high-frequency phase behavior of the ideal input admittance implies slightly increased sensitivity to the source interactions.

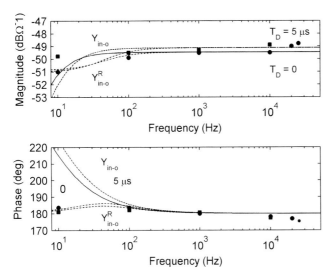

Figure 6.16 Predicted load-affected (dash-dotted lines) and internal (solid lies: $T_D = 0$, dashed lines: $T_D = 5$ μs) input admittances. The dots ($T_D = 0$) and squares ($T_D = 5$ μs) denote the simulated responses.

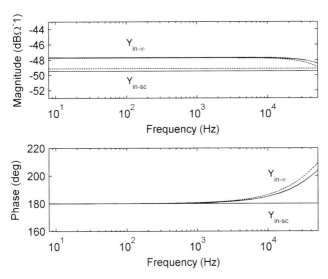

Figure 6.17 The predicted frequency responses of the ideal ($Y_{in-\infty}$) and short-circuit (Y_{in-sc}) input admittance, where the solid lines correspond to $T_D = 0$, and the dashed lines to $T_D = 5$ μs.

References

1. R. Redl, 'Power-factor correction in single-phase switching-mode power supplies – An overview,' *Int. J. Electron.*, vol. 77, no. 5, **1994**, pp. 555–582.
2. Y. Panov and M. Jovanovic, 'Performance evaluation of 70-W two-stage adapters for notebook computers,' in *Proc. IEEE Applied Power Electronics Conf.*, **1999**, pp. 1059–1065.
3. Use of rectified image voltage for controlling the switch on the primary side of a switched-mode power supply, U.S. Patent 6608768030B2, August 19, **2003**.
4. B. Irving and M. Jovanovic, 'Analysis and design of self-oscillating flyback converter,' in *Proc. IEEE Applied Power Electronics Conf.*, **2002**, pp. 897–903.
5. G. Spiazzi, D. Tagliavia, and S. Spampinato, 'DC–DC flyback converters in the critical conduction mode: A re-examination,' in *Proc. IEEE Industry Applications Society Annual Conf.*, **2000**, pp. 2426–2432.
6. B. Irving, Y. Panov, and M. Jovanovic, 'Small-signal model of variable-frequency flyback converter,' in *Proc. IEEE Applied Power Electronics Conf.*, **2003**, pp. 977–982.
7. J. Chen, R. Erickson, and D. Maksimovic, 'Averaged switch modeling of boundary conduction mode DC-to-DC converters,' in *Proc. IEEE Industrial Electronics Society Annual Conf.*, **2001**, pp. 844–849.
8. J. Lempinen and T. Suntio, 'Small-signal modeling for design of robust variable-frequency flyback battery chargers for portable device applications,' in *Proc. IEEE Applied Power Electronics Conf.*, **2001**, pp. 548–554.
9. T. Suntio, J. Lempinen, K. Hynynen, and P. Silventoinen, 'Analysis and small-signal modeling of self-oscillating converters with applied delay,' in *Proc. IEEE Applied Power Electronics Conf.*, **2002**, pp. 395–401.
10. T. Suntio, 'Average and small-signal modeling of self-oscillating flyback converter with applied switching delay,' *IEEE Trans. Power Electron.*, vol. 21, no. 2, **2006**, pp. 479–485.
11. T. Suntio, 'Unified average and small-signal modeling of direct-on-time control,' *IEEE Trans. Indust. Electron.*, vol. 53, no. 1, **2006**, pp. 287–295.
12. J. Sun, 'Small-signal modeling of variable-frequency pulsewidth modulators,' *IEEE Trans. Aerosp. Electron. Syst.*, vol. 38, no. 3, **2002**, pp. 1104–1108.

7
Dynamic Modeling and Analysis of Current-Output Converters

7.1
Introduction

Majority of the switched-mode converters operate at voltage-output mode providing constant output voltage. In the applications, where the storage batteries are used either to ensure uninterrupted operation of the end load as in the telecom DC UPS (uninterruptible power supply, Figure 7.1) systems [1] or as a standalone power source in mobile phones [2], the switched-mode converters recharging the storage batteries have to be provided with overcurrent or overload limiting for protecting them from damage. The over-current limiting changes the operation of the converter into the current-output mode.

The overload limiting is typically implemented using either constant-current or modified-constant-power limiting schemes as illustrated in Figure 7.2 [3–5]. The constant-current limiting can be accomplished either by using only the output current signal (i.e., single loop) or both the output voltage and the output current in cascade where the inner loop is the voltage loop and the outer loop the current loop. The modified constant-power limiting is typically implemented in such a way that the reference of the current loop is gradually increased along the decrease in the output voltage until the maximum defined output current is reached. After that the limiting follows the constant-current scheme as described in detail in [4].

The change of mode from the voltage output to the current output would also drastically change the dynamic characteristics of the converter from those observed in the voltage-output mode. It has been observed [6] that the crossover frequency of the current loop may increase even a decade or two compared to the measured frequency response at a resistive load, when the storage battery is connected as a load due to low internal impedance of the battery [7]. Several attempts have been made earlier to model the dynamics of the current-output converter [8–10], but the real dynamics were not found. The consistent methods to model the dynamics of the current-output converters are presented in [11–13] based on the models describing the dynamics of the voltage-output converter. The modeling technique applied in this chapter is extracted from [11].

Dynamic Profile of Switched-Mode Converter. Teuvo Suntio
© 2009 WILEY-VCH Verlag GmbH & Co. KGaA, Weinheim
ISBN: 978-3-527-40708-8

7 Dynamic Modeling and Analysis of Current-Output Converters

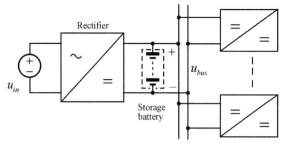

Figure 7.1 DC UPS system.

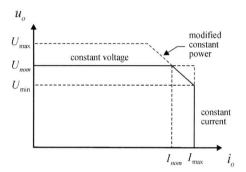

Figure 7.2 Typical output-voltage/current characteristics of a converter used in the battery-charging applications.

7.2
Dynamic Models for Current-Output Converter

The general load of a voltage-output converter consists of a parallel connection of an impedance (Z_L) and a constant-current sink (j_o) as depicted in Figure 7.3a. The current-output converter cannot operate properly with such a load. Therefore, its general load would consist of a series connection of an impedance (Z_L) and a constant-voltage source (e_o) as shown in Figure 7.3b. The internal dynamics of the current-output converter can be found assuming the load to be a pure constant-voltage source, which would be denoted by u_o. It may be obvious that the ideal load would effectively short-circuit the converter and, therefore, the converter cannot operate without damage at open loop. As a consequence, the internal dynamics is not fully directly measurable: near to ideal dynamic constant-voltage load can be implemented connecting a proper resistor and large capacitor with a low ESR in parallel, which would yield accurate enough frequency responses for the practical purposes. The dynamical models of the current-output converter can also be derived directly from the power stage changing the load as described above and by applying the modeling methods described in Chapters 3–6. The most convenient method would be, however, to derive the models based on the well-known transfer functions of the corresponding voltage-output converter.

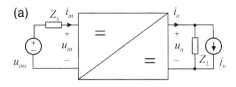

Figure 7.3 General load and supply conditions of (a) voltage-output converter and (b) current-output converter.

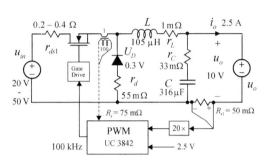

Figure 7.4 Experimental VM/PCM-controlled buck converter at current-output mode.

The application of the modified state-space-averaging method is demonstrated in Section 7.2.1 by using a CCM buck converter (Figure 7.4). The general dynamic models are derived in Section 7.2.2 based on the voltage-output converter.

7.2.1
Modified-State-Space-Averaging Technique

We demonstrate the application of the modified-state-space-averaging technique [14, 15] on the VMC buck converter as shown in Figure 7.4. The open-loop conditions we assume, when developing the models, are naturally fictive due to the reasons discussed earlier. We also assume that the output-current-sensing resistor (R_{s1}) is zero (note that the physical output-current-sensing resistor actually acts as a load and, therefore, it would be treated as a load resistor).

The variables of the current-output converter are basically same as the variables of the voltage-output converter except the output voltage (u_o) and the output current (i_o), which would change their roles from the output variable to the input variable (u_o) and from the input variable to the output variable (i_o). As a consequence, the small-signal state space can be given by (see Chapter 3, Section 3.3)

7 Dynamic Modeling and Analysis of Current-Output Converters

$$\frac{d\hat{i}_L}{dt} = -\frac{r_E}{L} \cdot \hat{i}_L + \frac{D}{L} \cdot \hat{u}_{in} - \frac{1}{L} \cdot \hat{u}_{in} + \frac{U_E}{L} \cdot \hat{d}$$

$$\frac{d\hat{u}_c}{dt} = -\frac{1}{r_c C} \cdot \hat{u}_c + \frac{1}{r_c C} \cdot \hat{u}_o$$

$$\hat{i}_{in} = D\hat{i}_L + I_L \hat{d} \quad (7.1)$$

$$\hat{i}_o = \hat{i}_L + \frac{1}{r_c} \cdot \hat{u}_c - \frac{1}{r_c} \cdot \hat{u}_o$$

where r_E and U_E are given by

$$r_E = r_L + Dr_{ds} + D'r_d$$

$$U_E = U_{in} + U_D + (r_d - r_{ds})I_o \quad (7.2)$$

The set of transfer functions defining the dynamics of the current-output converter can be obtained from (7.1) by first applying *Laplace* transform and secondly matrix algebra yielding

$$\begin{bmatrix} \hat{i}_{in} \\ \hat{i}_o \end{bmatrix} = \begin{bmatrix} Y^i_{in-o} & T^i_{oi-o} & G^i_{ci} \\ G^i_{io-o} & -Y^i_{o-o} & G^i_{co} \end{bmatrix} \begin{bmatrix} \hat{u}_{in} \\ \hat{u}_o \\ \hat{d} \end{bmatrix}$$

$$\begin{bmatrix} Y^i_{in-o} & T^i_{oi-o} & G^i_{ci} \\ G^i_{io-o} & -Y^i_{o-o} & G^i_{co} \end{bmatrix}$$

$$= \frac{\begin{bmatrix} D^2(1+sr_cC) & -D(1+sr_cC) & DU_E(1+sr_cC) \\ D(1+sr_cC) & -LC\left(s^2 + s\frac{r_E+r_c}{L} + \frac{1}{LC}\right) & U_E(1+sr_cC) \end{bmatrix}}{(r_E + sL)(1 + sr_cC)}$$

$$+ \begin{bmatrix} 0 & 0 & I_L \\ 0 & 0 & 0 \end{bmatrix} \quad (7.3)$$

According to (7.3), all the transfer functions except the output admittance (Y^i_{o-o}) are of the first order, which is as could be expected due to the dynamic short-circuit at the output of the converter. We also give for comparison the set of transfer functions defining the dynamics of the corresponding voltage-output converter as follows:

$$\begin{bmatrix} \hat{i}_{in} \\ \hat{u}_o \end{bmatrix} = \begin{bmatrix} Y^v_{in-o} & T^v_{oi-o} & G^v_{ci} \\ G^v_{io-o} & -Z^v_{o-o} & G^v_{co} \end{bmatrix} \begin{bmatrix} \hat{u}_{in} \\ \hat{i}_o \\ \hat{d} \end{bmatrix}$$

$$\begin{bmatrix} Y^v_{in-o} & T^v_{oi-o} & G^v_{ci} \\ G^v_{io-o} & -Z^v_{o-o} & G^v_{co} \end{bmatrix}$$

$$= \frac{\begin{bmatrix} \dfrac{D^2 s}{L} & \dfrac{D(1+sr_cC)}{LC} & \dfrac{DU_{ES}}{L} \\ \dfrac{D(1+sr_cC)}{LC} & -\dfrac{(r_E+sL)(1+sr_cC)}{LC} & \dfrac{U_E(1+sr_cC)}{LC} \end{bmatrix}}{s^2 + s\dfrac{r_E+r_c}{L} + \dfrac{1}{LC}} + \begin{bmatrix} 0 & 0 & I_L \\ 0 & 0 & 0 \end{bmatrix} \quad (7.4)$$

According to (7.2) and (7.4), the output admittance (Y^i_{o-o}) is exactly the inverse of the output impedance (Z_{o-o}) of the corresponding voltage-output converter. This means that the output admittance is the only directly measurable transfer function in a current-output converter.

The current-output mode has clearly and significantly changed the internal dynamics of the converter by removing its resonant nature from all the other transfer functions except from the output admittance. It may also be obvious that the gain of the internal current loop ($L_c(s)$) may be much higher than the corresponding gain of the voltage loop ($L_v(s)$) due to the changes in the control-to-output transfer function. We will later show in Section 7.3 that the load would easily recover the voltage-output nature of the transfer functions, which also explains the observed unexpected changes in the crossover frequency of the current loop observed in [6] as discussed earlier in Section 7.1. In short, the close-to-ideal voltage-type load such as storage batteries [4, 7] or a large capacitor would recover the internal dynamics of the converter into effect with the observed consequences.

The modified state-space-averaging technique can be equally applied to all the converters, but we have only demonstrated the use of the technique with the VMC buck converter. The more general and may be more convenient treatment of the modeling would be given in the next section.

7.2.2
General Dynamic Models

The general dynamic models of the current-output converter can be developed by applying the two-port model of the voltage-output converter (Figure 7.5a) and duality, which yield the two-port model as shown in Figure 7.5b.

The duality means that the Thevenin output port of the voltage-output converter would be substituted by means of the corresponding Norton equivalent circuit, and $\hat{i}_o$ in the input port by

$$\hat{i}_o = \frac{\hat{u}_T - \hat{u}_o}{Z_T} \quad (7.5)$$

When the described procedures are followed, the internal dynamic models of the current-output convert can be given based on the transfer functions of the corresponding voltage-output converter as follows:

7 Dynamic Modeling and Analysis of Current-Output Converters

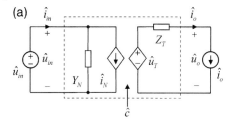

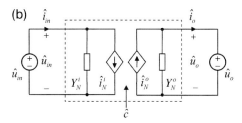

Figure 7.5 Two-port linear models of (a) voltage-output converter and (b) current-output converter.

$$\begin{bmatrix} \hat{i}_{in} \\ \hat{i}_o \end{bmatrix} = \begin{bmatrix} Y^v_{in-o} + \dfrac{G^v_{io-o} T^v_{oi-o}}{Z^v_{o-o}} & -\dfrac{T^v_{oi-o}}{Z^v_{o-o}} & G^v_{ci} + \dfrac{G^v_{co} T^v_{oi-o}}{Z^v_{o-o}} \\ \dfrac{G^v_{io-o}}{Z^v_{o-o}} & -\dfrac{1}{Z^v_{o-o}} & \dfrac{G^v_{co}}{Z^v_{o-o}} \end{bmatrix} \begin{bmatrix} \hat{u}_{in} \\ \hat{u}_o \\ \hat{c} \end{bmatrix} \quad (7.6)$$

According to (7.6), it is clear that the output admittance of the current-output converter is the inverse of the output impedance of the corresponding voltage-output converter as we observed already in the previous section. The order of the internal control-to-output (G^i_{co}), input-to-output (G^i_{io-o}), and output-to-input (T^i_{oi-o}) transfer functions are surely reduced compared to the order of the corresponding voltage-output converter, but the input admittance (Y^i_{in-o}) and the control-to-input (G^i_{ci}) may or may not have the same order and the features as the corresponding voltage-output converter. The internal input admittance (Y^i_{in-o}) is the same as the short-circuit input admittance (Y_{in-sc}) defined for the voltage-output converter, but the input admittance is not anymore load independent, which may be concluded according to Figure 7.3b.

7.3
Load and Supply Interactions

The effect of the nonideal load (Z_L, $\hat{e}_o$) can be found by computing $\hat{u}_o$ at the presence of the load from Figure 7.6, which yields

$$\hat{u}_o = \dfrac{Z_L(G^v_{io-o} \hat{u}_{in} + G^v_{co} \hat{c}) + Z^v_{o-o} \hat{e}_o}{Z_L + Z^v_{o-o}} \quad (7.7)$$

and substituting it in (7.6) with (7.7). These procedures yield

$$\begin{bmatrix} \hat{i}_{in} \\ \hat{i}_o \end{bmatrix} = \begin{bmatrix} Y_{in-o}^v + \dfrac{G_{io-o}^v T_{oi-o}^v}{Z_L + Z_{o-o}^v} & -\dfrac{T_{oi-o}^v}{Z_L + Z_{o-o}^v} & G_{ci}^v + \dfrac{G_{co}^v T_{oi-o}^v}{Z_L + Z_{o-o}^v} \\ \dfrac{\dfrac{G_{io-o}^v}{Z_{o-o}^v}}{1 + \dfrac{Z_L}{Z_{o-o}^v}} & -\dfrac{1}{Z_L + Z_{o-o}^v} & \dfrac{\dfrac{G_{co}^v}{Z_{o-o}^v}}{1 + \dfrac{Z_L}{Z_{o-o}^v}} \end{bmatrix} \begin{bmatrix} \hat{u}_{in} \\ \hat{e}_o \\ \hat{c} \end{bmatrix}$$

(7.8)

The effect of the nonideal source (Z_s, $\hat{u}_{ins}$) can be found by computing $\hat{u}_{in}$ at the presence of the source from Figure 7.6, which yields

$$\hat{u}_{in} = \dfrac{\hat{u}_{ins} + Z_s \left(\dfrac{T_{oi-o}^v}{Z_{o-o}^v} \hat{e}_o - \left(G_{ci}^v + \dfrac{G_{co}^v T_{oi-o}^v}{Z_{o-o}^v} \right) \hat{c} \right)}{1 + Z_s Y_{in-o}^i}$$

(7.9)

and substituting it in (7.6) with (7.9). These procedures yield

$$\begin{bmatrix} \hat{i}_{in} \\ \hat{i}_o \end{bmatrix} = \begin{bmatrix} \dfrac{Y_{in-o}^i}{1 + Z_s Y_{in-o}^i} & -\dfrac{\dfrac{T_{oi-o}^v}{Z_{o-o}^v}}{1 + Z_s Y_{in-o}^i} & \dfrac{G_{ci}^v + \dfrac{G_{co}^v T_{oi-o}^v}{Z_{o-o}^v}}{1 + Z_s Y_{in-o}^i} \\ \dfrac{\dfrac{G_{io-o}^v}{Z_{o-o}^v}}{1 + Z_s Y_{in-o}^i} & -\dfrac{1 + Z_s Y_{in-oc}^i}{1 + Z_s Y_{in-o}^i} \cdot \dfrac{1}{Z_{o-o}^v} & \dfrac{1 + Z_s Y_{in-\infty}^i}{1 + Z_s Y_{in-o}^i} \cdot \dfrac{G_{co}^v}{Z_{o-o}^v} \end{bmatrix} \begin{bmatrix} \hat{u}_{ins} \\ \hat{u}_o \\ \hat{c} \end{bmatrix}$$

(7.10)

where $Y_{in-\infty}^i$ (the ideal input admittance) is equal to the ideal input admittance of the voltage-output converter (i.e., $Y_{in-\infty}^i = Y_{in-\infty}^v = Y_{in-o}^v - \dfrac{G_{io-o}^v G_{ci}^v}{G_{co}^v}$), and Y_{in-oc}^i (the open-circuit input admittance) is equal to the internal open-loop input admittance of the voltage-output converter (i.e., $Y_{in-oc}^i = Y_{in-o}^v$). If the source interactions are to be defined for the load-affected converter, then all the parameters in (7.10) would be the corresponding load-affected transfer functions defined in (7.8). The ideal input admittance ($Y_{in-\infty}^v$) and the open-circuit input admittance (Y_{in-oc}^i) will be the only parameters, which will stay intact.

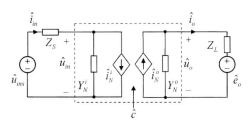

Figure 7.6 Two-port model of a current-output converter with nonideal source and load.

According to (7.6), the load-affected control-to-output transfer function (G_{co}^{i-L}) can be given by

$$\frac{\dfrac{G_{co}^v}{Z_{o-o}^v}}{1+\dfrac{Z_L}{Z_{o-o}^v}} \tag{7.11}$$

and the corresponding voltage-output transfer function (G_{co}^{v-L}) by

$$\frac{G_{co}^v}{1+\dfrac{Z_{o-o}^v}{Z_L}} \tag{7.12}$$

Equation (7.11) shows that the load does not affect the control-to-output transfer function, when $Z_L \ll Z_{o-o}^v$. Equation (7.12) shows that the load does not affect the control-to-output transfer function, when $Z_L \gg Z_{o-o}^v$. In both the cases, the resulting effect is typically a reduction of crossover frequency of the current or voltage loop.

According to (7.11), it may be obvious that the resistive load would recover the dynamical features of the voltage-output converter into the current-output converter, because $G_{co}^{i-L} \approx \frac{G_{co}^v}{Z_L}$, when $Z_L > Z_{o-o}^v$. Similarly, the low-impedance load would change the voltage-output converter to have the features resembling current-output converter, when $Z_L < Z_{o-o}^v$, because $G_{co}^{v-L} \approx Z_L \cdot \frac{G_{co}^v}{Z_{o-o}^v}$.

7.4
Cascaded Voltage-Current Loops

The rectifiers in the Telecom DC-UPS applications (Figure 7.1) [4] are normally operating at voltage-output mode, and when recharging the parallel connected battery, the current-output mode is automatically switched on. The required control system may be implemented either by using independent controllers for the voltage and current loops or using a cascaded configuration, where the voltage loop is the inner loop and the current loop the outer loop. The first case means that the voltage and current loops do not interact and, therefore, the basic current-output dynamical modeling presented in Section 7.2 applies. The second case means that the voltage and current loops are interconnected and, therefore, a different technique has to be applied to find the required dynamical models.

In the cascaded case, the closed-loop voltage-output converter has to be converted into the current-output mode. This can be accomplished by using similar methods as with the open-loop converter, but the open-loop transfer functions in (7.6) have to be replaced with the corresponding closed-loop transfer functions as follows:

$$\begin{bmatrix} \hat{i}_{in} \\ \hat{i}_o \end{bmatrix} = \begin{bmatrix} Y^v_{in-c} + \dfrac{G^v_{io-c} T^v_{oi-c}}{Z^v_{o-c}} & -\dfrac{T^v_{oi-c}}{Z^v_{o-c}} & G^v_{ci-c} + \dfrac{G^v_{co-c} T^v_{oi-c}}{Z^v_{o-c}} \\ \dfrac{G^v_{io-c}}{Z^v_{o-c}} & -\dfrac{1}{Z^v_{o-c}} & \dfrac{G^v_{co-c}}{Z^v_{o-c}} \end{bmatrix} \begin{bmatrix} \hat{u}_{in} \\ \hat{u}_o \\ \hat{c} \end{bmatrix} \quad (7.13)$$

where the subscript extension c denotes closed loop and the corresponding transfer functions are given in (7.14) (G_{se} = voltage sensing gain and $L_v(s)$ = voltage loop gain)

$$\begin{bmatrix} Y^v_{in-c} & T^v_{oi-c} & G^v_{ci-c} \\ G^v_{io-c} & -Z^v_{o-c} & G^v_{co-c} \end{bmatrix}$$

$$= \begin{bmatrix} Y^v_{in-o} - \dfrac{G^v_{io-o} G^v_{ci}}{G^v_{co}} \cdot \dfrac{L_v(s)}{1+L_v(s)} & T^v_{oi-o} + \dfrac{Z^v_{o-o} G^v_{ci}}{G^v_s} \cdot \dfrac{L_v(s)}{1+L_v(s)} & \dfrac{G^v_{ci}}{G_{se} G^v_{co}} \cdot \dfrac{L_v(s)}{1+L_v(s)} \\ \dfrac{G^v_{io-o}}{1+L_v(s)} & -\dfrac{Z^v_{o-o}}{1+L_v(s)} & \dfrac{1}{G_{se}} \cdot \dfrac{L_v(s)}{1+L_v(s)} \end{bmatrix}$$
$$(7.14)$$

7.5
Dynamic Review

The behavior of the control-to-output transfer function (G^i_{co}) and the current-loop gain ($L_c(c)$) of the current-output VMC and PCMC buck converter in CCM would be analyzed dynamically in this section. The experimental buck converter is shown earlier in Figure 7.4.

The predicted frequency responses of the internal (G^i_{co}) and resistive-load-affected (G^{i-R}_{co}) control-to-output transfer functions are shown in Figure 7.7 at the input voltage of 50 V and at the load of 2.5 A based on the analytical models derived earlier in Chapters 3 and 4.

The solid lines represent the internal dynamics and the dashed lines the 4-Ω-resistor-affected dynamics. The VMC control-to-output transfer function is multiplied by the modulator gain 1/3, and the PCMC control-to-output transfer function is divided by the inductor-current-sensing resistor of 75 mΩ.

According to Figure 7.7a, the current-loop crossover frequency has clear tendency to reduce approximately one decade, when the resistive load is connected compared to the internal crossover frequency. According to Figure 7.7b, the change of crossover frequency under PCM control could be approximately two decades or more. These are clear indications and explain the observed phenomena in [6]. This also means that the real reason for the observed problems has been the incorrect control design. It is discussed in [6] that the unexpected phenomenon can be avoided if the instantaneous current limiting of the PCM control is only applied. This is true because the internal instantaneous current limiting forces the converter to enter into the open-loop mode of operation and, therefore, no feedback loop from the output current does exist.

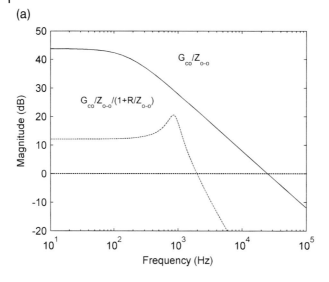
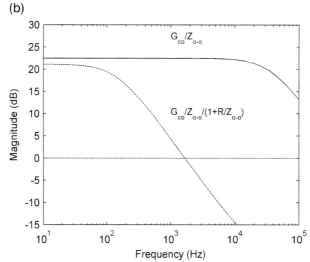

Figure 7.7 The frequency responses of the internal (solid lines) and resistive-load-affected (dashed lines) control-to-output transfer functions of the current-output buck converter at the input voltage of 50 V and at the load current of 2.5 A under (a) VM control and (b) PCM control.

The current-loop frequency responses were measured by injecting the excitation signal between the PWM block and the current-loop amplifier (20×) (Figure 7.4). The load was a parallel connection of a pure 4-Ω resistor and a 5-mF capacitor mimicking the dynamic short-circuit. The corresponding frequency responses are shown in Figure 7.8, which shows that the predictions

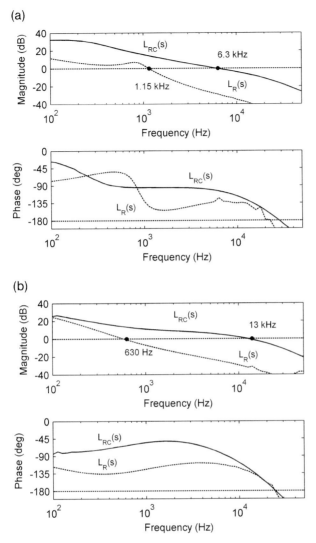

Figure 7.8 Measured frequency responses of the current-loop gain at the input voltage of 50 V and the output load current of 2.5 A using RC load (solid lines) and pure resistor load (R, dashed lines) of a current-output buck converter under (a) VM control and (b) PCM control.

made earlier in conjunction with Figure 7.7 are correct with respect to the crossover-frequency changes as well as the effect of a resistive load on the converter dynamics. The shown internal crossover frequencies are actually slightly load-affected due to the output-current-sensing resistor of 50 mΩ (Figure 7.4), which is higher than the high-frequency output impedance of the converters causing slight reduction of the loop crossover frequency.

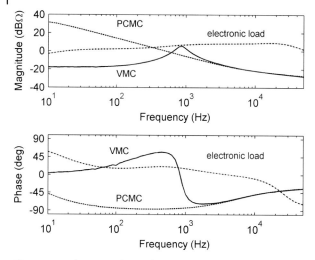

Figure 7.9 The internal impedance of the electronic load (dash-dot lines) and the output impedances of the experimental converters (VMC solid lines, PCMC dashed lines).

There are available electronic loads providing constant-voltage-type-load features. There are naturally high temptations to use them during the design phase due to the flexibility they provide. Figure 7.9 shows the internal impedance (dash-dot lines) of a certain electronic load compared to the output impedances of the experimental converters (VMC solid lines, PCMC dashed lines).

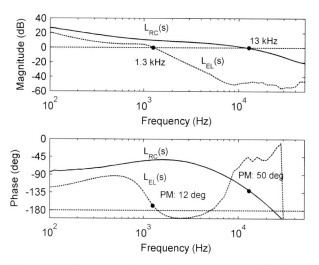

Figure 7.10 The measured current loop gains at the RC load (solid lines) and the electronic voltage-type load (dashed lines).

Figure 7.9 shows that the internal impedance of the electronic load is quite high and higher than the internal output impedances of the converters in the frequencies starting from 300 Hz. This means that the measured loop gain is load affected and does not give the correct information on the dynamics of the loop gain. The corresponding measured current-loop gains are shown in Figure 7.10, which confirms the conclusions made according to the impedances. The current loop indicates very low crossover frequency and phase margin (PM), which would lead to redesign of the controller. It may be obvious that the redesign could lead even to a worse design by extending the crossover frequency beyond the switching frequency. As a consequence, it is highly recommended to use an *RC* load with sufficiently large capacitor for the validation purposes in order to avoid the problems originating from the nonideal characteristics of the electronic loads.

References

1. T. Suntio and A. Glad, 'The batteries as a principal component in DC UPS systems,' in *Proc. IEEE International Telecommunications Energy Conf.*, **1990**, pp. 400–411.
2. J. Lempinen and T. Suntio, 'Small-signal modeling for design of robust variable-frequency flyback battery chargers for portable device applications,' in *Proc. IEEE Applied Power Electronics Conf.*, **2001**, pp. 548–554.
3. T. Suntio, A. Glad, and P. Waltari, 'Constant-current vs. constant-power protected rectifier as a DC UPS building block,' in *Proc. IEEE International Telecommunications Energy Conf.*, **1996**, pp. 227–233.
4. T. Suntio, I. Gadoura, J. Lempinen, and K. Zenger, 'Practical design issues of multi-loop controller for a telecom rectifier,' in *Proc. IEEE Telecommunications Energy Special Conf.*, **2000**, pp. 197–201.
5. I. Gadoura and T. Suntio, 'Implementation of optimal output characteristics for a telecom power supply: Fuzzy-logic approach,' *Int. J. Cont. Intell. Syst.*, vol. 30, no. 3, **2002**, pp. 134–139.
6. V.J. Thottuvelil, 'Modeling and analysis of power converter systems with batteries,' in *Proc. IEEE International Telecommunications Energy Conf.*, **1997**, pp. 517–522.
7. A. Tenno, R. Tenno, and T. Suntio, 'Battery impedance and its relation to battery characteristics,' in *Proc. IEEE International Telecommunications Energy Conf.*, **2002**, pp. 176–183.
8. T. Sato, T. Nakano, and K. Harada, 'The overload-protection characteristics of the current-mode DC-to-DC converter: Analysis and improvements,' in *Proc. IEEE Power Electronics Specialists Conf.*, **1988**, pp. 830–835.
9. H. Matsuo, F. Kurokawa, and T. Takeda, 'Analysis of the dynamic characteristics in the over-current limited mode of the DC–DC converter,' *IEEE Trans. Power Electron.*, vol. 4, no. 2, **1989**, pp. 175–180.
10. G.C. Hsieh, Y.H. Lin, and H.C. Tasi, 'Control strategy for constant voltage/constant current switching-mode instrumentation power supply,' in *Proc. IEEE Industrial Electronics Annual Conf.*, **2000**, pp. 983–988.
11. M. Hankaniemi and T. Suntio, 'Small-signal models for constant-current regulated converters,' in *Proc. IEEE Industrial Electronics Annual Conf.*, **2006**, pp. 2037–2042.

12. M. Hankaniemi, T. Suntio, and M. Sippola, 'Analysis of the load interactions in constant-current-controlled buck converter,' in *Proc. IEEE International Telecommunications Energy Conf.*, **2006**, pp. 343–348.
13. M. Hankaniemi and T. Suntio, 'Dynamical modeling and control of current-output converters,' *Int. Rev. Electr. Eng.*, vol. 2, no. 5, **2007**, pp. 671–680.
14. T. Suntio, 'Unified average and small-signal modeling of direct-on-time control,' *IEEE Trans. Indust. Electron.*, vol. 53, no. 1, **2006**, pp. 287–295.
15. R.D. Middlebrook and S. Cuk, 'A general unified approach in modeling switching-converter power stages,' *Int. J. Electron.*, vol. 42, no. 6, **1977**, pp. 521–550.

8
Interconnected Systems

8.1
Introduction

The switched-mode converter is usually a subsystem within a system either in a single equipment or in a wider constellation known as a distributed power system [1–3] forming a dynamically complicated interconnected overall system, which is prone to instability and performance degradation. The use of distributed power architectures is continuously growing and would become gradually the most popular way of applying the converters [3]. The distributed systems can be found, for example, on electronic circuit boards [4, 5], in space [6], and automotive [7] power applications, and so on. The stability and performance problems are actually general problems, which can also be encountered equally in the AC power systems [8, 9] and are supposed to be increasing when more and more power electronic equipments are deployed into the AC grid systems.

The switched-mode converters are a natural source of electromagnetic interference (EMI) due to its operating principles. Therefore, the converters are typically equipped with an input EMI filter to suppress the EMI noise to an acceptable level in order to comply with the stringent international EMI standards. The converter and the EMI filter form a complicated interconnected system, which may easily affect the stability and dynamic performance of the converter if not carefully designed [10–12]. Those effects are not very well understood [13–15]. Basically it is a question of the differences in the open-loop input admittance (Y_{in-o}), the ideal ($Y_{in-\infty}$) and short-circuit (Y_{in-sc}) input admittances, which dictate the way the converter would react on the source or EMI-filter interactions [16].

The stability analysis of the interconnected systems [17–20] is typically based on the application of the methods developed for the EMI-filter analysis and design [10, 11] as the impedance ratio consisting of the output impedance of the supplying subsystem and the input impedance of the load subsystem, which is commonly known as minor-loop gain according to [10]. It is shown in [21, 22] that the question is actually on the internal stability of the interconnected system, in which the stability can be related to the minor-loop gain with

Dynamic Profile of Switched-Mode Converter. Teuvo Suntio
© 2009 WILEY-VCH Verlag GmbH & Co. KGaA, Weinheim
ISBN: 978-3-527-40708-8

certain conditions. As a consequence, the internal stability can be concluded by applying the *Nyquist* stability criterion to the minor-loop gain. The results of the analysis are usually expressed as a certain forbidden region on the left-half plane of the complex plane out of which the minor-loop gain should stay in order to ensure stability to exist. It is natural that an internally unstable system is also unstable in the input–output sense, but the input–output instability may not be necessarily explicitly visible.

It is well known that the load [20, 23–26] and the source [27, 31, 32] as well as even the remote sensing [28, 29] may significantly change the dynamics of a converter, and lead to the degradation of the transient performance or even to instability. The combined effects of the load and source are usually most surprising [32].

It may be difficult to know accurately enough the state of the parameters, which may cause the stability and transient performance problems. Therefore, it is obvious that the methods by means of which the interactions can be reduced or totally eliminated are of importance [33–35]. It is known [36] that a high input-output-noise attenuation (i.e., $G_{io-o} \approx 0$) would make the converter highly invariant to the source interactions [37, 38] as well as would prevent the load interactions to propagate through the converter [32]. The load interactions would be greatly reduced if the open-loop output impedance (Z_{o-o}) can be made very small [36] as demonstrated in [38, 39], which naturally also gives very small closed-loop output impedance and highly improved load-transient dynamics. In theory, it is possible to develop methods based on the electrical interconnection of the subsystems [40], which would reduce the interactions, but in practice such methods would not work or are not applicable.

It may be obvious that most of the problems encountered in the practical applications are due to the system interactions, and cannot be easily reproduce in the laboratory for finding the reasons for the observed phenomena. Therefore, it is important that the possible causes for such phenomena are understood and recognized in advance. The knowing of the internal profile of the associated converters is the key for understanding and successful analyses. The internal dynamic profile can be found analytically by applying the modeling methods described in detail in Chapters 3–7 or by measuring the frequency responses of the defined transfer functions. The analytical models and the measured transfer functions can be combined by means of proper software packages such as Matlab™.

8.2
Theoretical Interaction Formulation

The internal dynamics of a voltage-output convert can be represented by means of a linear two-port model as shown in Figure 8.1 both at open and closed loop

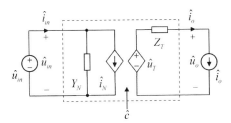

Figure 8.1 General internal linear two-port model of an electrical voltage-output system.

[41, 43]. The two-port model of Figure 8.1 can also be represented equally in a matrix form at open loop by

$$\begin{bmatrix} \hat{i}_{in} \\ \hat{u}_o \end{bmatrix} = \begin{bmatrix} Y_{in-o} & T_{oi-o} & G_{ci} \\ G_{io-o} & -Z_{o-o} & G_{co} \end{bmatrix} \begin{bmatrix} \hat{u}_{in} \\ \hat{i}_o \\ \hat{c} \end{bmatrix} \qquad (8.1)$$

and at closed loop by

$$\begin{bmatrix} \hat{i}_{in} \\ \hat{u}_o \end{bmatrix} = \begin{bmatrix} Y_{in-o} - \dfrac{G_{io-o}G_{ci}}{G_{co}} \cdot \dfrac{L_v(s)}{1+L_v(s)} & T_{oi-o} + \dfrac{Z_{o-o}G_{ci}}{G_{co}} \cdot \dfrac{L_v(s)}{1+L_v(s)} & \dfrac{G_{ci}}{G_{se}G_{co}} \cdot \dfrac{L_v(s)}{1+L_v(s)} \\ \dfrac{G_{io-o}}{1+L_v(s)} & -\dfrac{Z_{o-o}}{1+L_v(s)} & \dfrac{1}{G_{se}} \cdot \dfrac{L_v(s)}{1+L_v(s)} \end{bmatrix}$$
$$\times \begin{bmatrix} \hat{u}_{in} \\ \hat{i}_o \\ \hat{c} \end{bmatrix} \qquad (8.2)$$

where $L_v(s)$ denotes the output-voltage loop gain and G_{se} the output voltage sensing gain. If the electrical system is unregulated, the control variable ($\hat{c}$) and the voltage-loop gain $L_v(s)$ equals zero, and thus the dynamic description would consist of only four transfer functions instead of six. The control variable ($\hat{c}$) corresponds to the reference voltage ($\hat{u}_r$) at closed loop and may be usually zero for a constant-voltage converter. Therefore, the closed-loop system is most usually represented by using only four transfer functions (i.e., $Y_{in-\bar{c}}$, T_{oi-c}, G_{io-c}, and Z_{o-c}), which are the four left-most elements in (8.2).

Any electrical system can be represented by means of the two-port model of Figure 8.1. The models can be connected in series and/or parallel. We would only treat the systems comprising of series-connected subsystems of which the intermediate bus architecture (IBA, Figure 8.2) is a typical example and widely used in powering different electronics loads [3–5].

8.2.1
Load and Supply Interactions

The dynamic constellation of series-connected subsystems can be analyzed by considering the source or upstream subsystem represented by its output port

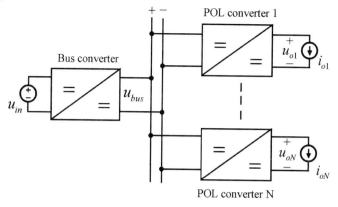

Figure 8.2 Intermediate-bus-architecture (IBA) based distributed power system.

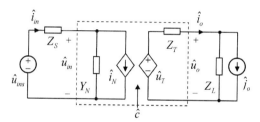

Figure 8.3 Two-port model with a nonideal input port (i.e., $\hat{u}_{ins}$, Z_s) and nonideal output port (i.e., $\hat{j}_o$, Z_L).

of the corresponding two-port model, and the load or downstream subsystem by its input port, as illustrated in Figure 8.3.

The load and source effect on the converter dynamics can be solved either by applying the extra-element-theorem method described in [42] or simply by computing $\hat{i}_o$ and $\hat{u}_{in}$ from Figure 8.3 in the presence of the nonideal source and load, and substituting them in (8.2) and (8.3) with the computed formulas. When following the described procedures, the load-affected sets of transfer functions become

$$\begin{bmatrix} \hat{i}_{in} \\ \hat{u}_o \end{bmatrix} = \begin{bmatrix} Y_{in-o} + \dfrac{G_{io-o} T_{oi-o}}{Z_{o-o} + Z_L} & \dfrac{Z_L T_{oi-o}}{Z_{o-o} + Z_L} & G_{ci} + \dfrac{G_{co} T_{oi-o}}{Z_{o-o} + Z_L} \\ \dfrac{G_{io-o}}{1 + \dfrac{Z_{o-o}}{Z_L}} & -\dfrac{Z_{o-o}}{1 + \dfrac{Z_{o-o}}{Z_L}} & \dfrac{G_{co}}{1 + \dfrac{Z_{o-o}}{Z_L}} \end{bmatrix}$$

$$\times \begin{bmatrix} \hat{u}_{in} \\ \hat{j}_o \\ \hat{c} \end{bmatrix} \quad (8.3)$$

and

$$\begin{bmatrix} \hat{i}_{in} \\ \hat{u}_o \end{bmatrix} = \begin{bmatrix} Y_{in-c} + \dfrac{G_{io-c}T_{oi-c}}{Z_{o-c}+Z_L} & \dfrac{Z_L T_{oi-c}}{Z_{o-c}+Z_L} & G_{ci-c} + \dfrac{G_{co-c}T_{oi-c}}{Z_{o-c}+Z_L} \\[2mm] \dfrac{G_{io-c}}{1+\dfrac{Z_{o-c}}{Z_L}} & -\dfrac{Z_{o-c}}{1+\dfrac{Z_{o-c}}{Z_L}} & \dfrac{G_{co-c}}{1+\dfrac{Z_{o-c}}{Z_L}} \end{bmatrix}$$

$$\times \begin{bmatrix} \hat{u}_{in} \\ \hat{j}_o \\ \hat{c} \end{bmatrix} \tag{8.4}$$

where the closed-loop internal transfer functions are defined explicitly in (8.2). Similarly, the source-affected sets of transfer functions can be given by

$$\begin{bmatrix} \hat{i}_{in} \\ \hat{u}_o \end{bmatrix} = \begin{bmatrix} \dfrac{Y_{in-o}}{1+Z_s Y_{in-o}} & \dfrac{T_{oi-o}}{1+Z_s Y_{in-o}} & \dfrac{G_{ci}}{1+Z_s Y_{in-o}} \\[2mm] \dfrac{G_{io-o}}{1+Z_s Y_{in-o}} & -\dfrac{1+Z_s Y_{in-sc}}{1+Z_s Y_{in-o}} \cdot Z_{o-o} & \dfrac{1+Z_s Y_{in-\infty}}{1+Z_s Y_{in-o}} \cdot G_{co} \end{bmatrix}$$

$$\times \begin{bmatrix} \hat{u}_{ins} \\ \hat{i}_o \\ \hat{c} \end{bmatrix} \tag{8.5}$$

and by

$$\begin{bmatrix} \hat{i}_{in} \\ \hat{u}_o \end{bmatrix} = \begin{bmatrix} \dfrac{Y_{in-c}}{1+Z_s Y_{in-c}} & \dfrac{T_{oi-c}}{1+Z_s Y_{in-c}} & \dfrac{G_{ci-c}}{1+Z_s Y_{in-c}} \\[2mm] \dfrac{G_{io-c}}{1+Z_s Y_{in-c}} & -\dfrac{1+Z_s Y_{in-sc}}{1+Z_s Y_{in-c}} \cdot Z_{o-c} & \dfrac{1+Z_s Y_{in-\infty}}{1+Z_s Y_{in-c}} \cdot G_{co-c} \end{bmatrix}$$

$$\times \begin{bmatrix} \hat{u}_{ins} \\ \hat{i}_o \\ \hat{c} \end{bmatrix} \tag{8.6}$$

where $Y_{in-\infty}$ is the ideal input admittance and Y_{in-sc} the short-circuit input admittance defined in (8.7). Both these special input admittances are load independent, and the ideal input admittance is operation- and control-mode independent as well [36]. The special admittances are explicitly presented for the basic converters in Chapter 3 in conjunction with the corresponding dynamical models, and naturally always computable by means of (8.7) for any converter. The short-circuit input admittance can be measured accurately enough by using the parallel connected resistor and large capacitor as a load as described in Chapter 7. The ideal input admittance cannot be measured directly because it is the internal feature of the converter.

$$\begin{aligned} Y_{in-\infty} &= Y_{in-o} - \dfrac{G_{io-o}G_{ci}}{G_{co}} = Y_{in-c} - \dfrac{G_{io-c}G_{ci-c}}{G_{co-c}} \\ Y_{in-sc} &= Y_{in-o} + \dfrac{G_{io-o}T_{oi-o}}{Z_{o-o}} = Y_{in-c} + \dfrac{G_{io-c}T_{oi-c}}{Z_{o-c}} \end{aligned} \tag{8.7}$$

If the open-loop input-to-output transfer function (G_{io-o}) can be made very small, then it may be obvious that the corresponding converter would be insensitive to the source interactions, because $Y_{in-o} = Y_{in-c} = Y_{in-sc} = Y_{in-\infty}$ and $Y_{in-\infty}$ is totally independent on the load impedance. The fact is, however, that the total independence cannot be achieved due to the possible changes in the output-to-input (T_{oi-o}) and control-to-input (G_{ci}) transfer functions. Sometimes the small value of T_{oi-o} or G_{ci} would yield similar results.

If the open-loop output impedance (Z_{o-o}) can be made very small, it is obvious that the load effects would be highly reduced. The load independence would not, however, be perfect, because the load may change the input admittance (Y_{in-o}) and thus also the control-to-output transfer function (G_{co}) in addition with the output-to-input (T_{oi-o}) and control-to-input (G_{ci}) transfer functions [32]. The small internal output impedance makes actually the short-circuit input admittance in (8.7) to become infinite, and consequently the source-interaction formulation given in (8.6) is not anymore valid. This also means that the extra-element-theorem-based method may not be either applicable. We have actually used the circuit theory in developing (8.6) and, therefore, the source-affected output impedance can be given by $Z^S_{o-o} = Z_s G_{io-o} T_{oi-o}/(1 + Z_s Y^L_{in-o})$, where the superscripts "S" and "L" denote the source and load, respectively. This means that the zero-output impedance can be ensured only if $G_{io-o} \approx 0$.

According to (8.3) and (8.4), the control-to-output transfer function (G_{co}) and thus the voltage-loop gain ($L_v(s)$) would be affected, if the open-loop output impedance (Z_{o-o}) is close to or greater than the load impedance (Z_L). The load-affected loop gain $L^L_v(s)$ can be given approximately by $Z_L L_v(s)/Z_{o-o}$, which can be used to approximate how the loop characteristics would change. It may be obvious that the output impedance having positive phase would increase the phase lag of the loop. If the phase of the output impedance is negative, then the phase lag will reduce. This means that the load effect may be quite easily predicted when the open-loop output impedance is known.

8.2.2
Internal and Input–Output Stabilities

If the input-to-output transfer function (G_{io-o}) is zero or very small, then the control-to-input transfer function (G_{co}) and the output impedance (Z_{o-o}) would stay intact under different source and load conditions as discussed in the previous section. This also means that the output-voltage loop gain ($L_v(s)$) and consequently the output transient dynamics would stay as they are designed to be. This does not, however, mean that the system using this kind of converters would be always stable. The same applies to the unregulated or open-loop converters, which are often considered to be always stable due to the lack of feedback.

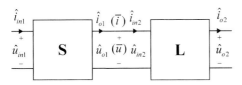

Figure 8.4 A cascaded system.

The concepts of internal and input–output stabilities would provide tools [21, 22] to analyze the stability of interconnected systems. Consider the cascaded system shown in Figure 8.4, where the source subsystem is denoted by **S** and the load subsystem by **L**. The subsystems may be either open- or closed-loop subsystems. Their internal dynamics can be represented by means of four transfer functions from the input variables (i.e., $\hat{u}_{\text{ini}}, \hat{i}_{\text{oi}}$) to the corresponding output variables (i.e., $\hat{u}_{\text{oi}}, \hat{i}_{\text{ini}}$). We assume without loss of generality that the subsystems are of voltage-output type. We denote the internal variables by $\bar{u}$ (i.e., $\hat{u}_{o1}, \hat{u}_{\text{in}2}$) and $\bar{i}$ (i.e., $\hat{i}_{o1}, \hat{i}_{\text{in}2}$), because they are same by definition.

The dynamic descriptions of the subsystems are given as follows:

$$\begin{bmatrix} \hat{i}_{\text{in}1} \\ \bar{u} \end{bmatrix} = \begin{bmatrix} S_{11} & S_{12} \\ S_{21} & S_{22} \end{bmatrix} \begin{bmatrix} \hat{u}_{\text{in}1} \\ \bar{i} \end{bmatrix}$$
$$\begin{bmatrix} \bar{i} \\ \hat{u}_{o2} \end{bmatrix} = \begin{bmatrix} L_{11} & L_{12} \\ L_{21} & L_{22} \end{bmatrix} \begin{bmatrix} \bar{u} \\ \hat{i}_{o2} \end{bmatrix} \qquad (8.8)$$

when the intermediate variables are taken into account. The internal and input–output stabilities can be studied by developing the mappings from the system input variables ($\hat{u}_{\text{in}1}, \hat{i}_{o2}$) to the intermediate variables ($\bar{u}, \bar{i}$) (i.e., internal stability) and to the system output variables ($\hat{u}_{o2}, \hat{i}_{\text{in}1}$) (i.e., input–output stability). Theses procedures yield

$$\begin{bmatrix} \bar{i} \\ \bar{u} \end{bmatrix} = \begin{bmatrix} \dfrac{S_{21} L_{11}}{1 - S_{22} L_{11}} & \dfrac{L_{12}}{1 - S_{22} L_{11}} \\ \dfrac{S_{21}}{1 - S_{22} L_{11}} & \dfrac{S_{22} L_{12}}{1 - S_{22} L_{11}} \end{bmatrix} \begin{bmatrix} \hat{u}_{\text{in}1} \\ \hat{i}_{o2} \end{bmatrix} \qquad (8.9)$$

and

$$\begin{bmatrix} \hat{i}_{\text{in}1} \\ \hat{u}_{o2} \end{bmatrix} = \begin{bmatrix} S_{11} + \dfrac{S_{12} S_{21} L_{11}}{1 - S_{22} L_{11}} & \dfrac{S_{12} L_{12}}{1 - S_{22} L_{11}} \\ \dfrac{S_{21} L_{21}}{1 - S_{22} L_{11}} & L_{22} + \dfrac{S_{22} L_{12} L_{21}}{1 - S_{22} L_{11}} \end{bmatrix} \begin{bmatrix} \hat{u}_{\text{in}1} \\ \hat{i}_{o2} \end{bmatrix} \qquad (8.10)$$

Some of the converters may have $S_{21} = 0$ or $L_{21} = 0$. This means that some of the entries in the matrices of (8.9) or (8.10) may be zero or equal to the internal transfer functions of the original subsystems (i.e., S_{11} or L_{22}) and consequently providing no additional information on the stability of the cascaded system. We assume that the subsystems are internally stable. Consequently, the interconnected system is stable if the characteristic polynomial $1 - S_{22} L_{11}$ does not have roots in the closed right-half plane of the complex plane. The stability can also be inferred applying the *Nyquist* stability criterion to $-S_{22}L_{11}$. $-S_{22}L_{11}$ is usually known as the minor-loop gain according to [10] and constitutes, in practice, of the output impedance of subsystem **S** and the input impedance of subsystem **L**, respectively. The presented system-theoretic approach is scientifically sound and proofs that the minor-loop gain can be used to study the stability of cascaded system, but it does not take any stand on the performance of the system (i.e., the gain and phase margins of the minor-loop gain do not necessarily comply to the same output-voltage-loop-gain margins of the supply or load converter) [30, 31].

In the case of supply converter (i.e., **S**), the minor-loop gain can be naturally given by Z_{o-c}/Z_L, which equals $Z_{o-o}/(1 + L_v(s))/Z_L$. If the voltage-loop gain is high, then the minor-loop gain can be approximated by $Z_{o-o}/L_v(s)/Z_L$, which has an inverse relation to the real load-affected loop gain as discussed in the previous section. If the loop gain is small, then the minor-loop gain would correspond to Z_{o-o}/Z_L, which does not have any relation to the real load-affected loop gain [30]. In practice, this means that the gain margin of the minor-loop gain corresponds to the gain margin of the voltage loop of the supply converter at low frequencies, but more gain margin is required at those frequencies close to and higher than the crossover frequency of the internal voltage loop in order to ensure that the dynamic characteristics related to the voltage loop would stay intact.

Similar explicit theoretical formulation between the minor-loop gain of the source impedance and the input impedance of the supply converter is difficult to develop, but quite the same as discussed above would apply [31].

The minor-loop gain is typically used to construct the forbidden region in the complex plane out of which the minor-loop gain should stay in order to ensure stability. The shaded are in Figure 8.5 indicates the forbidden region yielding from applying the ESAC criterion [18]. The area outside the circle

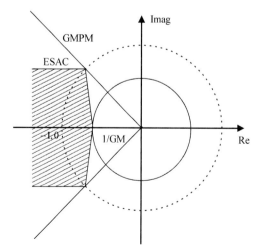

Figure 8.5 The forbidden region.

denoted by the radius 1/GM is the forbidden region when the input-filter design rules [10] are applied. This criterion is usually deemed to be too conservative [18] and requiring to using extra capacitors for ensuring stability. Therefore, new advanced criteria are developed to allow impedance overlap as well.

The system theoretical formulation given above indicates that the instability would take place, when the *Nyquist* stability criterion is violated regardless of the interface, where the impedances are defined [22]. This does not, however, ensure that the dynamic performance of the converter would be intact or sufficient.

Figure 8.6 shows an interconnected system, where we have different interfaces from A_1 to A_n. Let us assume that the interface A_1 describes a typical interface between an EMI filter having output impedance (Z_s) and a regulated converter consisting only of the pure power stage without any additional passive components added in its input with the closed-loop input impedance (Z_{in-c}). The closed-loop input impedance (Z_{in-c}) resembles typically an incremental negative resistor, especially at low frequencies but usually also at high frequencies. The *LC*-type input filter has its maximum impedance value at the resonant frequency. Its phase is also zero at the resonant frequency. This means that no impedance overlap is usually allowed at this interface in order to ensure stability. This also means that the input-filter design rules given in [10] are not conservative when considered in their correct context. Even if we do not have impedance overlap at the direct input of the converter, the dynamics of the converter may be degraded according to (8.5), because the open-loop input impedance would dictate the level of affection. Therefore, the minor-loop-gain margins should be sufficient as discussed in [10] in order to ensure proper dynamic performance.

If we assume that the interface A_4 is at the direct output of the corresponding converter, then the impedance overlap would always mean that the voltage-loop gain would be affected, and may not be desired. Even if we do not have impedance overlap, the loop gain may be affected because the level of affection would be determined by the open-loop output impedance as shown in (8.3) [30].

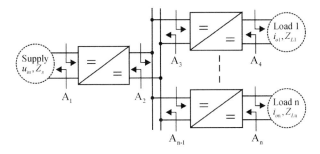

Figure 8.6 Practical interconnected system with several different interfaces.

If we assume that the load converters at the interfaces A_3 and A_{n-1} are equipped with input EMI filters, then the resonant nature of these filters would naturally cause impedance overlap at those interfaces. The EMI filters do easily mask the overlap from existing at the pure input of the corresponding converters and, therefore, the overlap would be allowed. If we assume that the interface A_2 is the direct output of the corresponding converter, then the impedance overlap would mean that the dynamics of the converter would change.

The engineers designing distributed interconnected systems should be concerned to ensure both stability and dynamic performance of the overall system. The presented stability-ensuring methods based on the nonconservative forbidden regions (Figure 8.5) do not necessarily ensure the dynamic quality of the system. Some practical evidence is provided in Section 8.4.

8.2.3
Output Voltage Remote Sensing

The output-voltage remote sensing (Figure 8.7) is sometimes used to improve the static voltage accuracy of the sensitive loads [28] or even to improve the load transient performance. A stability assessment tool to compute the effect of the connection impedance (Z_{con}, Figure 8.7) on the output-voltage loop gain of the converter is introduced in [28], but no analysis is provided to be able to conclude what the real dynamical effects of the remote sensing may actually be.

The effect of remote sensing [29] on the converter overall dynamics can be computed by modeling both the converter and the remote-sensing-impedance block (Z_{con}) by means of their corresponding two-port models as shown in Figure 8.8. The original dynamics of the converter is naturally activated when the output-voltage feedback is from the interface A_1 (Figure 8.8b). The remote-sensing-affected dynamics would be activated when the output-voltage feedback is moved to the interface A_2. The remote-sensing-affected dynamic representation can be solved by computing $\hat{i}_o$ and $\hat{u}_{inc}$ from Figure 8.8b and replacing them in the input port of the converter and in the output port of the impedance block with the computed formulas.

Following the defined procedures, the remote-sensing-affected dynamic representation of the corresponding open-loop system yields

$$\begin{bmatrix} \hat{i}_{in} \\ \hat{u}_{oc} \end{bmatrix} = \begin{bmatrix} Y_{in-o} + \dfrac{Y_{inc} G_{io-o} T_{oi-o}}{1 + Z_{o-o} Y_{inc}} & \dfrac{T_{oic} T_{oi-o}}{1 + Z_{o-o} Y_{inc}} & G_{ci} + \dfrac{Y_{inc} G_{co} T_{oi-o}}{1 + Z_{o-o} Y_{inc}} \\ \dfrac{G_{ioc} G_{io-o}}{1 + Z_{o-o} Y_{inc}} & -\dfrac{1 + Z_{o-o} Y^c_{in-sc}}{1 + Z_{o-o} Y_{inc}} \cdot Z_{oc} & \dfrac{G_{ioc} G_{co}}{1 + Z_{o-o} Y_{inc}} \end{bmatrix}$$

$$\times \begin{bmatrix} \hat{u}_{in} \\ \hat{i}_{oc} \\ \hat{c} \end{bmatrix} \quad (8.11)$$

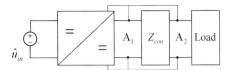

Figure 8.7 Application of output-voltage remote sensing.

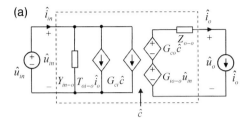

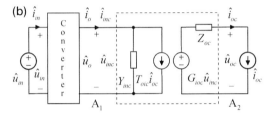

Figure 8.8 Output-voltage-remote-sensing constellation, where (a) defines two-port model of the converter and (b) the two-port model of the impedance block (Z_{con}).

where $Y^c_{\text{in-sc}}$ is the short-circuit input admittance of the impedance block given by

$$Y^c_{\text{in-sc}} = Y_{\text{inc}} + \frac{G_{\text{ioc}} T_{\text{oic}}}{Z_{\text{oc}}} \tag{8.12}$$

It may be obvious according to (8.11) that the dynamic effect of the remote sensing may be much more severe than expected if considering only the changes in the output-voltage loop gain (i.e., $G^{\text{RS}}_{\text{co}}$) as in [28]:

If we consider that the impedance block consists only of the connection cabling (Figure 8.9a), which can be modeled by means of a series connection of an inductor (L_c) and a resistor (r_{Lc}), then the impedance block model can be given by

$$\begin{bmatrix} 0 & 1 \\ 1 & -(sL_c + r_{Lc}) \end{bmatrix} \tag{8.13}$$

and the corresponding remote-sensing-affected dynamical models of the overall system by

$$\begin{bmatrix} \hat{i}_{\text{in}} \\ \hat{u}_{\text{oc}} \end{bmatrix} = \begin{bmatrix} Y_{\text{in-o}} & T_{\text{oi-o}} & G_{\text{ci}} \\ G_{\text{io-o}} & -(Z_{\text{o-o}} + sL_c + r_{Lc}) & G_{\text{co}} \end{bmatrix} \begin{bmatrix} \hat{u}_{\text{in}} \\ \hat{i}_{\text{oc}} \\ \hat{c} \end{bmatrix} \tag{8.14}$$

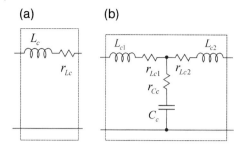

Figure 8.9 Impedance-block circuitries: (a) cabling connection and (b) T-type LCL circuit.

According to (8.14), the dynamic effect may be interpreted to be insignificant, but it is not true because the adding of an inductive component at the output would increase the load sensitivity of the converter: the capacitive load would easily reduce the phase margin of the voltage loop and deteriorate the load transient performance.

More complicated LC structures may change all the dynamical parameters and lead to a severe deterioration of the converter dynamics. Example of such a connection is a T or L_1CL_2 connection (Figure 8.9b), which can be characterized by

$$\begin{bmatrix} \dfrac{\dfrac{s}{L_{c1}}}{s^2 + s \cdot \dfrac{r_{Lc1} + r_{cc}}{L_{c1}} + \dfrac{1}{L_{c1}C_c}} & \dfrac{\dfrac{1 + s \cdot r_{cc}C_c}{L_{c1}C_c}}{s^2 + s \cdot \dfrac{r_{Lc1} + r_{cc}}{L_{c1}} + \dfrac{1}{L_{c1}C_c}} \\ \dfrac{\dfrac{1 + sr_{cc}C_c}{L_{c1}C_c}}{s^2 + s \cdot \dfrac{r_{Lc1} + r_{cc}}{L_{c1}} + \dfrac{1}{L_{c1}C_c}} & -(sL_{c2} + r_{Lc2} + \dfrac{\dfrac{(r_{Lc1} + s \cdot L_{c1})(1 + sr_{cc}C_c)}{L_{c1}C_c}}{s^2 + s \cdot \dfrac{r_{Lc1} + r_{cc}}{L_{c1}} + \dfrac{1}{L_{c1}C_c}}) \end{bmatrix}$$

(8.15)

According to (8.15), the converter-side LC circuit may cause peaking at the corresponding resonant frequency and the filter output-side inductor would increase the high-frequency output impedance, which implies sensitivity to capacitive load. The resonant frequency locates typically at high frequencies, and would increase the loop gain and decrease rapidly the phase. Such a phenomenon may easily lead to instability. Some practical evidence is provided in Section 8.4.

8.2.4
Input EMI Filter

The switched-mode converters are typically equipped with an input EMI filter (Figure 8.10a) in order to suppress the high-frequency current to a level complying with the relevant EMC standards. The EMI filter may affect the dynamics of the converter if not carefully designed [10]. The filter effect on

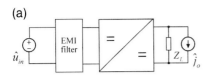
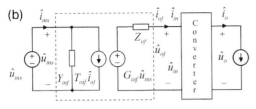

Figure 8.10 Input-EMI-filter constellation, where part (a) describes the physical connection and part (b) the dynamical model.

the converter dynamics may be analyzed by modeling the EMI filter with its two-port model (Figure 8.10b), and applying the same procedures as described in the previous section.

Following the proposed technique, the EMI-filter-affected dynamic open-loop representation of the converter yields

$$\begin{bmatrix} \hat{i}_{ins} \\ \hat{u}_o \end{bmatrix} = \begin{bmatrix} Y_{inf} + \dfrac{G_{iof} T_{oif} Y_{in-o}}{1 + Z_{of} Y_{in-o}} & \dfrac{T_{oif} T_{oi-o}}{1 + Z_{of} Y_{in-o}} & \dfrac{T_{oif} G_{ci}}{1 + Z_{of} Y_{in-o}} \\ \dfrac{G_{iof} G_{io-o}}{1 + Z_{of} Y_{in-o}} & \dfrac{1 + Z_{of} Y_{in-sc}}{1 + Z_{of} Y_{in-o}} \cdot Z_{o-o} & \dfrac{1 + Z_{of} Y_{in-\infty}}{1 + Z_{o-o} Y_{in-o}} \cdot G_{co} \end{bmatrix}$$

$$\times \begin{bmatrix} \hat{u}_{ins} \\ \hat{i}_o \\ \hat{c} \end{bmatrix} \tag{8.16}$$

where $Y_{in-\infty}$ and Y_{in-sc} are the ideal and short-circuit input admittances of the original converter, respectively. The stability of the EMI-filter-converter system has to be checked by applying the *Nyquist* stability criterion to the corresponding minor-loop gain, which usually consists of the filter output impedance (Z_{of}) and the closed-loop input impedance of the converter ($Z_{in-c} = 1/Y_{in-c}$). Sometimes, the converter is used at open loop, especially in the intermediate bus architectures [5]. In those cases, the minor-loop gain consists of the filter output impedance and the open-loop input impedance of the converter (i.e., $Z_{in-o} = 1/Y_{in-o}$).

According to (8.16), the EMI filter would always affect the dynamic profile of the associated converter regardless of the value of the internal input-to-output transfer function (G_{io-o}). The high input-to-output noise attenuation at open loop (i.e., $G_{io-o} \approx 0$) would make the output dynamics of the converter insensitive to the EMI filter providing that the EMI-filter-converter system is stable. If high input-to-output noise attenuation does not exist, then the behavior of the open-loop input impedance and the EMI-filter output impedance would determine the filter effects. Some experimental evidence is provided in Section 8.4.

8.3
Review of Methods to Reduce the Interactions

If the open-loop input-to-output transfer function (G_{io-o}) and the open-loop output impedance (Z_{o-o}) can be made very small, then the source and load interactions would be highly reduced according to the theoretical formulations derived in Section 8.2.1. The voltage-mode-controlled converters do not have such features as demonstrated in Chapter 3.

A CCM buck converter under optimally compensated peak-current-mode (Chapter 4), average-current-mode (Chapter 5), and self-oscillation (Chapter 6) controls would have high input-to-output noise reduction, which would reduce the source interactions and prevent the load interactions to propagate through the converter. The other basic converters do not posses such a feature. In addition to the above-named control techniques, the high input-noise attenuation can be accomplished with certain conditions applying input-voltage feedforward (IVFF) [37, 44, 45].

The control techniques introduced in Chapters 4–7 tend to increase the open-loop output impedance and, as a consequence, the load interactions, of which the reduction of the voltage-loop crossover frequency is the most common outcome. In order to reduce the basic load interactions, the open-loop output impedance shall be made small. In practice, this can be accomplished with certain conditions by applying output-current feedforward (OCFF) [38, 39, 46–48].

8.3.1
Input-Voltage Feedforward

The effect of IVFF on the converter dynamics can be analyzed by using the control engineering block diagrams given in Figure 8.11, which yield the dynamic description of the converter to be [37]

$$\begin{bmatrix} \hat{i}_{in} \\ \hat{u}_o \end{bmatrix} = \begin{bmatrix} Y_{in-o} - F_m q_i G_{ci} & T_{oi-o} & F_m G_{ci} \\ G_{io-o} - F_m q_i G_{co} & -Z_{o-o} & F_m G_{co} \end{bmatrix} \begin{bmatrix} \hat{u}_{in} \\ \hat{i}_o \\ \hat{u}_{co} \end{bmatrix} \quad (8.17)$$

It may be obvious according to (8.16) that the original dynamic nature of the converter would not be changed, that is, if the original converter has resonant nature, then the IVFF converter would have also the resonant nature. The IVFF control does not change the output impedance and the output-to-input transfer function.

The aim of the IVFF is to produce high input-noise attenuation and, therefore, G_{io-o}^{IVFF} has to be zero. According to (8.17), this would take place when $G_{io-o} - F_m q_i G_{co} = 0$, and as a consequence, $F_m q_i = G_{io-o}/G_{co}$. It may be obvious that the stated condition can be implemented properly only if G_{io-o} and G_{co} have the same zeros as in a buck converter. This means, in practice,

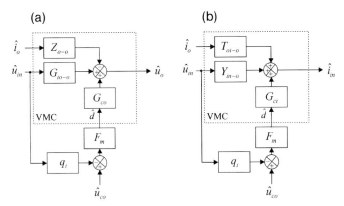

Figure 8.11 Control engineering block diagrams for assessing the effect of input-voltage feedforward on (a) output dynamics and (b) input dynamics.

that $F_m q_i \approx D/U_{in}$. The required condition can be accomplished making the PWM ramp slope directly dependent on the input voltage as illustrated in Figure 8.12. The comparator equation determining the duty ratio can be given by $u_{co} = k u_{in} d T_s$ according to Figure 8.12 from which the duty-ratio constraints can be derived by developing the proper partial derivatives as follows:

$$\hat{d} = \frac{1}{k U_{in} T_s}(\hat{u}_{co} - k D T_s \cdot \hat{u}_{co}) \tag{8.18}$$

which gives $F_m q_i = D/U_{in}$ as required for $G_{io-o}^{IVFF} \approx 0$.

The constant k (Figure 8.12) is designed so that the duty ratio is close to 100% at a certain minimum input voltage (U_{in-min}). According to this assumption, $k = V_M/U_{in-min}T_s$, where $V_M = U_{max} - U_{min}$ (Figure 8.12). Thus $F_m = U_{in-min}/V_M U_{in}$ and $q_i = D V_M/U_{in-min}$. This means that the control-to-output (G_{co}^{IVFF}) and the control-to-input (G_{ci}^{IVFF}) transfer functions would be near to independent of the input voltage and, therefore, the output-voltage-loop dynamics does not depend anymore on the input voltage.

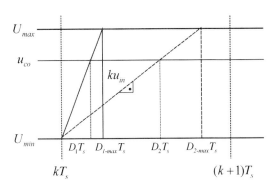

Figure 8.12 Duty-ratio generation under IVFF control.

The scheme of Figure 8.12 is typically implemented in the commercial PWM modulators by charging the timing capacitor via a resistor from the input voltage yielding an exponential curve instead of the linear curve. Despite the deviation from the theoretical ideal, the scheme would produce near to expected results [37].

It may be obvious from Figure 8.12 that the maximum pulsewidth of the converter would be limited. This means, in practice, that the load-transient dynamics of the converter would be deteriorated, because the derivative of the average inductor current would be limited according to $d_{max}m_1 - d'_{max}m_2$.

More detailed information on the IVFF scheme can be found from [37] and [45]. Experimental evidence would be provided in Section 8.4.

8.3.2
Output-Current Feedforward

It is well known [46] that the OCFF control would improve the load-transient dynamics of a converter. The improvement results from the reduction of the open-loop output impedance. It is claimed in [39] that the open-loop output impedance can be made zero for every converter, but it is not true because of the same reasons as discussed in the previous section in conjunction with the input-to-output transfer function.

The theoretical formulation defining the proper conditions for the zero open-loop output impedance can be derived from the control engineering block diagrams as shown in Figure 8.13, where R_{s2} is the output-current sensing resistor, H_i the output-current-loop sensing gain, and G_a the transfer function from the control voltage to the original control variable of the converter.

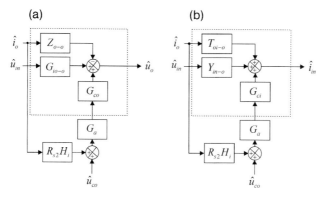

Figure 8.13 Control engineering block diagrams for assessing the effect of output-current feedforward on (a) output dynamics and (b) input dynamics.

According to Figure 8.13, the dynamic description of the converter can be given by

$$\begin{bmatrix} \hat{i}_{in} \\ \hat{u}_o \end{bmatrix} = \begin{bmatrix} Y_{in-o} & T_{oi-o} + R_{s2}H_iG_aG_{ci} & G_aG_{ci} \\ G_{io-o} & -(Z_{o-o} - R_{s2}H_iG_aG_{co}) & G_aG_{co} \end{bmatrix} \begin{bmatrix} \hat{u}_{in} \\ \hat{i}_o \\ \hat{u}_{co} \end{bmatrix} \quad (8.19)$$

which indicates that the OCFF control would change only the open-loop output impedance and the output-to-input transfer function. The other transfer functions would stay effectively intact, because the control-related transfer functions are actually same as in the original converter when the voltage-type control variable is considered.

The aim of the OCFF control is to produce zero open-loop output impedance. Therefore, $Z_{o-o}^{OCFF} = Z_{o-o} - R_{s2}H_iG_aG_{co} = 0$. As a consequence, the OCFF loop gain should be designed as

$$H_i = \frac{1}{R_{s2}G_a} \cdot \frac{Z_{o-o}}{G_{co}} \quad (8.20)$$

It may be obvious that the gain required for H_i in (8.20) may be implemented in reality only if the control-to-output transfer function (G_{co}) is of minimum phase, that is, it does not contain RHP zero. Thus the buck converter of the basic converters is the only prospective candidate. It is observed that the unity-gain OCFF control (i.e., $H_i = 1$) would provide highly improved load-transient response in a hysteretic current-mode [46] and peak-current-mode [38, 47] controlled buck converters. In order to improve the transient response of a voltage-mode-controlled buck converter, the OCFF loop gain should be effective at the resonant frequency of the converter, where the open-loop output impedance has the highest value. Such a principle has been tried in [48], but the deficiencies in the theoretical treatment have lead to poor performance. More detailed discussions on the subject can be found from [38]. Experimental evidence is provided in Section 8.4.

8.4
Experimental Dynamic Review

We show some practical evidence of the interactions, which may take place in the real interconnected systems comprising of switched-mode converters in the subsequent sections. The intention is to prove that different converters have their unique dynamic profiles determining how the converter would react to the external world. The dynamic profiles have been already disclosed in the previous chapters and some discussions made about the possible changes the external impedances may cause in the dynamic behavior of the converter. We also demonstrate that the severity of the interactions can be reduced by using

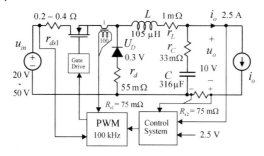

Figure 8.14 The experimental buck converter.

proper feedforward control. The experimental material is mainly based on the buck converter as shown in Figure 8.14 under different control methods.

The output-voltage loop gains of the VMC, PCMC, IVFF, and PCMC-OCF buck converters were designed to be close to similar at the input voltage of 50 V as shown in Figure 8.15.

The basic load effects are demonstrated connecting at the output of the converter a series resonant LC having resonant frequency close to that of the VMC converter. The converters are stable at the resonant load as shown in Figure 8.16, because the load impedance and the closed-loop output impedances of the converters do not overlap. The state of the dynamic profile of the converter cannot be concluded based on Figure 8.16.

The source effects are demonstrated connecting an EMI filter (Figure 8.17) at the input of the converter. The resonant frequency of the filter was chosen to be close to that of the VMC converter. The measured closed-loop input

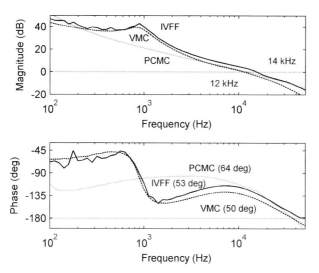

Figure 8.15 Measured frequency responses of the voltage-loop gains of the experimental converters at the input voltage of 50 V.

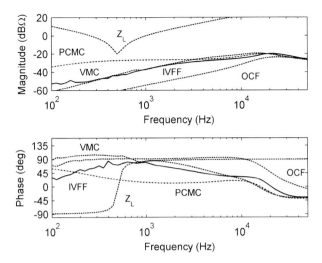

Figure 8.16 The measured closed-loop output impedances of the converters versus the load impedance.

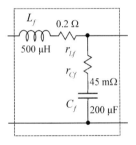

Figure 8.17 Input EMI filter.

impedances of the converters and the filter output impedance are shown in Figure 8.18 proving the stability of the cascaded system due to the lack of impedance overlap. The state of the dynamic profile of the converters cannot be concluded based on the information provided by Figure 8.18.

8.4.1
Load and Supply Interactions

A series-resonant load (i.e., 500 Hz) was connected at the output of the buck converters shown in Figure 8.14 at the input voltage of 20 V. The corresponding open-loop internal impedances (Z_{o-o}) and the load impedance (Z_L) are shown in Figure 8.19. According to (8.3), the load impedance would affect the voltage-loop gain if $Z_L \leq Z_{o-o}$. From Figure 8.19, it may be obvious that all the other voltage-loop gains except the loop gain of the PCMC-OCF converter would be

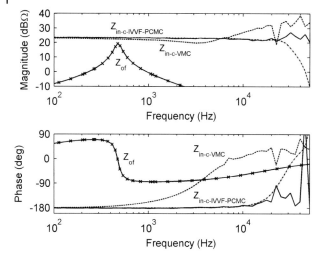

Figure 8.18 The closed-loop input impedances versus the output impedance of the EMI filter.

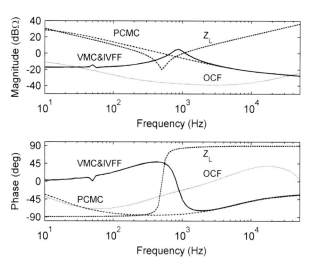

Figure 8.19 The open-loop output impedances versus the 500-Hz resonant load impedance.

affected. Figure 8.20 confirms the predictions. There are, however, no dynamic effects because the gain and phase margins are not changed.

The open-loop input impedances ($Z_{\text{in}-o}$) and the output impedance of the EMI filter (Figure 8.17) are shown in Figure 8.21. The impedance overlap in the VMC converter means that its voltage-loop gain would be affected. Figure 8.22 confirms the prediction. The other loop gains are intact.

8.4 Experimental Dynamic Review | 245

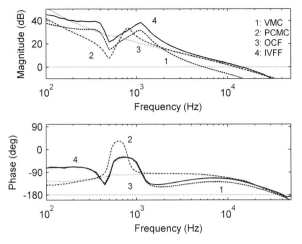

Figure 8.20 The resonant-load-affected output-voltage loop gains.

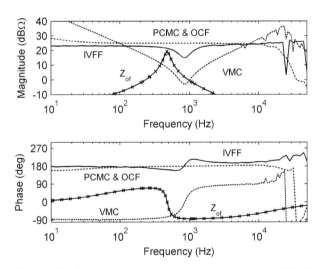

Figure 8.21 The open-loop input impedances versus the EMI-filter output impedance.

Figure 8.23 shows the combined effect of the resonant load and source. The VMC converter is very close to instability. The voltage-loop gains of the other converters are not changed.

The reason for the observed phenomenon is the behavior of the open-loop input-to-output transfer function (G_{io-o}) (Figure 8.24) and the output-to-input transfer function (T_{oi-o}) (Figure 8.25). According to (8.3), the load will affect the open-loop input admittance if the product $G_{io-o}T_{oi-o}$ is not zero. According to Figures 8.24 and 8.25, the product $G_{io-o}T_{oi-o}$ would amplify the load interactions especially at the frequencies close to the resonant frequencies

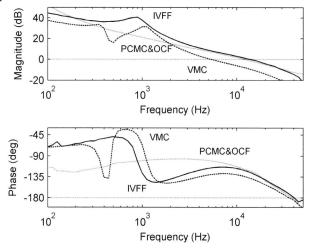

Figure 8.22 The EMI-filter affected output-voltage loop gains.

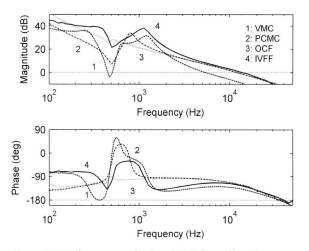

Figure 8.23 The resonant-load and EMI-filter-affected output-voltage loop gains.

as demonstrated in Figure 8.23. The PCMC and IVFF converters have rather high input-to-output noise attenuation, which also means reduced reflected load interactions.

A series resonant load (8 kHz) was connected at the output of the VMC and PCMC buck converters (Figure 8.14). The minor-loop gain (Z_{o-c}/Z_L) was measured and is shown in Figure 8.26. The cascaded systems are stable because of the positive phase margins. The minor-loop gain of the VMC converter shows slight impedance overlap, but the minor-loop gain of the PCMC converter is always less than unity and, therefore, no impedance overlap does exist. We

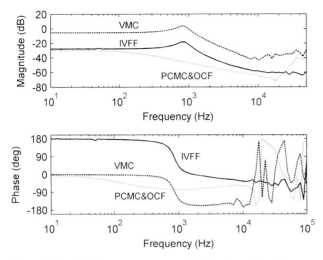

Figure 8.24 The input-to-output transfer functions (G_{io-o}).

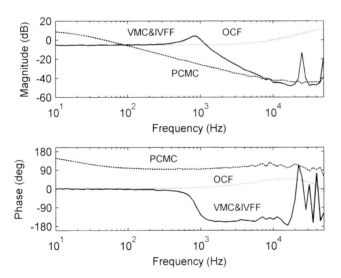

Figure 8.25 The output-to-input transfer functions (T_{oi-o}).

may conclude that the voltage-loop gain of the VMC converter may be slightly changed but the voltage-loop gain of the PCMC converter would be intact.

Figure 8.27 shows that the crossover frequencies of the converters are reduced equally from 11 to 6 kHz despite the information given by the minor-loop gain. The open-loop output impedances and the resonant-load impedance are shown in Figure 8.28, which indicates impedance overlap in both the converters and, consequently, a reduction of loop gain.

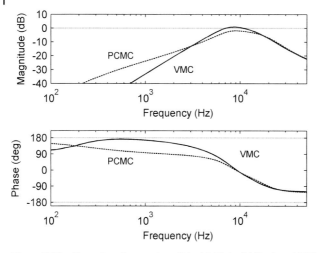

Figure 8.26 The minor-loop gains of the VMC (solid line) and PCMC (dashed line) converters at 8-kHz resonant load.

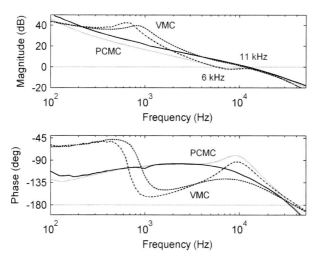

Figure 8.27 8-kHz-resonant-load-affected voltage-loop gains of the VMC and PCMC converters.

The load of the VMC buck converter was considered to be such that the minor-loop gain (Z_{o-c}/Z_L) has a continuous 6-dB gain margin. The corresponding phase margin was set to 0° and 60°. The original voltage-loop gain (solid line) and the load-affected loop gains with 0° phase margin (dashed line) and 60° phase margin (dotted line) are shown in Figure 8.29. At those frequencies, where the voltage-loop gain is high, the minor-loop-gain-related margins are also directly reflected to the voltage-loop gain. When approaching

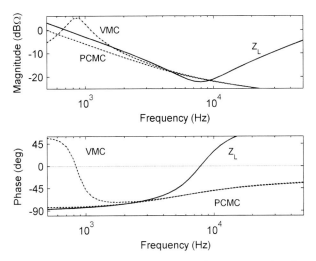

Figure 8.28 The open-loop output impedances versus the 8-kHz resonant-load impedance.

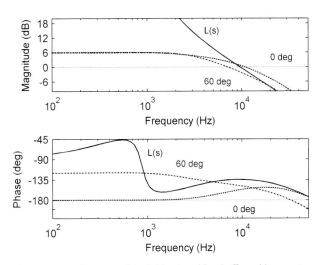

Figure 8.29 The internal (solid line) and load-affected loop gains of the VMC buck converter, where the minor-loop gain (Z_{oc}/Z_L) has continuous 6-dB gain margin and 0° (dashed line) or 60° (dash-dotted line) phase margin.

the crossover frequency, the 6-dB gain margin does not naturally hold any more, and the crossover frequency may either slightly increase (PM = 0°) or decrease (PM = 60°). In both the cases, the voltage-loop-gain phase margin reduces from the original phase margin. In practice, this means that the minor-loop-based margins have to be higher than that usually used in the voltage-loop gain for robust stability and in order to maintain adequate dynamics of the converter.

The source impedance of a VMC buck converter was considered to be such that the minor-loop gain (Z_s/Z_{in-c}) has a continuous 6-dB gain margin. The corresponding phase margins were set to 0° and 60°. The original voltage-loop gain (solid line) and the source-affected voltage-loop gains with 0° (dashed line) and 60° phase margin (dash-dotted line) are shown in Figure 8.30. At the low frequencies, the 6-dB minor-loop-gain-related margin would ensure intact dynamics. At the higher frequencies, the gain margin should be higher than the usual 6 dB in order to ensure robust stability and performance.

The closed-loop output impedance (Z_{o-c}) of the converter can be used to estimate the load sensitivity by considering that the instability boundary can be given according to the minor-loop gain (Z_{o-c}/Z_L) as $Z_L = -Z_{o-c}$. This means that $Z_L = |Z_{o-c}|\angle\varphi_{Z_{o-c}} - 180°$ [30]. According to this procedure, the load-impedance instability boundary can be given as shown in Figure 8.31 for VMC and PCM buck converters (Figure 8.14). It shows that the VMC converter is sensitive to the capacitive load at the frequencies lower than or equal to its resonant frequency (i.e., the phase is equal to or less than −90°). The PCMC converter can be unstable due to the capacitive load only at the low frequencies. The situation with the other types of loads can be similarly considered.

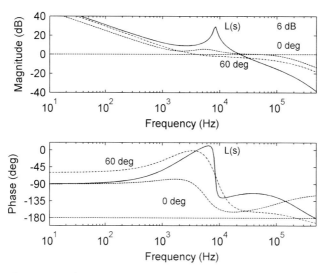

Figure 8.30 The internal (solid line) and source-affected loop gains of the VMC buck converter, where the minor-loop gain (Z_s/Z_{inc}) has continuous 6-dB gain margin and 0° (dashed line) or 60° (dotted line) phase margin.

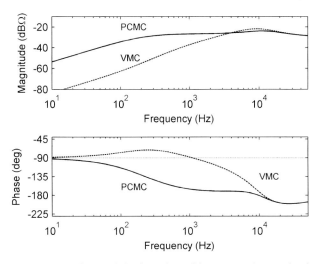

Figure 8.31 The instability boundary of the VMC and PCMC buck converters (Figure 8.14).

8.4.2
Remote Sensing

The impedance block (Figure 8.32) mimicking the impedance of a cabling was connected at the output of the VMC converter (Figure 8.14). The original and the cable-affected (RS) open-loop output impedances are shown in Figure 8.33.

The inductive nature of the high-frequency output impedance implies increased sensitivity to the capacitive load when the remote sensing is connected. The measured output-voltage loop gains at ideal constant current load ($L(s)$, solid line) and at capacitive load (i.e., $R_L = 4\Omega$, $C_L = 110\,\mu F$, ESR = 100 mΩ) with the remote sensing connected ($L^{RS}(s)$, dash-dot line) and without it ($L^C(s)$, dashed line) are shown in Figure 8.34: the remote sensing has clearly increased the phase lag and may easily lead to instability if the gain would increase.

The *LCL* circuit shown in Figure 8.36 was connected at the output of the synchronous buck converter shown in Figure 8.35. The measured original and the connection-block-affected output impedances are shown in Figure 8.37.

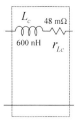

Figure 8.32 The cabling impedance.

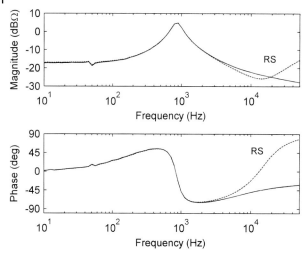

Figure 8.33 The original and cabling-affected (RS) open-loop output impedances.

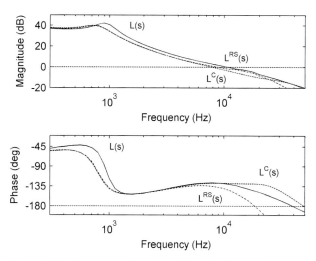

Figure 8.34 The measured internal ($L(s)$, solid line) the capacitive-load affected ($L^C(s)$, dashed lie) without remote sensing, and the capacitive-load-affected ($L^{RS}(s)$, dash-dot line) with remote sensing output-voltage loop gains.

The inductive nature of the high-frequency output impedance implies increased sensitivity to capacitive load: the measured voltage-loop gains in Figure 8.38 confirm the prediction, where the internal loop gain is denoted by means of solid line, the 4-μF-capacitive-load-affected loop gain ($L^C(s)$) by dashed line, and the 4-μF-capacitive-load-affected with remote sensing ($L^{RS}(s)$)

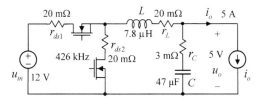

Figure 8.35 Experimental synchronous buck converter.

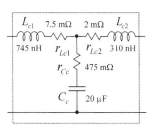

Figure 8.36 T-type connection impedance block.

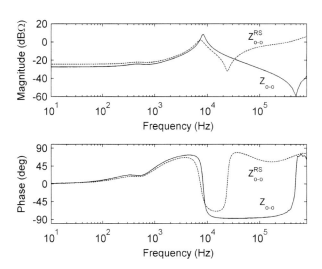

Figure 8.37 The measured internal (Z_{o-o}) and the connection-block-affected (Z_{o-o}^{RS}) output impedances.

by dash-dot line. It may be obvious that a slight increase in the loop gain would lead to instability.

The effect of the connection impedance (Figure 8.36) could have been more severe as discussed in Section 8.2.3, but the capacitor (C_c) had to be selected to be an electrolytic capacitor with rather high ESR in order to keep the converter stable. Therefore, the resonant peaking in the *LCL* circuit is very low as shown in Figure 8.39. Consequently, the interactions are also reduced. In the practical applications, the resonant peaking could

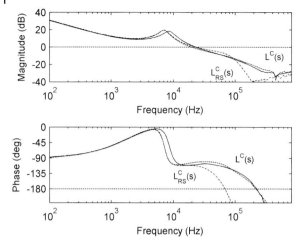

Figure 8.38 The measured internal (solid line), the capacitive-load-affected ($L^C(s)$, dashed line), and the capacitive-load-affected loop gain with remote sensing ($L_{RS}^C(s)$, dash-dot line).

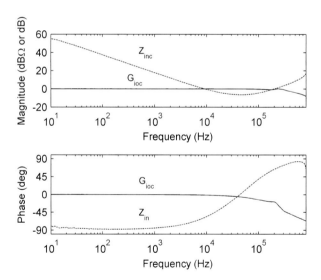

Figure 8.39 The internal input impedance (Z_{inc}) and the input-to-output transfer function (G_{ioc}) of the LCL circuit.

be high and the interactions also more severe due to the nature of the circuitry.

The synchronous buck converter was subjected to a constant-current-type load change from 1 to 5 A with slew rate of 2.5A/μs. The corresponding

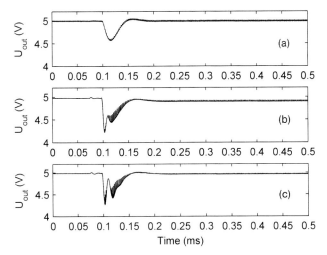

Figure 8.40 The output-voltage transient responses of the synchronous buck converter, when the constant-current-type load changes from 1 to 5 A (2.5 A/μs): (a) original response, (b) response with the LCL circuit connected between the converter and load, (c) response, when the remote sensing is connected over the LCL circuit.

output-voltage transient responses without the LCL circuit (a), with the LCL connected between the converter and load (b), and applying remote sensing over the LCL circuit (c) are shown in Figure 8.40: the remote sensing obviously improves the static accuracy but not the transient response.

8.4.3
System Stability

A small-scale IBA system was constructed based on nonisolated synchronous buck converters as shown in Figure 8.41. There are numerous different interfaces (A_i) from which the stability information can be extracted in terms of different minor-loop gains. The stated connection impedances are cabling impedances. The buck converters have to be equipped with an input capacitor in order to work well. Therefore, the plain input of the converter is never measurable.

Figure 8.42 shows the measured minor-loop gains at the interfaces A_{P4} and A_{P3} as well as the computed minor-loop gain existing at the direct input of the corresponding buck converter. The minor-loop gain at A_{P4} (solid line) shows that the system is stable although slight impedance overlap exists. The minor-loop gain at A_{P3} indicates high impedance overlap and phase close to 180°. The *Nyquist* plot shows that the minor-loop gain at A_{P3} also indicates stable system, because it does not encircle the point $(-1, 0)$ in clockwise direction. The information the minor-loop gain provides about the robust stability is,

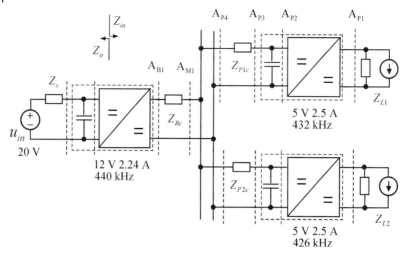

Figure 8.41 Experimental IBA system.

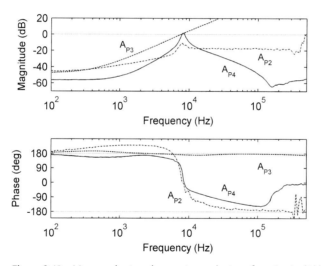

Figure 8.42 Measured minor-loop gains at the interface A_{P4} (solid line) and A_{P3} as well as the computed minor-loop gain at A_{P2}.

however, very vague. The computed minor-loop gain at A_{P2} confirms the situation. The message is that every interface has its own minor-loop gain giving information, which may or may not be useful. The dynamic performance of the system can be inferred only if the minor-loop gain represents the direct input or output of the converter.

The time-domain information may often be very confusing as shown in Figure 8.43, where two different voltage measurements are shown at the

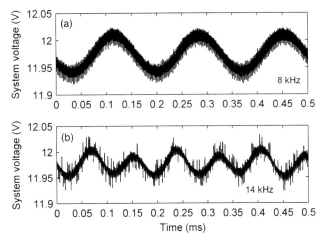

Figure 8.43 The voltage of the main bus (A_{M1}): (a) the beat frequency of the bus converter and the first POL converter, and (b) the beat frequency of the bus converter and the second POL converter.

interface A_{M1} (i.e., the main bus). The curves show sinusoidal signals at different frequencies depending on the loading condition of the system and the POL converters. Figure 8.43a shows a sinusoidal signal of 8 kHz, which is actually the switching-frequency difference (beat frequency) of the bus converter and the first POL converter. Figure 8.43b shows similar beat-frequency signal of the bus converter and the second POL converter. These signals can be easily interpreted as a sign of instability when they appear along the increase of the loading.

References

1. S. Luo and I. Battarseh, 'A review of distributed power systems Part I: DC distributed power system,' *IEEE Aerosp. Electron. Syst. Mag.*, vol. 20, no. 8, **2005**, pp. 5–15.
2. S. Luo and I. Battarseh, 'A review of distributed power systems Part II: High frequency AC distributed power systems,' *IEEE Aerosp. Electron. Syst. Mag.*, vol. 21, no. 6, **2006**, pp. 5–14.
3. L. Brush, 'Distributed power architecture demand characteristics,' in *Proc. IEEE Applied Power Electronics Conf.*, **2004**, pp. 342–345.
4. R.V. White, 'Using on-board power systems,' in *Proc. IEEE International Telecommunications Energy Conf.*, **2004**, pp. 234–240.
5. H. Huang, 'Coordination of design issues in the intermediate bus architecture,' in *Proc. IEEE Applied Power Electronics Conf.*, **2005**, pp. 169–175.
6. J.R. Lee, B.H. Cho, S.J. Kim, and F.C. Lee, 'Modeling and simulation of spacecraft power systems,' *IEEE Trans. Aerosp. Electron. Syst.*, vol. 24, no. 3, **1988**, pp. 295–303.
7. A. Emadi, A. Khaligh, C.H. Rivetta, and G.A. Williamson, 'Constant power loads and negative impedance instability in automotive systems:

Definition, modeling, stability, and control of power electronic converters and motor drives,' *IEEE Trans. Veh. Technol.*, vol. 55, no. 4, **2006**, pp. 1112–1125.

8. C.D. Davidson and R. Scasz, 'Compatibility of switched-mode rectifiers with engine generators,' in *Proc. IEEE International Telecommunications Energy Conf.*, **2000**, pp. 626–631.

9. C.M. Hoff and S. Mulukutla, 'Analysis of the instability of PFC power supplies with various AC sources,' in *Proc. IEEE Applied Power Electronics Conf.*, **1994**, pp. 696–702.

10. R.D. Middlebrook, 'Input filter considerations in design and application of switching regulators,' in *Proc. IEEE Industry Applications Society Annual Meeting*, **1976**, pp. 366–382.

11. R.D. Middlebrook, 'Design techniques for preventing input-filter oscillations in switched-mode regulators,' in *Proc. National Solid-State Power Conversion Conf.* **1978**, pp. A3.1–A3.16.

12. F.C. Lee and Y. Yu, 'Input-filter design for switching regulators,' *IEEE Trans. Aerosp. Electron. Syst.*, vol. AES-15, no. 5, **1979**, pp. 627–634.

13. S.Y. Erich and W.M. Polivka, 'Input filter design criteria for current-programmed regulators,' *IEEE Trans. Power Electron.*, vol. 7, no. 1, **1992**, pp. 143–151.

14. Y. Jang and R.W. Erickson, 'Physical origin of input filter oscillations in current programmed converters,' *IEEE Trans. Power Electron.*, vol. 7, no. 4, **1992**, pp. 725–733.

15. S. Ang and A. Oliva, *Power-Switching Converters*, Taylor & Francis, Boca Raton, FL, USA, **2005**, 2nd Edition.

16. T. Suntio, I. Gadoura, and K. Zenger, 'Input filter interactions in peak-current-mode controlled buck converter operating in CICM,' *IEEE Trans. Indust. Electron.*, vol. 49, no. 1, **2002**, pp. 76–86.

17. C.M. Wildrick, F.C. Lee, B.H. Cho, and B. Choi, 'A method of defining the load impedance specification for a stable distributed power system,' *IEEE Trans. Power Electron.*, vol. 10, no. 3, **1995**, pp. 280–285.

18. S.D. Sudhoff, S.F. Glover, P.T. Lamm, D.H. Schmucker, and D.E. Delisle, 'Admittance space stability analysis of power electronic systems,' *IEEE Aerosp. Electron. Syst.*, vol. 36, no. 3, **2000**, pp. 965–973.

19. J.M. Zhang, X.G. Xie, D.Z. Jia, and Z. Qian, 'Stability problems and input impedance improvement for cascaded power electronic systems,' in *Proc. IEEE Applied Power Electronic Conf.*, **2004**, pp. 1018–1024.

20. S. Abe, H. Nakagawa, M. Hirokawa, T. Zaitsu, and T. Ninomiya, 'System stability of full-regulated bus converter in distributed power system,' in *Proc. IEEE International Telecommunications Energy Conf.*, **2005**, pp. 563–568.

21. K. Zenger, A. Altowati, and T. Suntio, 'Dynamic properties of interconnected power systems – A system theoretic approach,' in *Proc. IEEE Industrial Electronics and Applications Conf.*, **2006**, pp. 835–840.

22. K. Zenger, A. Altowati, and T. Suntio, 'Stability and performance analysis of regulated converter systems,' in *Proc. IEEE Industrial Electronics Society Annual Conf.*, **2006**, pp. 1975–1980.

23. P. Li and B. Lehman, 'Performance prediction of DC–DC converters with impedances as loads,' *IEEE Trans. Power Electron.*, vol. 19, no. 1, **2004**, pp. 201–209.

24. B. Choi, J. Kim, B.H. Cho, S. Choi, and C.M. Wildrick, 'Designing control loop for DC-to-DC converters with unknown AC dynamics,' *IEEE Trans. Indust. Electron.*, vol. 49, no. 4, **2002**, pp. 925–932.

25. B. Choi, B.H. Cho, and S.-S. Hong, 'Dynamics and control of DC-to-DC converters driving other converters down stream,' *IEEE Trans. Circuits Syst. – I: Fundam. Theory Appl.*, vol. 46, no. 10, **1999**, pp. 1240–1248.

26. D. Lee, B. Choi, J. Sun, and B.H. Cho, 'Interpretation and prediction of loop gain characteristics for switching power converters loaded with general load subsystem,' in *Proc. IEEE Power*

Electronic Specialist. Conference, **2005**, pp. 1024–1029.

27. P. Li and B. Lehman, 'Accurate loop gain prediction for DC–DC converters due to impact of source/input filter,' *IEEE Trans. Power Electron.*, vol. 20, no. 4, **2005**, pp. 754–1017.

28. C. Gezgin, W.C. Bowman, and V.J. Thottuvelil, 'A stability assessment tool for DC-DC converters,' in *Proc. IEEE Applied Power Electronics Conf.*, **2002**, pp. 367–373.

29. M. Karppanen, T. Suntio, and M. Sippola, 'Impact of output-voltage remote sensing on converter dynamics,' *Int. Rev. Electr. Eng.*, vol. 2, no. 2, **2007**, pp. 196–202.

30. M. Hankaniemi, M. Karppanen, and T. Suntio, 'Load imposed instability and performance degradation in a regulated converter,' *IEE Proc. Electr. Power Appl.*, vol. 153, no. 6, **2006**, pp. 781–786.

31. M. Karppanen, M. Sippola, and T. Suntio, 'Source-imposed instability and performance degradation in a regulated converter,' in *Proc. IEEE Power Electronics Specialists Conf.*, **2007**, pp. 194–200.

32. M. Hankaniemi, M. Karppanen, T. Suntio, A. Altowati, and K. Zenger, 'Source-reflected load interactions in a regulated converter,' in *Proc. IEEE Industrial Electronics Society Annual Conf.*, **2006**, pp. 2893–2898.

33. T. Suntio, K. Kostov, T. Tepsa, and J. Kyyrä, 'Using input invariance as a method to facilitate system design DPS applications,' *J. Circuits, Syst. Comput.*, vol. 13, no. 4, **2004**, pp. 707–723.

34. T. Suntio, 'Input invariance as a method to reduce EMI filter interactions in telecom DPS systems,' in *Proc. IEEE International Telecommunications Energy Conf.*, **2003**, pp. 592–597.

35. M. Karppanen, M. Hankaniemi, and T. Suntio, 'Load and supply invariance in a regulated converter,' in *Proc. IEEE Power Electronics Specialist Conference*, **2006**, pp. 2663–2668.

36. M. Hankaniemi, T. Suntio, and M. Sippola, 'Characterization of regulated converters to ensure stability and performance in distributed power supply system,' in *Proc. IEEE International Telecommunications Energy Conf.*, **2005**, pp. 533–538.

37. M. Karppanen, T. Suntio, and M. Sippola, 'Dynamical characterization of input-voltage-feedforward-controlled buck converter,' *IEEE Trans. Indust. Electron.*, vol. 54, no. 2, **2007**, pp. 1005–1013.

38. M. Karppanen, M. Hankaniemi, T. Suntio, and M. Sippola, 'Dynamical characterization of peak-current-mode controlled buck converter with output-current feedforward,' *IEEE Trans. Power Electron.*, vol. 22, no. 2, **2007**, pp. 444–451.

39. L.D. Varga and N.A. Losic, 'Synthesis of zero-impedance converter,' *IEEE Trans. Power Electron.*, vol. 7, no. 1, **1992**, pp. 152–170.

40. X. Wang, R. Yao, and F. Rao, 'Subsystem-interaction restraint in the two-stage DC distributed power systems with decoupling-controlled-integration structure,' *IEEE Trans. Indust. Electron.*, vol. 52, no. 6, **2005**, pp. 1555–1563.

41. P.G. Maranesi, V. Tavazzi, and V. Varoli, 'Two-port modeling of PWM voltage regulators at low frequencies,' *IEEE Trans. Indust. Electron.*, vol. 35, no. 3, **1988**, pp. 444–450.

42. R.D. Middlebrook, 'Null double injection and the extra element theorem,' *IEEE Trans. Edu.*, vol. 32, no. 3, **1989**, pp. 167–180.

43. T. Suntio, and I. Gadoura, 'Use of unterminated two-port modeling technique in analysis of input filter interactions in telecom DPS systems,' in *Proc. IEEE International Telecommunications Energy Conf.*, **2002**, pp. 560–565.

44. D.V. Otto, 'Reduction of switching regulator audiosusceptibility to zero,' *IEE Electron. Lett.*, vol. 22, no. 8, **1986**, pp. 441–442.

45. J.P. Sjöroos, T. Suntio, K. Kostov, and J. Kyyrä, 'Dynamic performance

of the buck converter with input voltage feed-forward control,' in *Proc. European Power Electronics and Applications Conf.*, **2005**, paper no. 630, pp. 1–10.

46. R. Redl, and N.O. Sokal, 'Near-optimum dynamic regulation of DC–DC converters using feedforward of output current and input voltage with current-mode control,' *IEEE Trans. Power Electron.*, vol. PE-1, no. 3, **1986**, pp. 181–191.

47. G.K. Schoneman and D.M. Mitchell, 'Output impedance considerations for switching regulators with current-injected control,' *IEEE Trans. Power Electron.*, vol. 4, no. 1, **1989**, pp. 25–35.

48. S. Kanemaru, T. Hamada, T. Nabeshima, T. Sato, and T. Nakano, 'Analysis and optimum design of a buck-type DC-to-DC converter employing load current feedforward,' in *Proc. IEEE Power Electronic Specialist Conf.*, **1998**, pp. 309–314.

9
Control Design Issues

9.1
Introduction

The switched-mode converters are usually used to supply power for different electronic loads, where the transient-performance requirements can be very stringent due to low supply voltages and rapidly changing load currents as in powering the microcomputers [1–3]. Such applications are usually based on the use of the intermediate bus architecture (IBA) shown in Figure 9.1a [4], where the converters operate as voltage-output converters. Although the load-transient requirements are generally stringent, a part of the converters (i.e., the bus converter) may be operating even without feedback from the output voltage for reducing the costs of implementation.

In some applications, the system may incorporate elements, which would require the use of output-current limiting to protect the converters from damage as in Telecom uninterruptible power supply (UPS) systems shown in Figure 9.1b [5]. In such a system, the rectifiers providing the recharging of the storage batteries have to be able to operate both as a voltage-output and current-output converter requiring the use of multiloop-control arrangement as depicted in Figure 9.2 [6]. The cascaded nature of the control as well as the varying dynamic features of the voltage-output and current-output operations would complicate the design of the required controllers as we have discussed in Chapter 7.

Basically it is always the question of maintaining robust stability and achieving adequate transient dynamics [7–20] within the certain constraints stipulated by the application. In practice, the internal dynamic profile of the converter would determine what the real transient dynamics would be in the operational environment [21]. The practical dynamic behavior of the converter would depend on the voltage and/or current-loop design but also on the other factors related to the internal dynamics and external interactions [22–33] as we have discussed especially in Chapters 2 and 8. Of course, the fact is that a poor control-loop design would yield poor performance but an excellent design would not necessarily yield the opposite.

Dynamic Profile of Switched-Mode Converter. Teuvo Suntio
© 2009 WILEY-VCH Verlag GmbH & Co. KGaA, Weinheim
ISBN: 978-3-527-40708-8

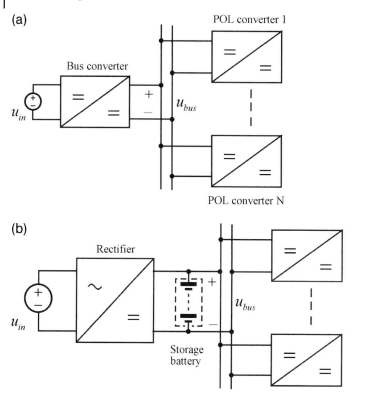

Figure 9.1 Typical distributed power architectures: (a) IBA system and (b) DC UPS system.

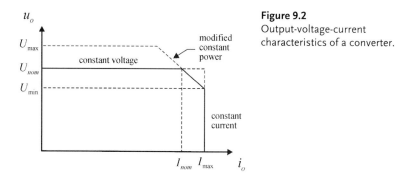

Figure 9.2
Output-voltage-current characteristics of a converter.

Excellent input and output transient performance can be achieved without feedback control by providing proper feedforward from the input voltage and output current as implied and demonstrated in [34]. Actually the feedforward schemes were applied in [23] and [35] as introduced in Chapter 8, Section 8.3, but the goals of the feedforward design were to obtain invariance to the input and output interactions. Figure 9.3a shows the control block diagram for the output dynamics including the feedforward gains (q_i, q_o). The output voltage can be solved to be

$$\hat{u}_o = \left(G_{io-o} + q_i G_a G_{co}\right) \hat{u}_{in} - \left(Z_{o-o} - q_o G_a G_{co}\right) \hat{i}_o \tag{9.1}$$

which is zero when the feedforward gains are chosen to be

$$q_i = -\frac{G_{io-o}}{G_a G_{co}}$$
$$q_o = \frac{Z_{o-o}}{G_a G_{co}} \tag{9.2}$$

The corresponding control block diagram is presented in Figure 9.3b from which we can solve the input current with the defined feedforward gains (9.2) to be as follows:

$$\hat{i}_{in} = \left(Y_{in-o} - \frac{G_{io-o} G_{ci}}{G_{co}}\right) \hat{u}_{in} + \left(T_{oi-o} + \frac{Z_{o-o} G_{ci}}{G_{co}}\right) \hat{i}_o \tag{9.3}$$

Equation (9.3) can also be given as

$$\hat{i}_{in} = Y_{in-\infty} \hat{u}_{in} + T_{oi-\infty} \hat{i}_o \tag{9.4}$$

where $Y_{in-\infty}$ is the familiar ideal input admittance defined earlier in Chapter 2 and discussed throughout the book. $T_{oi-\infty}$ is actually also a special parameter – ideal reverse transfer ratio – not observed earlier but actually characterizing the low frequency behavior of the closed-loop T_{oi-c} as will be shown later. In the case of the second-order converters, the ideal reverse transfer ratio corresponds to $M(D)$ (see Chapter 3, Section 3.4.3). Both of these special parameters are independent of control mode, load, and state of output feedback. They can be always solved according to (9.3), when the transfer functions are known. When $Z_{o-o} \approx 0$ then $T_{oi-o} = T_{oi-c} = T_{oi-\infty}$, which will change the propagation of the output-current noise into the input current profoundly.

The feedback-amplifier design is most often based on the use of proportional-integral-derivative (PID) or proportional-integral (PI) controls [36, 37]. More advanced linear and nonlinear control methods [38–44] are available and suited well for the multivariable control applications [41] especially when digital controllers are in use [45]. Usually the PID-type controllers would provide adequate dynamic performance and robustness of stability. In analog control, the constraints of operational amplifiers such as the limited gain-bandwidth (GBW) product would usually dictate the maximum high-frequency gain of the controller [20] in low-cost applications but the constraints can be somewhat relaxed by using several discrete operational amplifiers in cascade as demonstrated in [10].

The control systems in the extremely low-cost and high-volume switched-mode converters, for example in the mobile-phone-battery-charger applications

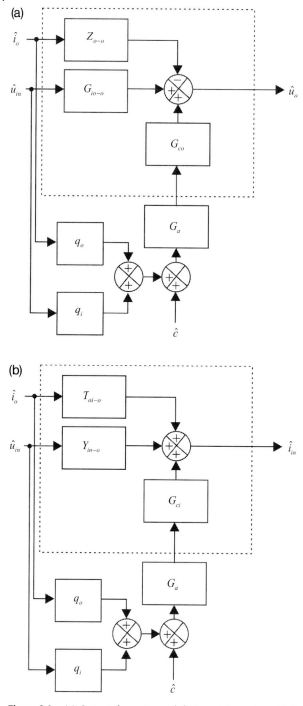

Figure 9.3 (a) Output dynamics and (b) input dynamics with feedforward gains.

[55], are implemented by using a shunt regulator TL 431 [53–62] designed originally for providing a stable voltage reference [56]. The overall low cost has made the shunt regulator a popular building block of a simple control system. Its dynamic properties are not usually what the designer assumes [58, 60]. Therefore, the dynamics and stability of the associated converters may not be satisfactory [59].

The intention of the chapter is not to provide comprehensive control-design methods but to highlight the constraints the design would incorporate as well as to introduce an effective way to design the required loop characteristics. In addition, we will discuss the problems associated with the simple control systems based on the use of shunt regulators. The voltage-output converters are treated only. The same methods and constraints also apply to the current-output converters treated in Chapter 7. The comprehensive list of the references would provide hints for further reading if more thorough knowledge and more advanced control methods are of interest.

9.2
Feedback-Loop-Design Constraints

An optimum feedback-loop design according to [8] is the one with the highest gain below the loop crossover frequency and lowest gain above the crossover frequency. According to [9], the optimal controller would be quickly responsive to substantial changes in output voltage and also provide precise steady-state control. These properties are also inherent in the definition provided by [8].

The first definition of the optimum design [8] is actually in line with the requirements of robust stability and fast transient response even if the external interactions were quite unknown during the writing of [8]: The optimum design might be expressed as illustrated in Figure 9.4, where the loop gain is as high as possible up to the loop crossover and as small as possible after the loop crossover with constant phase margin for all the frequencies.

The reasoning behind the optimality is the fact that such a converter would be invariant to load and would have fast transient response: The high-loop gain means that the closed-loop output impedance is extremely low up to the loop crossover frequency, which means that the load impedance would not easily affect the internal dynamics [22] and the transient setup time is very short [29–33]. In practice, the load transient is dictated by the properties of the output capacitors and the other parasitic circuit elements at the output of the converter and along the path between the load and the converter. The load impedance does not affect the input impedance of the converter due to the extremely high attenuation capability of the closed-loop input-to-output transfer function [22]. Therefore, the stability of the converter due to the source interactions would be dictated by the ideal input impedance in such a way that no impedance overlap is tolerated at the direct input of the converter.

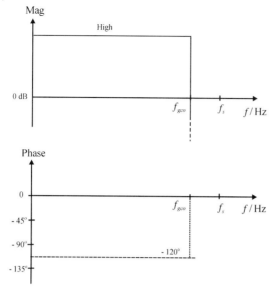

Figure 9.4 Optimal loop shape.

The reality is, however, that such a loop behavior cannot be accomplished due to the practical constraints involved in the components such as operational amplifiers, and A/D converters as well as to the duty-ratio generation. The same constraints would also dictate that the internal disturbance-input transfer functions (i.e., Z_{o-o}, G_{io-o}) would actually dominate the transient responses [48–52] as illustrated in [22] and, therefore, the controllers cannot be tuned based on the time-domain transient responses even if stated so for example in [15–17].

9.2.1
Phase and Gain Margins

The output-voltage transient (Δu_o) can be mathematically presented in frequency domain [29–32] by

$$\Delta u_o(s) = Z_{o-c}\Delta i_o(s) \tag{9.5}$$

where Z_{o-c} is the closed-loop output impedance of the converter and $\Delta i_o(s)$ the *Laplace* transformation of the excitation signal (i.e., the shape of output current change). The time-domain responses can be naturally computed by means of (9.5). Similar formulation can also be derived for the input-voltage transient behavior. The value of the output impedance at the voltage-loop crossover frequency corresponds to the amount of undershoot in the output-voltage transient [29]. This means that a low value of phase

(a)

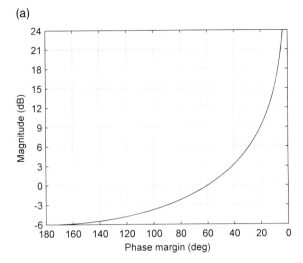

(b)

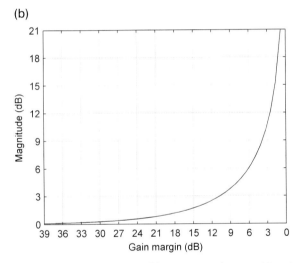

Figure 9.5 The peak value of the sensitivity function (a) at the gain crossover frequency (f_{gco}) as a function of phase margin and (b) at the phase crossover frequency (f_{phco}) as a function of gain margin.

(PM) or gain (GM) margin would increase the undershoot by causing peaking in the output impedance or the input-to-output transfer function due to the sensitivity function S (i.e., $Z_{o-c} = Z_{o-o}S$, $G_{io-c} = G_{io-o}S$; $S = (1 + L(s))^{-1}$): The magnitude of the sensitivity function can be given at the loop crossover frequency by $|S|_{|L|=1} = 1/\sqrt{2(1-\cos(\mathrm{PM}))}$ and at the phase crossover frequency by $|S|_{\varphi=-180°} = |\mathrm{GM}/(\mathrm{GM}-1)|)$. Therefore, the output impedance or the input-to-output transfer function would experience peaking if PM < 60°

(see Figure 9.5a) or GM < 18 dB (see Figure 9.5b). The typically used gain margin of 6 dB would actually cause the peaking of 6 dB. As a consequence, the loop-gain related margins should be maintained adequate for preventing the excess peaking and the deterioration of transient performance.

9.2.2
RHP Zeros and Poles

The boost and buck–boost converters would exhibit nonminimum-phase behavior, which means that their control-to-output transfer functions contain right-half-plane (RHP) zero. This also applies for the fourth-order and higher order converters. The existence of the RHP zero means actually [46] that the maximum control bandwidth has to be limited to the frequency of the RHP zero and usually below it. The situation is worst in the CCM converters where the RHP zero locates at low frequencies and thereby also effectively deteriorates the corresponding transient performance. The control-bandwidth limitation cannot be removed even if such claims have been presented in [47]. In the DCM converters, the RHP zero typically locates at a rather high frequencies and does not deteriorate the transient performance similarly to the CCM mode but it would limit the maximum achievable control bandwidth.

Sometimes, the open-loop converter may even contain an unstable pole (i.e., RHP pole) or the load may impose it. This means that the control bandwidth has to be designed to be higher than the RHP pole in order to maintain stability [46]. An illustrative example is the PCM-controlled superboost converter (Figure 9.6a) [63] sometimes used in the space applications. Such a fourth-order converter contains usually a resonant RHP zero as shown in Figure 9.6b (solid line) in addition to the resonant LHP pole. It may be obvious that the control bandwidth has to be designed shorter than the RHP zero. When the input voltage is further decreased, the resonant LHP pole moves to the RHP (dashed-line) in such a way that it locates at the higher frequency than the RHP zero. As a consequence, the converter cannot be controlled but the duty ratio has to be limited in order that the described situation does not activate. There are also methods to design the converter in such a way that the appearance of the RHP pole is shifted to higher duty ratios. Figure 9.6c shows the output-voltage waveform when the RHP pole is present at open loop, proving that the unstable pole really exists.

9.2.3
Minimum and Maximum Loop Crossover Frequencies

In practice, the usable loop crossover frequencies (f_{gco}) without the limitations caused by the RHP zeros and poles are defined according to [18] to be as follows:

1. $f_{gco} > 3f_n$, where f_n is the resonant frequency of the converter: The resonant pole will cause ringing and the control has to eliminate this ringing. For doing so the control has to have adequate gain.
2. $f_{gco} < \frac{f_s}{5}$, where f_s is the switching frequency: The high gain at the high frequencies would also amplify noise, which may affect the pulsewidth generation and lead to instability (i.e., ripple effects, see e.g., [19]).

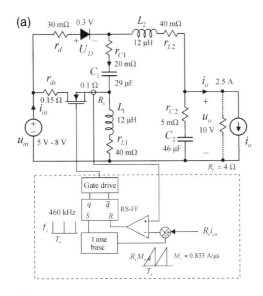

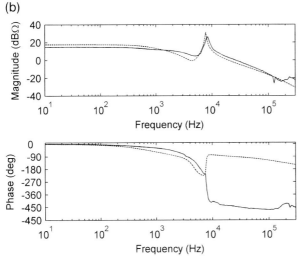

Figure 9.6 PCM-controlled superboost converter: (a) schematics, (b) control-to-output transfer function, and (c) output-voltage waveform due to the unstable pole at open loop.

270 | 9 Control Design Issues

(c)

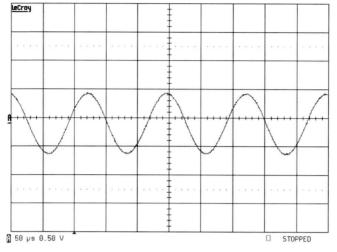

Figure 9.6 (Continued).

3. $f_{gco} < f_{C r_C}$, where $f_{C r_C}$ is the zero caused by the output capacitor: The output capacitor usually dominates the high-frequency transient behavior and, therefore, there is no benefit raising the crossover frequency above the output-capacitor ESR frequency.

9.2.4
Internal Gain of an Operational Amplifier

The internal gain of the operational amplifier (op amp) [18, 20] may also limit the achievable crossover frequency due to limiting the gain and phase of the controller. Typical transfer function of an op amp with the gain-bandwidth (GBW) product of 1.5 MHz is shown in Figure 9.7.

If the gain of the controller exceeds the internal gain of the op amp, the controller gain and phase would follow the gain and phase of the op amp with surprising result. This can be easily concluded from (9.6), where the open-loop gain of the op amp is denoted by G_{OPA} and the ideal gain of the controller by Z_f/Z_{in} (see Figure 9.8b).

$$G_{cc}^{OPA} = -\frac{Z_f}{Z_{in}} \cdot \frac{G_{OPA}}{1 + \frac{Z_f}{Z_{in}} + G_{OPA} + \frac{Z_f}{R_b}} \quad (9.6)$$

At low frequencies, where the controller gain is usually very high due to the integral-control action, the phase of the actual controller would not start from $-90°$ but from zero. This would sometimes remove the conditional-stability condition to exist even if the theory predicts that to exist.

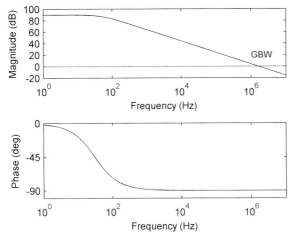

Figure 9.7 The open-loop gain of an op amp with 1.5-MHz GBW-product.

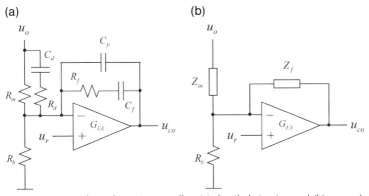

Figure 9.8 Simple analog PID controller: (a) detailed circuitry and (b) general circuitry.

9.3
Controller Implementations

The most common feedback-loop controller is a modified proportional-integral-derivative (PID) controller [8, 17, 36, 37], which usually yields sufficient dynamic properties. The simple analog solution of such a controller also known as type-3 controller [8] is shown in Figure 9.8a. The resistor R_b contributes only to the steady-state value of the output voltage when the reference is not used for control purposes (i.e., $\hat{u}_r$). The corresponding transfer function (i.e., $\hat{u}_{co}/\hat{u}_o$) can be given by

$$G_{cc} = \frac{\left(1 + s \cdot R_f C_f\right)\left(1 + s \cdot (R_{in} + R_d)C_d\right)}{R_{in}\left(C_f + C_p\right)s \cdot (1 + s \cdot R_d C_d)\left(1 + s \cdot \dfrac{R_f C_f C_p}{C_f + C_p}\right)} \tag{9.7}$$

where C_f is typically much larger than C_p. As a consequence, the transfer function simplifies to

$$G_{cc} = \frac{(1 + s \cdot R_f C_f)(1 + s \cdot (R_{in} + R_d) C_d)}{R_{in} C_f s \cdot (1 + s \cdot R_d C_d)(1 + s \cdot R_f C_p)} \qquad (9.8)$$

The PI-type or type-2 controller can be obtained by removing R_d and C_d from Figure 9.8a. The corresponding transfer function can be obtained from (9.7) by setting them to zero.

The transfer function from the reference to the control voltage (i.e., $\hat{u}_{co}/\hat{u}_r$) can be given by

$$G_{cc-r} = 1 + \left(1 + \frac{Z_{in}}{R_b}\right) G_{cc} \qquad (9.9)$$

where Z_{in} is defined in Figure 9.6b and G_{cc} in (9.7). As a consequence, the controller can be defined by

$$\hat{u}_{co} = G_{cc}(\hat{u}_r - \hat{u}_o) + \left(1 + \frac{Z_f}{R_b}\right) \cdot \hat{u}_r \qquad (9.10)$$

where Z_f is defined in Figure 9.7b. The reference is usually embedded in the corresponding PWM microcircuit and, therefore, it is not available and does not contribute to the controller behavior. If the controller is implemented by using discrete components then the reference section should be noise free due to the amplifying effects of G_{cc-r} causing easily unexpected noise problems.

For the control-design purposes, the controller transfer function (G_{cc}) is most convenient to be transformed into the form

$$G_{cc} = \frac{K(1 + s/\omega_{z1})(1 + s/\omega_{z2})}{s \cdot (1 + s/\omega_{p1})(1 + s/\omega_{p2})} \qquad (9.11)$$

The contributions of the gain factor (K), the different zeros (ω_{zi}) and poles (ω_{pi}) to the physical circuitry in Figure 9.8a can be easily deduced comparing (9.7) and (9.11) to each other. The typical frequency responses of the PID and PI controllers are shown in Figures 9.9 and 10, respectively, where f_{zi} and f_{pi} are the frequencies in Hz corresponding to the zeros and poles given in angular frequencies (rad/s) in (9.11).

9.4
Optocoupler Isolation

The feedback loop of the transformer isolated converters has to be also isolated. The most common isolating component is an optocoupler [53, 54]. The optocoupler would be always a critical component in the loop, because its

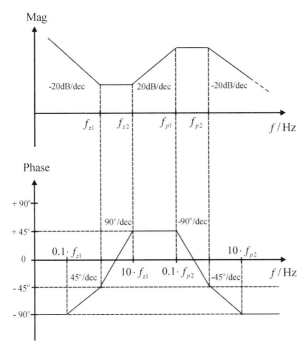

Figure 9.9 The frequency response of a PID controller.

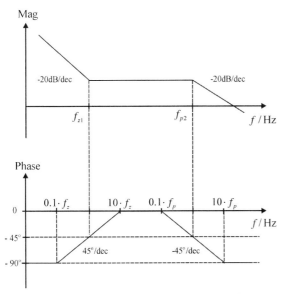

Figure 9.10 The frequency response of a PI controller.

current transfer ratio (CTR) can vary substantially depending on temperature, the level of diode current, the type and manufacturer, it may have a delay, and an *RC*-circuit-like behavior highly dependent on the collector/emitter resistance. The transfer function describing its dynamical behavior [55] can be given by

$$G_{\text{opto}} \approx \text{CTR} \cdot \frac{\left(1 - s \cdot \frac{t_d}{2}\right)}{(1 + s/\omega_{-3\,\text{dB}})\left(1 + s \cdot \frac{t_d}{2}\right)} \quad (9.12)$$

where the effect of the delay (t_d) is approximated by using the first-order Padé approximation of e^{-st_d}. In theory, the delay would only increase the phase lag by $-\omega t_d$ but does not affect the magnitude (i.e., $e^{-st_d} \triangleq 1 \angle - \omega t_d$). The cut-off frequency ($\omega_{-3\,\text{dB}}$) can be found from frequency responses typically given by the manufacturer at different collector resistors or from the time-domain step response given at a certain collector resistor (R_{c1}): The rise time (t_r) and the corresponding cut-off frequency are related by $\omega_{-3\,\text{dB}} \approx 2.2/t_r$, from which the equivalent $C_{\text{opto}} \approx t_r/2.2 R_{c1}$. The corresponding $\omega_{-3\,\text{dB}}$ at the used collector resistor (R_{c2}) equals $R_{c1}/R_{c2} \cdot 2.2/t_r$, respectively.

9.5
Shunt-Regulator-Based Control Systems

The control system of a low-cost converter is usually built around the shunt regulator TL431 developed in the late 1970s for providing stable reference voltage [56]. In principle, the shunt regulator is an excellent device for implementing a simple control system due to its ability to drive rather large currents [53–61] when the optocoupler is needed in the control loop as illustrated in Figure 9.11.

9.5.1
Dynamic Model

The shunt regulator is, however, quite a complicated device dynamically and its properties are not usually understood [58, 60]. The first mistake is to assume that the shunt regulator provides the same properties as an op amp [60, 62]. The reality is, however, that it is a transconductance amplifier (i.e., voltage-to-current amplifier) having current as an output signal and no negative feedback present as shown in Figure 9.12a [59]. The corresponding dynamic equivalent circuit is shown in Figure 9.12b. The op-amp-like features are valid only if the transconductance gain (g_m) and the output impedance (Z_o) are high (Figure 9.12b).

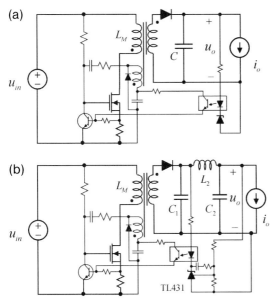

Figure 9.11 Application of TL431 in controlling self-oscillating flyback converter with (a) combined feedback loops and (b) separate feedback loops.

The shunt regulator also provides two feedback loops (Figure 9.12c) – one via the reference input (slow loop) and the other via the cathode (fast loop) [57]. The loops may be connected together (Figure 9.11a) or separately (Figure 9.11b). In the case of Figure 9.11b, the overall loop gain is difficult to measure and analyze. Therefore, the effect of the fast loop in those cases is most often just forgotten but may have profound effect on the loop behavior.

The dynamic properties of TL431 are usually defined by means of the slow-loop open-loop frequency response at the bias current of 10 mA by using the measurement setup shown in Figure 9.13a and based on the cathode voltage (i.e., $\hat{u}_{ca}/\hat{u}_{Ref}$) [56]. The DC specifications are stated to be valid down to the collector current of 1 mA. The open-loop output impedance is not usually shown in the manufacturers' specifications but can be measured using the measurement setup shown in Figure 9.13b.

It should be noted that the open-loop response incorporates the effects of the transconductance gain (g_m) and the output impedance (Z_o) as follows:

$$\frac{\hat{u}_{ca}}{\hat{u}_{Ref}} = -\frac{R_d Z_o}{R_d + Z_o} \cdot g_m \tag{9.13}$$

The real output signal is, however, the cathode current (i_{ca}) and, therefore, the desired transfer function can be given by

$$\frac{\hat{i}_{ca}}{\hat{u}_{Ref}} = \frac{Z_o}{R_d + Z_o} \cdot g_m \tag{9.14}$$

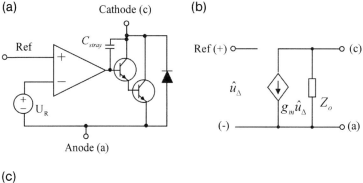

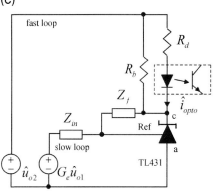

Figure 9.12 Shunt regulator TL431: (a) high-level equivalent circuit, (b) dynamical equivalent circuit, and (c) the two feedback loops.

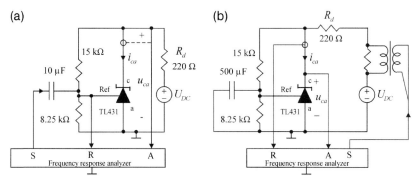

Figure 9.13 The frequency-response measurement setups for (a) open-loop gain and (b) open-loop output impedance.

which shows the influence of different circuit elements as well as the existence of the positive feedback. It is obvious that the transfer functions in (9.13) and (9.14) equals $R_d g_m$ and g_m, when $Z_o \gg R_d$. Typically g_m is assumed to be constant and Z_o a capacitor [58, 59], which can be estimated from the given

9.5 Shunt-Regulator-Based Control Systems

frequency responses: the low-frequency gain corresponds to $R_d g_m$ and the cut-off frequency to $1/2\pi R_d C_o$.

The fast and slow-loop transfer functions can be given from Figure 9.12c as follows:

$$\frac{\hat{i}_{opto}}{\hat{u}_{o-f}} = \frac{1}{(Z_{in} + Z_f)(R_d + Z_o) + R_d Z_o} \left(Z_{in} + Z_f + Z_o \right.$$
$$\left. + \frac{Z_{in} Z_o^2 (Z_{in} + Z_f) g_m}{(Z_{in} + Z_f)(R_d + Z_o) + R_d Z_o (1 + Z_{in} g_m)} \right) \quad (9.15)$$

$$\frac{\hat{i}_{opto}}{\hat{u}_{o-s}} = \frac{G_e Z_o (Z_f g_m - 1)}{(Z_{in} + Z_f)(R_d + Z_o) + R_d Z_o (1 + Z_{in} g_m)}$$

where the subscript extensions f and s denote fast and slow, respectively, and the termination resistor R_d contains the effect of the corresponding discrete resistor and the dynamical resistance of the diode, which may be large at the low diode current. It is also assumed that the bias resistor (R_b) is much larger than R_d. If $g_m \gg 1$ then (9.15) becomes

$$\frac{\hat{i}_{opto}}{\hat{u}_{o-f}} = \frac{1}{R_d}$$
$$\frac{\hat{i}_{opto}}{\hat{u}_{o-s}} = \frac{G_e Z_f}{R_d Z_{in}} \quad (9.16)$$

which is the form typically used in the analyses as, for example, in [60] except that there is clearly no negative feedback. If we consider the transfer functions in (9.16) with respect to the cathode voltage (u_{ca}) then the negative sign would appear but the cathode voltage is not the signal transforming the control information through the optocoupler.

The dynamic parameters of the shunt regulator (i.e., g_m and Z_o) can be extracted by measuring the open-loop response (4) from Figure 9.13a, and the output impedance $\left(Z_o = \frac{\hat{u}_{ca}}{\hat{i}_{ca}} \right)$ from Figure 9.13b: The transconductance (g_m) can be computationally solved by means of the measured responses from

$$g_m = \left(1 + \frac{R_d}{Z_o} \right) \cdot \frac{\hat{i}_{ca}}{\hat{u}_{Ref}} \quad (9.17)$$

A certain shunt regulator was analyzed experimentally by measuring its open-loop gain ($\hat{i}_{ca}/\hat{u}_{Ref}$) and the output impedance (Z_o) as well as extracting the transconductance gain computationally from the measured responses from (9.17).

It was noticed that the shunt regulator does not function dynamically properly at the specified minimum cathode current of 1 mA: The measured cathode voltage (u_{ca}) is shown in Figure 9.14 during the measurement of the open-loop gain indicating clearly that the cathode current of 1 mA is not sufficient for the proper dynamic operation. The proper operation was recovered when the

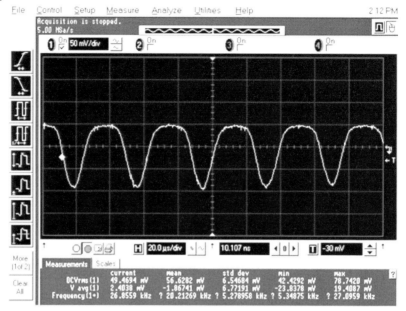

Figure 9.14 The corrupted cathode-voltage response at the cathode current of 1 mA.

cathode current was increased to 2 mA although the excitation level had to be kept rather low compared to the excitation at the cathode current of 10 mA in order to avoid the corrupted response.

The measured open-loop gain and the output impedance at the cathode current of 10 mA (solid line) and 2 mA (dashed line) are shown in Figures 9.15 and 16, respectively. Figure 9.16 shows that the output impedance corresponds to a capacitor, which is expected due to the stray capacitances in the microelectronic circuit (Figure 9.12a) [56]. The level of cathode current affects the ESR of the capacitor. The high-frequency responses defines the value of the ESR (R_o) and the corresponding value of the capacitor (C_o) can be solved from the corner frequency ($f_{-3\text{ dB}}$) by $C_o = \frac{1}{2\pi f_{-3\text{ dB}} R_o}$.

The transconductance gain (g_m) was computationally solved by means of the responses in Figures 9.15 and 16. The corresponding responses are shown in Figure 9.17 indicating that the high-frequency gain depends on the level of cathode current.

The shape of g_m implies that it can be modeled by means of a second-order transfer function as

$$g_m \approx \frac{g_{m-o}(1 + s/\omega_{z1})}{s^2 + s \cdot 2\zeta\omega_n + \omega_n^2} \tag{9.18}$$

where g_{m-o} is the low-frequency transconductance, ζ the damping factor, and ω_n the undamped natural frequency. At the low cathode current, the resonant behavior is well damped. The zero (ω_{z1}) can be found at the frequency where

9.5 Shunt-Regulator-Based Control Systems | 279

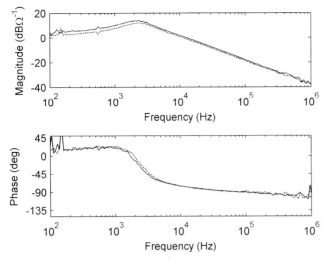

Figure 9.15 The open-loop gain ($\hat{i}_{ca}/\hat{u}_{\text{Ref}}$) at the cathode current of 10 mA (solid line) and 2 mA (dashed line).

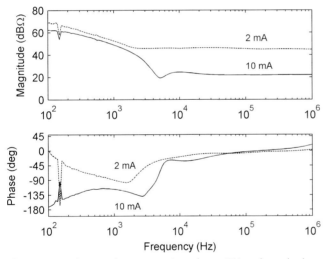

Figure 9.16 The open-loop output impedance (Z_o) at the cathode current of 10 mA (solid line) and 2 mA (dashed line).

the phase is close to 45° and the resonant pole (ω_n) at the frequency where the phase is close to 0°.

The actual behavior of such a control system shown in Figure 9.18 was measured by injecting the excitation signal into both of the loops (see Figure 9.12). The bias current was adjusted by means of the bias resistor (R_b) to 2 and 10 mA. The corresponding responses are shown in Figure 9.19.

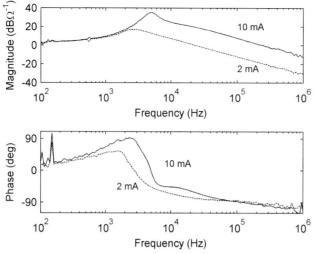

Figure 9.17 The computed transconductance gain (g_m) at the cathode current of 10 mA (solid line) and 2 mA (dashed line).

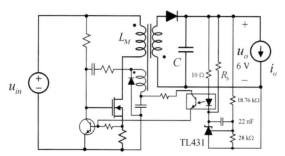

Figure 9.18 A flyback converter with the TL-431-based control system.

The total response of the control system is the sum of the slow- and fast-loop responses [59] defined in (9.15). The slow-loop response dominates at the low frequencies and the fast-loop response at the high frequencies. From Figure 9.18, the slow loop forms basically an integral controller where the maximum gain is defined by the transconductance gain. Therefore, it is natural that the low-frequency gain of the control system is very low due to the behavior of g_m shown in Figure 9.17. It may be obvious that the steady-state voltage accuracy can be poor. The high-frequency response should correspond to $1/R_d$ (i.e., −20 dB) but is actually lower. The reason may be that the dynamic resistance of the optocoupler diode increases the value of R_d.

The models shown may not be generalized because the properties of various shunt regulators are known to vary significantly [59]. The methods to extract the models are naturally generally applicable. The important message is

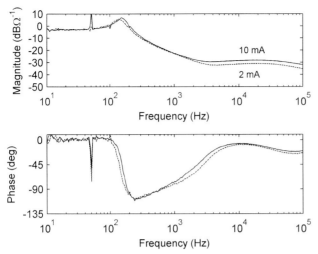

Figure 9.19 The control-system frequency responses at the cathode current of 10 mA (solid line) and 2 mA (dashed line).

that the properties of the shunt regulator and the control systems based on them are not what usually are assumed based on the information the open literature has provided. In addition, the minimum bias current should be higher than the specified minimum, and preferably close to 10 mA. If the overall current consumption has to be kept low then the shunt regulators having much lower minimum currents should be used as discussed in [61].

9.5.2
Two-Loop Control System

The slow and fast loops of the shunt regulator can be connected into the output of the converter with several ways:

1. Both of the loops are connected together at the direct output of the converter as illustrated in Figure 9.11a. The overall output-voltage loop gain of the converter can be easily measured injecting the excitation signal into both of the loops simultaneously. The overall control-system transfer function is the sum of the slow- and fast-loop transfer functions defined in (9.15).
2. The fast loop may be connected directly to an auxiliary voltage supply and the slow loop at the output of the converter as shown in [58]. The fast loop does not contribute dynamically and, therefore, the overall control-system transfer function is the slow-loop transfer function defined in (9.15).

3. The converters may be equipped with an extra output filter. It is customary to connect the fast loop before the filter and the slow loop directly at the output of the converter as illustrated in Figure 9.11b. As a consequence, the voltage-loop gain is difficult to measure by means of a single measurement but it can be found by using the methods defined below [57].

In the case of separately connected fast and slow loops, the overall voltage-loop gain can be found by measuring separately the responses of those loops when the other loop is connected. If we denote the measured fast loop by $L_f^M(s)$ and the measured slow loop by $L_s^M(s)$ then the overall voltage-loop gain ($L(s)$) can be given by

$$L(s) = \frac{L_f^M(s) + L_s^M(s) + 2L_f^M(s)L_s^M(s)}{1 - L_f^M(s)L_s^M(s)} \tag{9.19}$$

where $L_f^M(s)$ and $L_s^M(s)$ can be given as a function of the original fast ($L_f(s)$) and slow ($L_s(s)$) loops, respectively, by

$$\begin{aligned} L_f^M &= \frac{L_f(s)}{1 + L_s(s)} \\ L_s^M &= \frac{L_s(s)}{1 + L_f(s)} \end{aligned} \tag{9.20}$$

The validity of (9.19) assumes that the measured loops are given extracting the effect of negative feedback as is usual in control engineering. If the measurement data are the authentic data from the frequency-response analyzer (i.e., the phase margin is usually observable with respect to zero) then the sign of the product of the loop gains in the numerator has to be changed to minus for obtaining correct overall loop gain [57].

In order to analyze the dynamics of the two-loop system shown in Figure 9.11b, the corresponding transfer functions containing the effect of the extra filter have to be determined: Those transfer functions can be found by utilizing the linear two-port model of the converter shown in Figure 9.20, where the open-loop parameters constitute the G-parameter set defined in (9.21).

$$\begin{bmatrix} \hat{i}_{in} \\ \hat{u}_o \end{bmatrix} = \begin{bmatrix} Y_{in-o} & T_{oi-o} & G_{ci} \\ G_{io-o} & -Z_{0-o} & G_{co} \end{bmatrix} \begin{bmatrix} \hat{u}_{in} \\ \hat{i}_{o1} \\ \hat{c} \end{bmatrix} \tag{9.21}$$

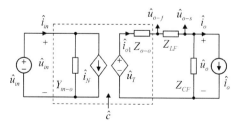

Figure 9.20 Two-port model of a converter with the added secondary filter.

The necessary transfer functions can be solved from Figure 9.20 by means of pure circuit theory yielding the control-block diagram shown in Figure 9.21, which represents the output dynamics of the converter. The shunt-regulator-based control system is also added into the block diagram, where G_{cc-f} and G_{cc-s} are the fast and slow-loop transfer functions defined in (9.15), respectively, Z_{LF} and Z_{CF} the inductive and capacitive impedances of the filter, respectively, and $Z_{SF} = Z_{LF} + Z_{CF}$, G_{a-i} the combined transfer function of the other dynamical elements along the path from the shunt regulator to the control variable ($\hat{c}$) as well as G_{io-o}, Z_{o-o}, and G_{co} the internal transfer functions of the original converter defined in the bottom row of (9.21).

From Figure 9.21, the fast ($L_f(s)$) and slow ($L_s(s)$) loops can be, respectively, given by

$$L_f(s) = -G_{cc-f} G_{a-i} \cdot \frac{G_{co}}{1 + \dfrac{Z_{o-o}}{Z_{SF}}}$$

$$L_s(s) = -G_{cc-s} G_{a-i} \cdot \frac{G_{co}}{1 + \dfrac{Z_{o-o}}{Z_{SF}}} \cdot \frac{Z_{CF}}{Z_{SF}} \quad (9.22)$$

and the overall loop gain ($L(s)$) by

$$L(s) = L_f(s) + L_s(s) \quad (9.23)$$

This means that the closed-loop input-to-output transfer function (G_{io-c}) and the closed-loop output impedance (Z_{o-c}) can be given by

$$G_{io-c} = \frac{\dfrac{Z_{CF}}{Z_{o-o} + Z_{SF}}}{1 + L_f(s) + L_s(s)} \cdot G_{io-o}$$

$$Z_{o-c} = \frac{(Z_{o-o} + Z_{LF}) Z_{CF}}{(Z_{o-o} + Z_{SF})(1 + L_f(s) + L_s(s))} \quad (9.24)$$

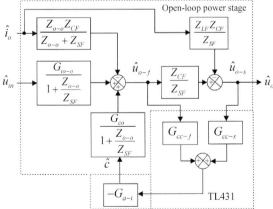

Figure 9.21 The control-block diagram for the overall system representing the output dynamics.

9.6
Simple Control-Design Method

The internal open-loop dynamics of the voltage-output converter can be represented by means of a set of G-parameters by

$$\begin{bmatrix} \hat{i}_{in} \\ \hat{u}_o \end{bmatrix} = \begin{bmatrix} Y_{in-o} & T_{oi-o} & G_{ci} \\ G_{io-o} & -Z_{o-o} & G_{co} \end{bmatrix} \begin{bmatrix} \hat{u}_{in} \\ \hat{i}_o \\ \hat{c} \end{bmatrix} \quad (9.25)$$

where the upper row represents the input dynamics and the bottom row the output dynamics. The matrix form can be equally given as the linear two-port model shown in Figure 9.22.

The closed-loop dynamics can be most conveniently found by utilizing the control block diagrams shown in Figure 9.23. The voltage-loop gain ($L(s)$) is defined from the variables shown in Figure 9.23a by

$$L(s) = G_{se} G_{cc} G_a G_{co} \quad (9.26)$$

where G_{se} is the output-voltage sensing gain, G_{cc} the controller transfer function, G_a the transfer function of all the other dynamical elements within the loop from the controller to the control input ($\hat{c}$), and the control-to-output transfer function (G_{co}). The inputs where the input variables ($\hat{u}_{in}, \hat{i}_o$) are connected (Figure 9.23a) are known as disturbance inputs.

The closed-loop dynamic representation can be solved from Figure 9.23 to be as follows:

$$\begin{bmatrix} \hat{i}_{in} \\ \hat{u}_o \end{bmatrix} =$$

$$\begin{bmatrix} \dfrac{Y_{in-o}}{1+L(s)} + Y_{in-\infty} \cdot \dfrac{L(s)}{1+L(s)} & \dfrac{T_{oi-o}}{1+L(s)} + T_{oi-\infty} \cdot \dfrac{L(s)}{1+L(s)} & \dfrac{G_{ci}}{G_{se}G_{co}} \cdot \dfrac{L(s)}{1+L(s)} \\ \dfrac{G_{io-o}}{1+L(s)} & -\dfrac{Z_{o-o}}{1+L(s)} & \dfrac{1}{G_{se}} \cdot \dfrac{L(s)}{1+L(s)} \end{bmatrix}$$

$$\times \begin{bmatrix} \hat{u}_{in} \\ \hat{i}_o \\ \hat{u}_r \end{bmatrix} \quad (9.27)$$

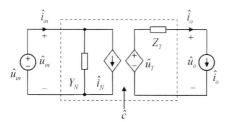

Figure 9.22 Two-port model of a voltage-output converter.

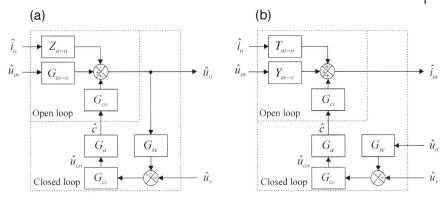

Figure 9.23 Control block diagrams representing closed-loop (a) output dynamics and (b) input dynamics.

where $Y_{\text{in}-\infty}$ is the ideal input admittance (i.e., $Y_{\text{in}-o} - G_{\text{io}-o}G_{\text{ci}}/G_{\text{co}}$) and $T_{\text{oi}-\infty}$ is the ideal reverse transfer ratio (i.e., $T_{\text{oi}-o} + \frac{Z_{o-o}G_{\text{ci}}}{G_{\text{co}}}$). The goal of the control design is to obtain robust stability in terms of the phase (PM) and gain (GM) margins as well as adequate transient performance associated with the load-current (i.e., $\frac{Z_{o-o}}{1+L(s)}$) and input-voltage (i.e., $\frac{G_{\text{io}-o}}{1+L(s)}$) disturbances as discussed earlier in Section 9.2. The control design examples in the subsequent subsections are given based on the buck and boost converters defined in Figure 9.24 and operating in CCM.

9.6.1
Control Design Example: VMC Buck Converter

The buck converter under VM control is shown in Figure 9.25 and the associated circuit elements are defined in Figure 9.24a. The dynamic modeling and characterization of the VMC buck converter is previously presented in Chapter 3. The goal of the control design is to limit the maximum loop crossover frequency to 10 kHz and obtain the phase margin of 60°. The resonant nature of the converter requires the use of PID controller in order to obtain enough phase boost for the stable operation.

The output-voltage loop gain ($L(s)$) can be given by

$$L(s) = \frac{1}{V_M} \cdot G_{\text{cc}} \cdot G_{\text{co}} \qquad (9.28)$$

where V_M is the peak-to-peak voltage of the PWM ramp (Figure 9.25), G_{cc} the error-amplifier transfer function, and G_{co} the control-to-output transfer function of the buck converter. It is obvious that V_M (3 V) and G_{co} (Chapter 3) are known, and we have to choose G_{cc} such that the goals are met. Therefore,

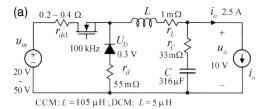

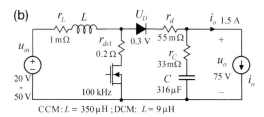

Figure 9.24 The schematics of (a) buck converter and (b) boost converter.

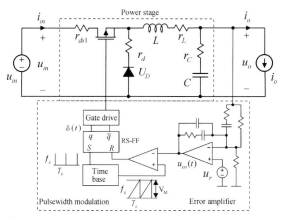

Figure 9.25 VM-controlled buck converter.

the loop gain (9.28) can be given by

$$L(s) = \frac{1}{3} \cdot \frac{K(1 + s/\omega_{z1})(1 + s/\omega_{z2})}{s \cdot (1 + s/\omega_{p1})(1 + s/\omega_{p2})}$$

$$\times \frac{(U_{\text{in}} + U_D + (r_d - r_{\text{ds1}})I_L)(1 + sr_C C)}{LC}{s^2 + s\frac{r_L + Dr_{\text{ds1}} + D'r_d + r_C}{L} + \frac{1}{LC}} \quad (9.29)$$

G_{co} of the VMC buck converter has the highest gain at the maximum input voltage, which is 50 V. Therefore, we first plot the frequency response of the known part of the loop gain at the input voltage of 50 V as shown in Figure 9.26 (solid line). The next task is to shape the phase of the loop such that the phase

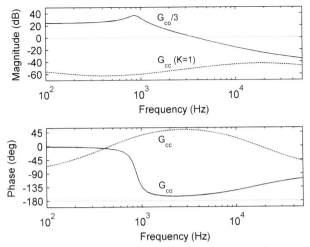

Figure 9.26 The first phase of the control design: VMC buck converter.

margin of 60° is obtained. This actually means that we assume $K = 1$. The maximum phase boost is obtained if the zeros (ω_{zi}) are placed at the same frequency. The resonant nature of the converter ($f_n \approx 900$ Hz) requires that the zeros have to be placed before the resonant frequency. It is usually advised that the last pole (ω_{p2}) should be placed to compensate the ESR zero of the output capacitor but due to the high uncertainty of the value of the zero the last pole is recommended to be placed at half the switching frequency or closer. The frequency response of the controller ($K = 1$) with the zero-pole placements defined in (9.30) is shown in Figure 9.26 (dashed line).

$$\omega_{z1,2} = 0.5 \cdot \frac{1}{\sqrt{LC}}$$
$$\omega_{p1} = 30 \cdot \omega_{z1} \quad (9.30)$$
$$\omega_{p2} = \frac{2\pi f_s}{2}$$

In the next phase, the voltage-loop gain with $K = 1$ should be plotted as shown in Figure 9.27. According to it, the gain of the controller shall be increased by 61 dB in order to obtain the crossover frequency of 10 kHz, which gives $K = 1122$. The resulting predicted loop gains at the input voltages of 50 V (solid line) and 20 V (dashed line) are shown in Figure 9.28. The reduction of the crossover frequency due to the decrease of the input voltage is typical to the voltage-mode control. The crossover-frequency reduction may be eliminated by using several cascaded controllers as in [10] to make the slope of the loop gain steeper in the vicinity of the crossover frequency.

The same design procedures were also applied to the experimental converter shown in Figure 9.24a. The measured output-voltage loop gains are shown

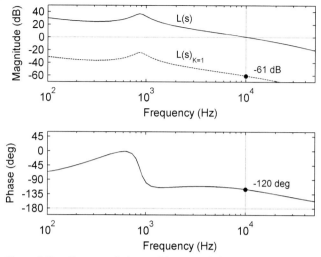

Figure 9.27 The second phase of the control design: VMC buck converter.

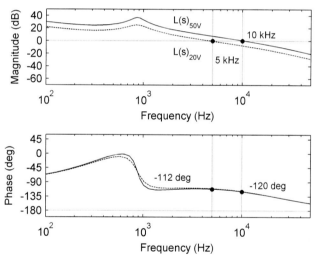

Figure 9.28 The predicted voltage-loop gains of the VMC buck converter at the input voltages of 50 V (solid line) and 20 V (dashed line).

in Figure 9.29. The reduction of the phase margins from 60° is caused by the higher phase lag in the real converter compared to the predictions. More accurate design would have been obtained if the measured control-to-output transfer function had been used as the base of the design.

The open-loop (dashed line) and closed-loop (solid line) output impedances of the practical converter at the input voltage of 50 V are shown in Figure 9.30. The magnitude of the closed-loop output impedance at the voltage-loop crossover

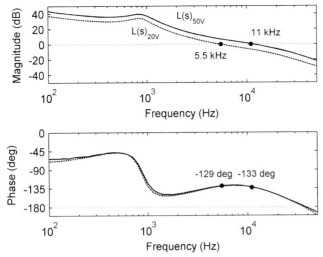

Figure 9.29 The measured output-voltage loop gains at the input voltages of 50 V (solid line) and 20 V (dashed line).

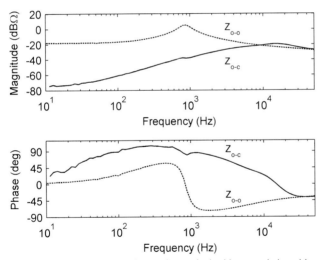

Figure 9.30 The measured open-loop (dashed line) and closed-loop (solid line) output impedances at the input voltage of 50 V.

frequency (11 kHz) is approximately 80 mΩ. The converter was subjected to a constant-current-type load change with the slew rate of 250 mA/μs from 0.2 to 2.5 A. The corresponding output-voltage response is shown in Figure 9.31 where the voltage dip of 170 mV corresponds well with the magnitude of the load change and the magnitude of the closed-loop output impedance at the voltage-loop crossover frequency. The setup time is also rather quick due to the small output impedance at the low frequencies.

9.6.2
Control Design Example: PCMC Buck Converter

The buck converter under PCM control is shown in Figure 9.32 and the associated circuit elements are defined in Figure 9.24a. The equivalent inductor-current sensing resistor (R_s) is 75 mΩ. The dynamic modeling and characterization of the PCMC buck converter is previously presented in Chapter 4. The goal of the control design is to limit the maximum voltage-loop crossover frequency to 10 kHz and obtain the phase margin of 60°. The resonant-free nature of the converter allows the use of PI controller. The PWM modulator used in the experimental converter is UCC 3842, where the error-amplifier section contains an extra gain of 1/3.

The output-voltage loop gain ($L(s)$) can be given by

$$L(s) = \frac{1}{3R_s} \cdot G_{cc} \cdot G_{co} \tag{9.31}$$

where R_s (Figure 9.32) is the equivalent inductor-current sensing resistor, G_{cc} the error-amplifier transfer function, and G_{co} the control-to-output transfer function of the PCMC buck converter. It is obvious that R_s(75 mΩ) and G_{co} (Chapter 4) are known and we have to choose G_{cc} such that the goals are met.

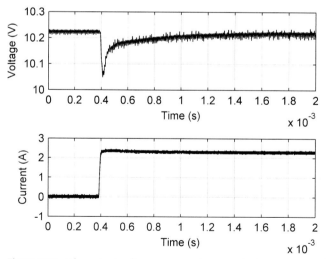

Figure 9.31 The output-voltage response to a constant-current step change with slew rate of 250 mA/µs at the input voltage of 50 V.

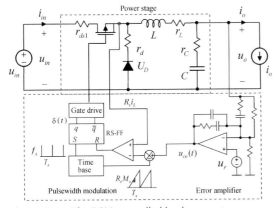

Figure 9.32 The PCM-controlled buck converter.

Therefore, the loop gain (9.31) can be given by

$$L(s) = \frac{1}{3 \cdot 75 \times 10^{-3}} \cdot \frac{K(1 + s/\omega_{z1})}{s \cdot (1 + s/\omega_{p2})}$$

$$\times \frac{\dfrac{F_m U_E (1 + s r_C C)}{LC}}{s^2 + s \dfrac{r_E + F_m q_c U_E + r_C}{L} + \dfrac{1}{LC}}$$

$$U_E = U_{\text{in}} + U_D + (r_d - r_{\text{ds1}}) I_L, \; r_E = r_L + D r_{\text{ds1}} + D' r_d \quad (9.32)$$

The converter is supposed to be compensated such that $M_c \approx 1.45 \cdot U_o/2L$ (see Chapter 4, Section 4.5.1). G_{co} of the PCMC buck converter has the highest phase lag at the highest input voltage but its magnitude stays virtually intact when the input voltage changes as demonstrated in Chapter 4. Therefore, we plot the known part of the loop gain as shown in Figure 9.33 (solid line) at the input voltage of 50 V. The next task is to shape the phase of the loop such that the phase margin of 60° is obtained. Similarly to the VMC converter the zero (ω_{z1}) can be placed before the virtual resonant frequency ($f_n \approx 900$ Hz) and the last pole (ω_{p2}) to obtain the phase margin of 60°. The frequency response of the controller ($K = 1$) with the zero-pole placements defined in (9.33) is shown in Figure 9.34 (dashed line) indicating that the phase margin would be 68°. In order to obtain 60°, the last pole should be placed at slightly lower frequencies than defined in (9.33). In the practical converter, the phase margin would be lower than predicted and, therefore, we do not make the correction.

$$\omega_{z1} = 0.5 \cdot \frac{1}{\sqrt{LC}}$$
$$\omega_{p2} = \frac{2\pi f_s}{8} \quad (9.33)$$

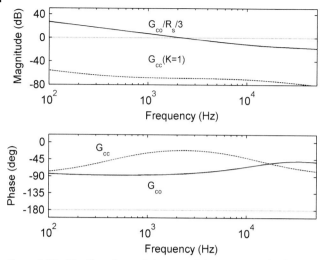

Figure 9.33 The first phase of the control design: PCMC buck converter.

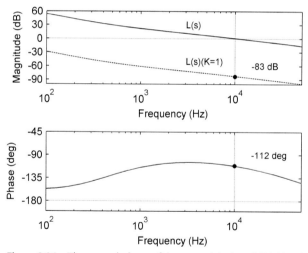

Figure 9.34 The second phase of the control design: PCMC buck converter.

In the next phase, the voltage-loop gain with $K = 1$ should be plotted as shown in Figure 9.34. According to it, the gain of the controller shall be increased by 83 dB to obtain the crossover frequency of 10 kHz, which gives $K = 14125$. The resulted predicted loop gains at the input voltages of 50 V (solid line) and 20 V (dashed line) are shown in Figure 9.35 indicating that the magnitude of the voltage-loop gain does not change, when the input voltage changes.

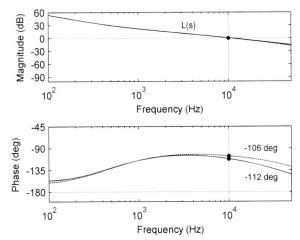

Figure 9.35 The predicted voltage-loop gains at the input voltages of 50 V (solid line) and 20 V (dashed line).

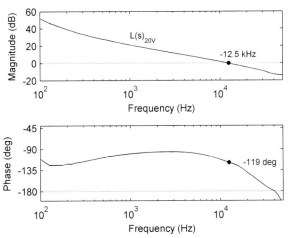

Figure 9.36 The measured voltage-loop gain of the PCMC buck converter at the input voltage of 20 V.

The similar control-design procedures were applied to the experimental converter. The measured voltage-loop gain at the input voltage of 20 V is shown in Figure 9.36. The reduction of the phase margin from the predicted phase margin is caused by the higher phase lag in the real converter compared to the predictions. More accurate design would have been obtained if the measured control-to-output transfer function had been used as a base of the design.

The open-loop (dashed line) and closed-loop (solid line) output impedances at the input voltage of 50 V are shown in Figure 9.37. The magnitude of

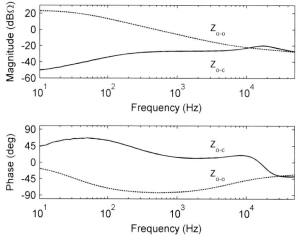

Figure 9.37 The measured open-loop (dashed line) and closed-loop (solid line) output impedances at the input voltage of 50 V.

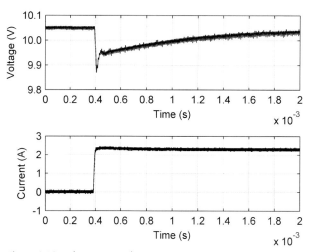

Figure 9.38 The output-voltage response to a constant-current-type step load change from 0.2 to 2.5 A with the slew rate of 250 mA/μs at the input voltage of 50 V.

the closed-loop output impedance at the voltage-loop crossover frequency (13 kHz) is approximately 80 mΩ. The converter was subjected to a constant-current-type load change from 0.2 to 2.5 A with the slew rate of 250 mA/μs. The corresponding output-voltage response is shown in Figure 9.38, where the voltage dip of 180 mV corresponds well to the magnitude of the load change and the magnitude of the closed-loop output impedance at the voltage-loop crossover frequency. The setup time is rather sluggish compared to the response of the VMC converter (Section 9.6.1, Figure 9.31) due to the larger output impedance at the low frequencies.

9.6.3
Control Design Example: VMC Boost Converter

The boost converter under VM control is shown in Figure 9.39 and the associated circuit elements are defined in Figure 9.24b. The dynamic modeling and characterization of the VMC boost converter is previously presented in Chapter 3. G_{co} of the VMC boost converter contains a RHP zero, which automatically limits the achievable control bandwidth to the frequency of the RHP zero or below it. The goal of the control design is to maximize the crossover frequency with a satisfactory phase margin. The resonant nature of the converter requires the use of PID controller in order to obtain enough phase boost for the stable operation.

Similarly to the VMC buck converter, the output-voltage loop ($L(s)$) can be given by

$$L(s) = \frac{1}{V_M} \cdot G_{cc} \cdot G_{co} \tag{9.34}$$

where V_M (3 V) and G_{co} (Chapter 3) are known and we have to choose G_{cc} such that the design goals are met. Therefore, the loop gain (9.34) can be given by

$$L(s) = \frac{1}{3} \cdot \frac{K(1 + s/\omega_{z1})(1 + s/\omega_{z2})}{s \cdot (1 + s/\omega_{p1})(1 + s/\omega_{p2})}$$

$$\times \frac{\left(D'(U_o + U_D) - (r_L + r_{ds1} + D'^2 r_c)I_L - sLI_L\right)(1 + sr_C C)}{LC} \tag{9.35}$$
$$s^2 + s\frac{r_L + Dr_{ds1} + D'(r_d + r_C)}{L} + \frac{D'^2}{LC}$$

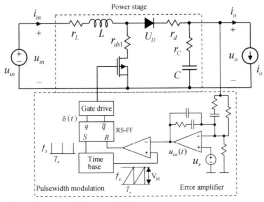

Figure 9.39 VM-controlled boost converter.

The RHP zero (i.e., $\frac{D'(U_o+U_D)-(r_L+r_{ds1}+D'^2 r_c)I_L}{LI_L}$) locates closest to the origin when the input voltage is lowest and the output load highest. It should be also observed that the resonant frequency (i.e., $D'/\sqrt{LC}$) moves toward the origin when the input voltage decreases. Therefore, the known part of the loop gain has to be plotted at the input voltage of 20 V and at the maximum load power as shown in Figure 9.40 (solid line). The RHP zero locates at 1.369 kHz at the defined condition and the resonant frequency at 121 Hz. According to the phase behavior of G_{co}, it may be obvious that the crossover frequency has to be limited to approximately half the frequency of the RHP zero. The first zero (ω_{z1}) has to be placed before the resonant frequency. The phase boost is maximized when the zeros ($\omega_{z1,2}$) are placed at the same frequency. The last zero should be placed below half the switching frequency in order to maintain sufficient gain margin due to the shape of the high-frequency magnitude of G_{co}. The frequency response of the controller ($K = 1$) is shown in Figure 9.40 (dashed line) with the zero-pole placements defined in (9.36).

$$\omega_{z1,2} = 0.5 \cdot \frac{D'}{\sqrt{LC}}$$
$$\omega_{p1} = 30 \cdot \omega_{z1} \qquad (9.36)$$
$$\omega_{p2} = \frac{2\pi f_s}{2}$$

In the next design phase, the voltage-loop gain with $K = 1$ should be plotted as shown in Figure 9.41. The phase behavior of the loop gain reveals that the worst-case phase margin has to be chosen rather low in order to obtain a high enough crossover frequency: the crossover frequency of 750 Hz means that the gain of the controller has to be increased by 21.3 dB. This gives $K = 11.6$

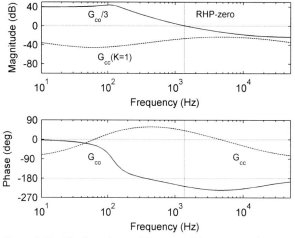

Figure 9.40 The first phase of the control design: VMC boost converter.

9.6 Simple Control-Design Method | 297

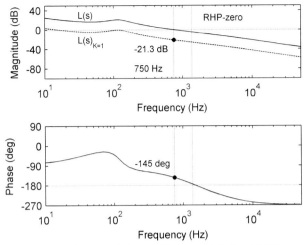

Figure 9.41 The second phase of the control design: VMC boost converter.

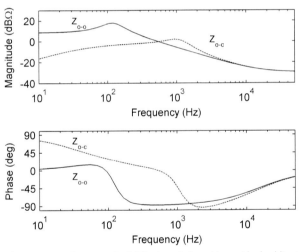

Figure 9.42 The open (solid line) and closed-loop (dashed line) output impedances of the VMC boost converter at the input of 20 V.

and the phase margin of 25°. The corresponding open- and closed-loop output impedances are shown in Figure 9.42 explaining clearly the reason for the known inferior load-transient response of a VMC boost converter operating in CCM.

The predicted voltage-loop gains at the input voltage of 20 V (solid line) and 50 V (dashed line) are shown in Figure 9.43. It may be obvious from Figure 9.43 that the controller design at the high input voltage would have yielded unstable

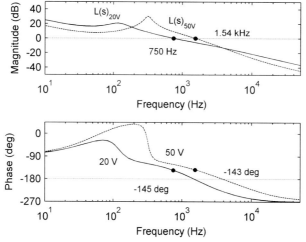

Figure 9.43 The predicted voltage-loop gains of the VMC boost converter at the input voltages of 20 V (solid line) and 50 V (dashed line).

operation at the low input voltage. Even if the crossover frequency has doubled due to the increase of the input voltage, the load-transient response would not improve because of the output impedances (i.e., the phase margin is still low causing peaking in the sensitivity function, the resonant frequency has moved and the damping has reduced).

9.6.4
Control Design Example: PCMC Boost Converter

The boost converter under PCM control is shown in Figure 9.44 and the associated circuit elements are defined in Figure 9.24b. The equivalent inductor-current sensing resistor (R_s) is 150 mΩ. The dynamic modeling and characterization of the PCMC boost converter is previously presented in Chapter 4. G_{co} of the PCMC boost converter contains the same RHP zero as the corresponding VMC boost converter, which automatically limits the achievable control bandwidth to the frequency of the RHP zero or below it. The goal of the control design is to maximize the crossover frequency with a satisfactory phase margin.

The resonant-free nature of the converter allows the use of PI controller. The PWM modulator used in the experimental converter is UCC 3842, where the error-amplifier section contains an extra gain of 1/3.

The output-voltage loop gain ($L(s)$) can be given by

$$L(s) = \frac{1}{3R_s} \cdot G_{cc} \cdot G_{co} \qquad (9.37)$$

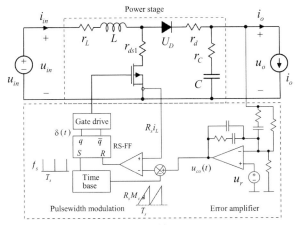

Figure 9.44 The PCM-controlled boost converter.

similarly to the corresponding PCM-controlled buck converter. It is obvious that R_s (150 mΩ) and G_{co} (Chapter 4) are known, and we have to choose G_{cc} such that the design goals are met. Therefore, the loop gain (9.37) can be given by

$$L(s) = \frac{1}{3 \cdot 0.15} \cdot G_{cc}$$

$$\times \frac{\frac{F_m^{sp}(D'U_E - I_L(r_E + D'^2 r_C))}{LC}\left(1 - s \cdot \frac{LI_L}{D'U_E - I_L(r_E + D'^2 r_C)}\right)(1 + s \cdot r_C C)}{\Delta}$$

$$\Delta = s^2 + s \cdot \left(\frac{r_E + D'^2 r_C + F_m^{sp} U_E q_c^{sp}}{L} - \frac{F_m^{sp} I_L q_o^{sp}}{C}\right) +$$

$$\frac{D'^2 + D' F_m^{sp}(U_E q_o^{sp} + I_L q_c^{sp}) - F_m^{sp} I_L q_o^{sp}(r_E + D'^2 r_C)}{LC}$$

$$U_E = U_o + U_D + (r_d + r_C - r_{ds1})I_L - r_C I_o$$

$$r_E = r_L + D r_{ds1} + D'(r_d + D r_C) \tag{9.38}$$

The converter is supposed to be compensated such that $M_c \approx U_o/2L$ for providing the duty ratio of 100% (see Chapter 4, Section 4.5.2). The design of the controller shall be carried out at the minimum input voltage and maximum output power similarly to the corresponding VMC converter. Therefore, we plot the known part of the loop gain as shown in Figure 9.45 at the input voltage of 20 V. The next task is to shape the phase of the loop gain, which can be done similarly to the corresponding VMC converter (i.e., $\omega_{z1} < 2\pi f_n$, $\omega_{p1} < 0.5 \cdot 2\pi f_s$). The resulting frequency responses of the controller ($K = 1$) with the pole-zero placements defined in (9.35)

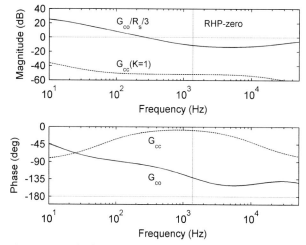

Figure 9.45 The first phase of the control design: PCMC boost converter.

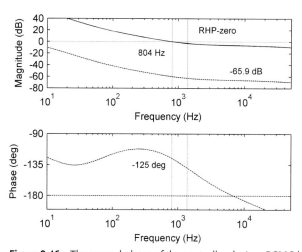

Figure 9.46 The second phase of the controller design: PCMC boost converter.

(dashed line) and the corresponding loop gain (solid line) are shown in Figures 9.45 and 9.46, respectively.

$$\omega_{z1} = 0.5 \cdot \frac{D'}{\sqrt{LC}}$$
$$\omega_{p2} = \frac{2\pi f_s}{8}$$
(9.39)

The high-frequency shape of the loop gain ($K = 1$) in Figure 9.46 indicates that the limiting factor in the design is maintaining of the adequate gain margin.

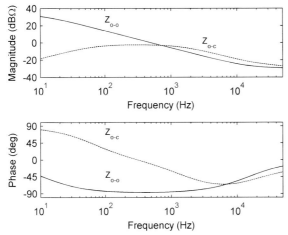

Figure 9.47 The open (solid line) and closed-loop (dashed line) output impedances of the PCMC boost converter at the input voltage of 20 V.

Therefore, the controller gain can be increased only by 65.9 dB − GM. If GM of 6 dB is adequate then the maximum gain increase is 59.9 dB corresponding to $K = 988$, the crossover frequency equals 804 Hz, and the phase margin equals 55°. It may be obvious that the PCM control does not remove the effect of the RHP zero as claimed in [47].

The open- and closed-loop output impedances of the PCMC boost converter at the input voltage of 20 V are shown in Figure 9.47. The closed-loop output impedance implies that the load-transient response would

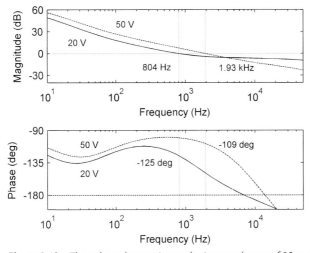

Figure 9.48 The voltage-loop gains at the input voltages of 20 and 50 V.

be quite similar to the corresponding VMC boost converter discussed earlier.

The predicted voltage-loop gains at the input voltages of 20 V (solid line) and 50 V (dashed line) are shown in Figure 9.48. The shape of the loop gains implies that the load-transient behavior would be slightly improved when the input voltage increases compared to the low-line response.

9.7
Conclusions

Basically the control design is trivial when knowing the behavior of the dynamic elements contributing to the loop gain and having an ability to interpret the information incorporated into the frequency responses. In practice, the control design is difficult because of the uncertainties incorporated into different passive and active circuit elements. The possible effect of the uncertainties can be naturally studied effectively only if the analytical models in the symbolic form are available. The most accurate control design would be achieved if the control-to-output transfer function is experimentally measured at the operating point most crucial to the design but naturally the other uncertainties will stay.

As high loop gain as possible before the crossover frequency and as low loop gain as possible after the crossover frequency would provide the best condition the control design can ever provide. The single-op-amp-based controller cannot provide such features but several cascaded controllers or digital control has to be used for obtaining such a goal.

The simple shunt-regulator-based controllers frequently utilized especially in the low-cost and high-volume applications deserve much greater concern than typically given in order to achieve high product quality.

An excellent control design would not necessarily yield robust stability and excellent transient response, because the external interactions can change the dynamics of the converter profoundly. The peak-current-mode control and input-current-feedforward in a buck-type converter would make the converter quite invariant to those external interactions and also improve substantially the input-voltage and load transient responses.

References

1. A. Lidow, D. Kinzer, G. Sheridan, and D. Tam, 'The semiconductor roadmap for power management in the new millennium,' *Proc. IEEE*, vol. 89, no. 6, **2001**, pp. 803–812.
2. A. Lidow and G. Sheridan, 'Defining future for microprocessor power delivery,' in P *roc. IEEE Applied Power Electronics Conf.*, **2003**, pp. 3–9.
3. R. Miftakhutdinov, 'An analytical comparison of alternative control techniques for powering next generation microprocessors,' in *Proc. Power Supply Design Seminar (SEM*

1400), Texas Instruments, Inc., Dallas, TX, USA, **2001**, pp. 1–1–1–39.
4. H. Huang, 'Coordination of design issues in the intermediate bus architecture,' in *Proc. IEEE Applied Power Electronics Conf.*, **2005**, pp. 169–175.
5. T. Suntio and A. Glad, 'The batteries as a principal component in DC UPS systems,' in *Proc. IEEE Int. Telecommunications Energy Conf.*, **1990**, pp. 400–411.
6. T. Suntio, I. Gadoura, J. Lempinen, and K. Zenger, 'Practical design issues of multiloop controller for a telecom rectifier,' in *Proc. Telecommunications Energy Special Conf.*, **2000**, pp. 197–201.
7. D. Venable, 'Testing power supplies for stability,' Technical Report 1, Venable Industries, Inc., Austin, TX, USA (online: www.venable.biz).
8. D. Venable, 'Optimum feedback amplifier design for control systems,' Technical Report 3, Venable Industries, Inc., Austin, TX, USA, (online: www.venable.biz).
9. M.E. Jacobs, 'Optimal feedback control of switch-mode power converters,' in *Proc. IEEE International Telecommunications Energy Conf.*, **2006**, pp. 508–517.
10. E. Figueres, G. Garcera, J.M. Benavent, M. Pascal, and J.A. Martinez, 'Adaptive two-loop voltage-mode control of DC–DC switching converters,' *IEEE Trans. Indust. Electron.*, vol. 53, no. 1, **2006**, pp. 239–253.
11. L. Dixon, 'Switching power supply control loop design,' in *Proc. Power Supply Design Seminar (SEM-1000)*, Unitrode Corp, Merrimack, NH, USA, **1994**, pp. C1–1–C1–10.
12. L. Dixon, 'Control loop cookbook,' in *Proc. Power Supply Design Seminar (SEM-1100)*, Unitrode Corp, Merrimack, NH, USA, **1996**, pp. 5–1–5–26.
13. D. Mitchell and L. Dixon, 'Designing stable control loops,' in *Proc. Power Supply Design Seminar*, Texas Instruments, Inc., Dallas, TX, USA, **2001**, pp. 5–1–5–30.
14. J. Sun, 'Control design considerations for voltage regulator modules,' in *Proc. IEEE International Telecommunications Energy Conf.*, **2003**, pp. 84–91.
15. W.H. Tuttle, 'Relating converter transient response characteristics to feedback loop control design,' in *Proc. 11th National Power Conversion Conf.*, **1984**, pp. 10–1–10–12.
16. J.C. Basio and S.R. Matos, 'Design of PI and PID controllers with transient performance specifications,' *IEEE Trans. Edu.*, vol. 45, no. 4, **2002**, pp. 364–370.
17. K. Ogata, *Modern Control Engineering*, Prentice-Hall, Upper Saddle River, NJ, USA, **1997**.
18. R. Ridley, 'Loop gain crossover frequency,' Switching Power Magazine, **2006** (on-line: www.switchingpowermagazine.com).
19. S. Banerjee and G.C. Verghese, *Nonlinear Phenomena in Power Electronics*, IEEE Press, New York, USA, **2001**.
20. R. Miftakhutdinov, 'Compensating DC/DC converters with ceramic output capacitors,' in *Proc. Power Supply Design Seminar (SEM 1600)*, Texas Instruments, Dallas, TX, USA, **2004**, pp. 7–1–7–15.
21. T. Roinila, M. Hankaniemi, T. Suntio, M. Sippola, and M. Vilkko, 'Dynamical profile of a switched-mode converter – Reality or imagination,' in *Proc. IEEE International Telecommunications Energy Conf.*, **2007**, pp. 420–427.
22. T. Suntio, M. Hankaniemi, and M. Karppanen, 'Analysing the dynamics of regulated converters,' *IEE Proc. Electr. Power Appl.*, vol. 153, no. 6, **2006**, pp. 905–910.
23. M. Karppanen, M. Hankaniemi, T. Suntio, and M. Sippola, 'Dynamical characterization of peak-current-mode controlled buck converter with output-current feedforward,' *IEEE Trans. Power Electron.*, vol. 22, no. 2, **2007**, pp. 444–451.
24. M. Hankaniemi, T. Suntio, and M. Sippola, 'Characterization of regulated converters to ensure stability and performance,' in *Proc. IEEE*

International Telecommunications Energy Conf., **2005**, pp. 533–538.

25. M. Hankaniemi, M. Karppanen, and T. Suntio, 'Load-imposed instability and performance degradation in a regulated converter,' *IEE Proc. Electr. Power Appl.*, vol. 153, no. 6, **2006**, pp. 781–786.

26. M. Karppanen, M. Sippola, and T. Suntio, 'Source-imposed instability and performance degradation in a regulated converter,' in *Proc. IEEE Power Electronics. Specialists Conf.*, **2007**, pp. 194–200.

27. M. Karppanen, T. Suntio, and M. Sippola, 'Impact of output-voltage remote sensing on converter dynamics,' *Int. Rev. Electr. Eng.*, vol. 2, no. 2, **2007**, pp. 196–202.

28. C. DeVries, 'A novel technique for compensating POL converters with variable output capacitance,' in *Proc. IEEE Applied Power Electronics Conf.*, **2006**, pp. 474–479.

29. B. Choi, 'Step load response of a current-mode-controlled DC-to-DC converter,' *IEEE Trans. Aerosp. Electron. Syst.*, vol. 33, no. 4, **1997**, pp. 1115–1121.

30. C. Gezgin, 'Predicting load transient response of output voltage in DC–DC converters,' in *Proc. IEEE Applied Power Electronics Conf.*, **2004**, pp. 1339–1343.

31. D. Morrison, 'Modeling DC–DC converter transient response,' *Power Electronics Technology*, **2004**, p. 56 (online: www.powerelectronics.com).

32. J. Betten and R. Kollman, 'Easy calculation yields load transient response,' *Power Electronics Technology*, 2005, pp. 40–48 (online: www.powerelctronics.com).

33. R. Redl, B.P. Erisman, and Z. Zansky, 'Optimizing the load transient response of the buck converter,' in *Proc. IEEE Applied Power Electronics Conf.*, **1998**, pp. 170–176.

34. R. Redl and N.O. Sokal, 'Near-optimum dynamic performance of the buck converter using feed-forward of output current and input voltage with current-mode control,' *IEEE Trans. Power Electron.*, vol. PE-1, no. 3, **1986**, pp. 186–191.

35. M. Karppanen, T. Suntio and M. Sippola, 'Dynamical characterization of input-voltage-feedforward-controlled buck converter,' *IEEE Trans. Indust. Electron.*, vol. 54, no. 2, **2007**, pp. 1005–1013.

36. K.J. Åström and T. Hägglund, *Advanced PID Control*, Instrumentation, Systems, and Automation Society (ISA), Research Triangle Park, NC, USA, **2006**.

37. Y. Li, K.H. Ang and G.C.Y. Chong, 'PID control system analysis and design,' *IEEE Control Syst. Mag.*, vol. 26, no. 1, **2006**, pp. 32–41.

38. H. Kwakernaak, 'Robust control and H_∞-optimization – Tutorial paper,' *Automatica*, vol. 29, no. 2, **1993**, pp. 255–273.

39. K. Zhou and J.C. Doyle, *Essentials of Robust Control*, Prentice-Hall, Upper Saddle River, NJ, USA, **1998**.

40. I. Gadoura and T. Suntio, 'Practical robust control of paralleled DC/DC converters using H_∞ loop-shaping technique,' in *Proc. Power Conversion and Intelligent Motion Conf.* (Europe), **2002**, pp. 335–340.

41. J.M. Maciejowski, *Multivariable Feedback Design*, Addison-Wesley, Reading, MA, USA, **1989**.

42. F.H. Leung, P.K.S. Tam, and C.K. Li, 'The control of switching dc–dc converters – A general LQR problem,' *IEEE Trans. Indust. Electron.*, vol. 38, no. 1, **1991**, pp. 65–71.

43. G. Escobar. R. Ortega, H. Sira-Ramirez, J.-P. Villain, and I. Zein, 'An experimental comparison of several nonlinear controllers for power converters,' *IEEE Control Syst. Mag.*, vol. 19, no. 1, **1999**, pp. 66–82.

44. P. Mattavelli, L. Rossetto, and G. Spiazzi, 'Small-signal analysis of DC–DC converters with sliding mode control,' *IEEE Trans. Power Electron.*, vol. 12, no. 1, **1997**, pp. 96–102.

45. H. Al-Atrash and I. Batarseh, 'Digital controller design for a practicing power electronics engineer,' in *Proc.*

IEEE Applied Power Electronics Conf., **2007**, pp. 34–41.
46. J.S. Freudenberg and D.P. Looze, 'Right half plane poles and zeros and design tradeoffs in feedback systems,' *IEEE Trans. Autom. Cont.*, vol. AC-30, no. 6, **1985**, pp. 555–565.
47. D.M. Mitchell, 'Tricks of the Trade: Understanding the right-half-plane zero in small-signal DC-DC converter models,' *IEEE Power Electronics Society Newsletter*, **2001**, pp. 5–6.
48. K.D.T. Ngo, S. Kirachaiwanich, and M. Walters, 'Buck modulator with improved large-power bandwidth,' *IEEE Trans. Aerosp. Electron. Syst.*, vol. 38, no. 4, **2002**, pp. 1335–1343.
49. K. Yao, Y. Meng, and F.C. Lee, 'Control bandwidth and transient response of buck converters,' in *Proc. IEEE Power Electronics Specialists Conf.*, **2002**, pp. 137–142.
50. Y. Qiu, J. Sun, M. Xu, K. Lee, and F.C. Lee, 'High-bandwidth designs for voltage regulators with peak-current control,' in *Proc. IEEE Applied Power Electronics Conf.*, **2006**, pp. 24–30.
51. S.A. Chickamenahalli, S. Mahadevan, E. Standford, and K. Merley, 'Effect of target impedance and control loop design on VRM stability,' in *Proc. IEEE Applied Power Electronics Conf.*, **2002**, pp. 196–202.
52. K. Yao, M. Xu, Y. Meng, and F.C. Lee, 'Design considerations for VRM transient response based on the output impedance,' *IEEE Trans. Power Electron.*, vol. 18, no. 6, **2003**, pp. 1270–1277.
53. R. Mammano, 'Isolating the control loop,' in *Proc. Power Supply Design Seminar (SEM-1000)*, Unitrode Corp, Merrimack, NH, USA, **1994**, pp. C2–1–C2–15.
54. Y. Panov and M.M. Jovanovic, 'Small-signal analysis and control design of isolated power supplies with optocoupler feedback,' *IEEE Trans. Power Electron.*, vol. 20, no. 4, **2005**, pp. 823–832.
55. J. Lempinen and T. Suntio, 'Small-signal modeling for design of robust variable-frequency flyback battery charger,' in *Proc. IEEE Applied Power Electronics Conf.*, **2001**, pp. 548–554.
56. Texas Instruments, Inc., Dallas (**2003**, February.), Precision adjustable shunt regulator, focus.ti.com/lit/ds/symlink/tl431a.pdf.
57. D. Venable, 'Testing and stabilizing feedback loops in today's power supplies,' Technical Report 17, Venable Industries, Inc. Austin, TX, USA, (online: www.venable.biz).
58. R. Kollman and J. Betten, 'Closing the loop with a popular shunt regulator,' *Power Electronics Technology*, **2003**, pp. 30–36 (online: www.powerelctronics.com).
59. T. Tepsa and T. Suntio, 'Adjustable shunt regulator based control systems,' *IEEE Power Electron. Lett.*, vol. 1, no. 4, **2003**, pp. 93–96.
60. R. Ridley, 'Designing with the TL431,' *Switching Power Magazine*, vol. 5, no. 2, **2004**, pp. 20–26 (online: www.switchingpowermagazine.com).
61. C. Basso, 'Biasing the TL431 for improved output impedance,' *Power Electronics Technology*, **2005**, pp. 56–57 (online: www.powerelectronics.com).
62. B.T. Irving and M.M. Jovanovic, 'Analysis and design of self-oscillating flyback converter,' in *Proc. IEEE Applied Power Electronics Conf.*, **2002**, pp. 897–903.
63. F. Tonicello, 'The control problem of maximum point power tracking in power systems,' in *Proc. 7th European Space Power Conf.*, **2005**, CD-ROM publication, pp. 7.

10
The Fourth-Order Converter – Superbuck

10.1
Introduction

The second-order converters – buck, boost, and buck–boost – treated in Chapters 3–6 form the basis for implementing majority of the converters used in the practical applications [1]. In some applications such as power-factor correction [2, 3] and interfacing of solar arrays [4–8], a highly desired feature of the converter is the continuous input current. The continuous current means also reduced EMI noise and current stresses in the associated capacitors. Only the boost converter of the basic converters can offer such a feature but its other properties may not be always proper for the intended application. The continuous input current exists only in the converters having an inductor placed at the input current path. This means the necessity to use higher order converters such as introduced in [9].

The Cuk converter [10] (Figure 10.1a) invented in the late 1970s is one of the best known higher order converters along with the SEPIC converter (Figure 10.1b) introduced in [1]. From Figure 10.1, the input current in both of the converters is continuous due to the placement of L_1 but the output capacitor is supplied with continuous current only in the Cuk converter due to the placement of L_2. The inverse polarity of the output voltage in the Cuk converter makes its direct use limited and questionable as an optimal converter.

Several other topological alternatives to implement the desired feature are available as described in [9] of which the buck-like converter known as step-down, current-sourced buck, two-inductor buck or superbuck has become a popular choice especially in the solar energy applications [4, 5, 11] (Figure 10.2). Its static and dynamic features greatly resemble those of the conventional buck converter introduced earlier in Chapters 3 and 4. It may, however, have certain dynamic anomalies such as RHP zeros and poles not present in the conventional converter. The appearance of those anomalies would greatly affect the controller design and obtainable transient dynamics [16, 17], but they can be controlled by means of the power-stage design.

Dynamic Profile of Switched-Mode Converter. Teuvo Suntio
© 2009 WILEY-VCH Verlag GmbH & Co. KGaA, Weinheim
ISBN: 978-3-527-40708-8

10 The Fourth-Order Converter – Superbuck

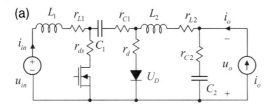

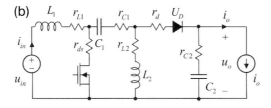

Figure 10.1 Fourth-order converters known as (a) Cuk and (b) SEPIC.

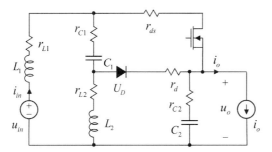

Figure 10.2 Superbuck converter.

The input-current ripple can be further reduced by using coupled-inductor design [12–15] as demonstrated in [2–4]. The coupling of inductors may further complicate the dynamics of the converter by either introducing anomalies such as RHP poles and zeros or advancing their appearance. Usually the simultaneous appearance of RHP poles and zeros makes the converter inapplicable as demonstrated in [18] (see Chapter 9, Section 9.2.2, Figure 9.6).

Actually the superbuck converter incorporates features making it a good candidate for an optimal converter.

The rest of the chapter is dedicated for the modeling and analysis of the superbuck converter as a representative of the fourth-order converters, as an example of the anomalies they can introduce, for demonstrating the applicability of the methods introduced in the previous chapters, and for illustrating the extent of analysis needed for understanding fully the converter dynamics. Experimental evidence is provided from two different PCM-controlled superbuck designs. The study of the existence of the RHP poles requires the use of *Routh–Hurwitz* analysis methods [19], which are shortly introduce. The methods for analyzing the dynamics

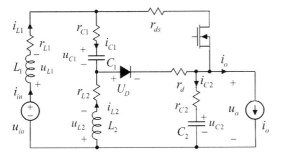

Figure 10.3 Superbuck converter with the direction of currents and the polarity of voltages.

related to the coupled-inductor design [20] are also shortly introduced and applied to the superbuck converter under VM and PCM control. Design considerations are provided for avoiding the appearance of the RHP zeros and poles and designing the optimal compensation for the PCM control.

10.2
Basic Dynamics

The basic dynamics of a certain converter is related to the direct-duty ratio or voltage-mode control as stated in Chapter 2, which has to be modeled first: The direction of the currents and polarity of the voltages constituting the state variables in a superbuck converter are shown in Figure 10.3. During the on-time, the main switch (MOSFET) is controlled on. As a consequence, the circuit structure of the converter becomes as shown in Figure 10.4a. During the off-time, the main switch is turned off and the diode conducts. As a consequence, the circuit structure of the converter becomes as shown in Figure 10.4b. From Figure 10.4, we can construct the state-space representations for the on- and off-times from which the average and small-signal state spaces and corresponding models can be solved by applying the methods introduced in Chapter 3. During the on-time the state space can be given by

$$\frac{di_{L1}}{dt} = -\frac{r_{L1} + r_{ds} + r_{C2}}{L_1} i_{L1} - \frac{r_{ds} + r_{C2}}{L_1} i_{L2} - \frac{u_{C2}}{L_1} + \frac{u_{in}}{L_1} + \frac{r_{C2}}{L_1} i_o$$

$$\frac{di_{L2}}{dt} = -\frac{r_{ds} + r_{C2}}{L_2} i_{L1} - \frac{r_{L2} + r_{ds} + r_{C1} + r_{C2}}{L_2} i_{L2} + \frac{u_{C1}}{L_2} - \frac{u_{C2}}{L_2} + \frac{r_{C2}}{L_2} i_o$$

$$\frac{du_{C1}}{dt} = -\frac{i_{L2}}{C_1}$$

$$\frac{du_{C2}}{dt} = \frac{i_{L1}}{C_2} + \frac{i_2}{C_2} - \frac{i_o}{C_2}$$

$$i_{in} = i_{L1}$$
$$u_o = r_{C2}i_{L1} + r_{C2}i_{L2} + u_{C2} - r_{C2}i_o \tag{10.1}$$

and during the off-time by

$$\frac{di_{L1}}{dt} = -\frac{r_{L1} + r_{C1} + r_d + r_{C2}}{L_1}i_{L1} - \frac{r_d + r_{C2}}{L_1}i_{L2} - \frac{u_{C1}}{L_1}$$
$$- \frac{u_{C2}}{L_1} + \frac{u_{in}}{L_1} + \frac{r_{C2}}{L_1}i_o - \frac{U_D}{L_1}$$
$$\frac{di_{L2}}{dt} = -\frac{r_d + r_{C2}}{L_2}i_{L1} - \frac{r_{L2} + r_d + r_{C2}}{L_2}i_{L2} - \frac{u_{C2}}{L_2} + \frac{r_{C2}}{L_2}i_o - \frac{U_D}{L_2}$$
$$\frac{du_{C1}}{dt} = \frac{i_{L1}}{C_1} \tag{10.2}$$
$$\frac{du_{C2}}{dt} = \frac{i_{L1}}{C_2} + \frac{i_{L2}}{C_2} - \frac{i_o}{C_2}$$
$$i_{in} = i_{L1}$$
$$u_o = r_{C2}i_{L1} + r_{C2}i_{L2} + u_{C2} - r_{C2}i_o$$

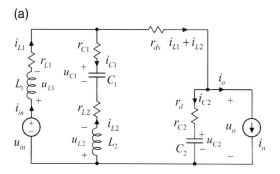

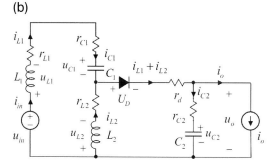

Figure 10.4 Superbuck structures during the (a) on time and (b) off time.

10.2.1
Averaged Models

10.2.1.1 Averaged State Space

The averaged state space can be obtained by multiplying (10.1) with d and (10.2) with d' and summing them together. These procedures yield

$$\frac{d\langle i_{L1}\rangle}{dt} = -\frac{r_{L1} + r_{C2} + dr_{ds} + d'r_d + d'r_{C1}}{L_1} \cdot \langle i_{L1}\rangle - \frac{r_{C2} + dr_{ds} + d'r_d}{L_1} \cdot \langle i_{L2}\rangle$$

$$- \frac{d'\langle u_{C1}\rangle}{L_1} - \frac{\langle u_{C2}\rangle}{L_1} + \frac{\langle u_{in}\rangle}{L_1} + \frac{r_{C2}}{L_1} \cdot \langle i_o\rangle - \frac{d'U_D}{L_1}$$

$$\frac{d\langle i_{L2}\rangle}{dt} = -\frac{r_{C2} + dr_{ds} + d'r_d}{L_2}\langle i_{L1}\rangle - \frac{r_{L2} + r_{C2} + dr_{ds} + dr_{C1} + d'r_d}{L_2}\langle i_{L2}\rangle$$

$$+ \frac{d\langle u_{C1}\rangle}{L_2} - \frac{\langle u_{C2}\rangle}{L_2} + \frac{r_{C2}}{L_2}\langle i_o\rangle - \frac{d'U_D}{L_2} \qquad (10.3)$$

$$\frac{d\langle u_{C1}\rangle}{dt} = \frac{d'\langle i_{L1}\rangle}{C_1} - \frac{d\langle i_{L2}\rangle}{C_1}$$

$$\frac{d\langle u_{C2}\rangle}{dt} = \frac{\langle i_{L1}\rangle}{C_2} + \frac{\langle i_{L2}\rangle}{C_2} - \frac{\langle i_o\rangle}{C_2}$$

$$\langle i_{in}\rangle = \langle i_{L1}\rangle$$

$$\langle u_o\rangle = r_{C2}\langle i_{L1}\rangle + r_{C2}\langle i_{L2}\rangle + \langle u_{C2}\rangle - r_{C2}\langle i_o\rangle$$

10.2.1.2 Steady-State Operating Point

The steady-state operating point can be solved from the averaged state space (10.3) by setting the derivatives to zero. These procedures yield

$$U_o = DU_{in} - D'U_D - \left(Dr_{ds} + DD'r_{C1} + D'r_d + D^2 r_{L1} + D'^2 r_{L2}\right) I_o$$

$$U_{C2} = U_o$$

$$U_{C1} = U_{in} - D'U_D - \left(Dr_{L1} - D'R_{L2}\right) I_o \qquad (10.4)$$

$$I_{L1} = DI_o$$

$$I_{L2} = D'I_o$$

$$I_{in} = DI_o$$

The duty ratio as a function of the input and output voltages, the load current, and the parasitic elements can be solved from

$$D^2 - \frac{U_{in} + U_D + (r_{C1} + r_{ds} - r_d - 2r_{L2})I_o}{(r_{L1} + r_{L2} - r_{C1})I_o} \cdot D + \frac{U_o + U_D + (r_{L2} + r_d)I_o}{(r_{L1} + r_{L2} - r_{C1})I_o} = 0$$

$$(10.5)$$

10.2.1.3 Boundary Conduction Mode

The peak-to-peak ripple currents of L_1 and L_2 can be computed according to the on-time derivatives of the corresponding currents from (10.1) at a certain operating point (10.4) to be as follows:

$$\Delta i_{L1-pp} = \frac{DT_s \left(U_{in} - U_o - (Dr_{L1} + r_{ds}) I_o\right)}{L_1}$$

$$\Delta i_{L2-pp} = \frac{DT_s \left(U_{in} - U_o - (Dr_{L1} + D'r_{C1}) I_o\right)}{L_2} \quad (10.6)$$

The mode boundary between the CCM and DCM conduction modes would take place [4], when the output current equals the half the sum of the inductor-current peak-to-peak ripples (10.6) or $(\Delta i_{L1-pp} + \Delta i_{L2-pp})/2 = I_o$, because the sum of the inductor currents is supplied to the output capacitor (Figure 10.4b) via a diode during the off-time. The critical K-value based on the sum of inductor currents yields $K_{crit} = D'$, which is the same as defined for the conventional buck converter in [1]. The K-value for analyzing the operation mode (i.e., CCM: $K > K_{crit}$, DCM: $K < K_{crit}$, and BCM: $K = K_{crit}$) can be given by $2L_p/T_s R_{eq}$, where $L_p = L_1 L_2/(L_1 + L_2)$ and $R_{eq} = U_o/I_o$.

10.2.2
Small-Signal Models

10.2.2.1 Small-Signal State Space

The small-signal state space can be obtained from (10.3) by developing the proper partial derivatives as introduced in Chapter 2. This procedure yields when the steady-state operating point (10.4) is also taken into account:

$$\frac{d\hat{i}_{L1}}{dt} = -\frac{R_1}{L_1}\hat{i}_{L1} - \frac{R_2}{L_1}\hat{i}_{L2} - \frac{D'}{L_1}\hat{u}_{C1} - \frac{1}{L_1}\hat{u}_{C2} + \frac{1}{L_1}\hat{u}_{in} + \frac{r_{C2}}{L_1}\hat{i}_o + \frac{U_1}{L_1}\hat{d}$$

$$\frac{d\hat{i}_{L2}}{dt} = -\frac{R_2}{L_2}\hat{i}_{L1} - \frac{R_3}{L_2}\hat{i}_{L2} + \frac{D\hat{u}_{C1}}{L_2} - \frac{\hat{u}_{C2}}{L_2} + \frac{r_{C2}}{L_2}\hat{i}_o + \frac{U_2}{L_2}\hat{d}$$

$$\frac{d\hat{u}_{C1}}{dt} = \frac{D'}{C_1}\hat{i}_{L1} - \frac{D}{C_1}\hat{i}_{L2} - \frac{I_o}{C_1}\hat{d} \quad (10.7)$$

$$\frac{d\hat{u}_{C2}}{dt} = \frac{1}{C_2}\hat{i}_{L1} + \frac{1}{C_2}\hat{i}_{L2} - \frac{1}{C_2}\hat{i}_o$$

$$\hat{i}_{in} = \hat{i}_{L1}$$

$$\hat{u}_o = \hat{u}_{C2} + r_{C2}C_2\frac{d\hat{u}_{C2}}{dt}$$

which can also be given in a matrix form as

$$\begin{bmatrix} \dfrac{d\hat{i}_{L1}}{dt} \\ \dfrac{d\hat{i}_{L2}}{dt} \\ \dfrac{d\hat{u}_{C1}}{dt} \\ \dfrac{d\hat{u}_{C2}}{dt} \end{bmatrix} = \begin{bmatrix} -\dfrac{R_1}{L_1} & -\dfrac{R_2}{L_1} & -\dfrac{D'}{L_1} & -\dfrac{1}{L_1} \\ -\dfrac{R_2}{L_2} & -\dfrac{R_3}{L_2} & \dfrac{D}{L_2} & -\dfrac{1}{L_2} \\ \dfrac{D'}{C_1} & -\dfrac{D}{C_1} & 0 & 0 \\ \dfrac{1}{C_2} & \dfrac{1}{C_2} & 0 & 0 \end{bmatrix} \begin{bmatrix} \hat{i}_{L1} \\ \hat{i}_{L2} \\ \hat{u}_{C1} \\ \hat{u}_{C2} \end{bmatrix} + \begin{bmatrix} \dfrac{1}{L_1} & \dfrac{r_{C2}}{L_1} & \dfrac{U_1}{L_1} \\ 0 & \dfrac{r_{C2}}{L_2} & \dfrac{U_2}{L_2} \\ 0 & 0 & -\dfrac{I_o}{C_1} \\ 0 & -\dfrac{1}{C_2} & 0 \end{bmatrix} \begin{bmatrix} \hat{u}_{in} \\ \hat{i}_o \\ \hat{d} \end{bmatrix}$$

$$\begin{bmatrix} \hat{i}_{in} \\ \hat{u}_o \end{bmatrix} = \begin{bmatrix} 1 & 0 & 0 & 0 \\ 0 & 0 & 0 & 1+r_{C2}C_2\dfrac{d}{dt} \end{bmatrix} \begin{bmatrix} \hat{i}_{L1} \\ \hat{i}_{L2} \\ \hat{u}_{C1} \\ \hat{u}_{C2} \end{bmatrix} + \begin{bmatrix} 0 & 0 & 0 \\ 0 & 0 & 0 \end{bmatrix} \begin{bmatrix} \hat{u}_{in} \\ \hat{i}_o \\ \hat{d} \end{bmatrix}$$

(10.8)

where

$$R_1 = r_{L1} + r_{C2} + Dr_{ds} + D'(r_d + r_{C1})$$
$$R_2 = r_{C2} + Dr_{ds} + D'r_d$$
$$R_3 = r_{L2} + r_{C2} + D(r_{ds} + r_{C1}) + D'r_d$$
$$U_1 = U_{in} + U_D + (r_d - r_{ds} + Dr_{C1} - Dr_{L1} + D'r_{L2})I_o$$
$$U_2 = U_1 - r_{C1}I_o$$

10.2.2.2 Transfer Functions

The corresponding transfer functions (i.e., G-parameters) can be solved from (10.8) most conveniently by using proper software packages such as Matlab™ with Symbolic Toolbox yielding

Input dynamics:

$$\Delta Y_{in-o} = \dfrac{s}{L_1}\left(s^2 + s\dfrac{R_3}{L_2} + \dfrac{C_1 + D^2 C_2}{L_2 C_1 C_2}\right)$$

$$\Delta T_{oi-o} = \dfrac{1}{L_1 C_2}\left(s^2 + s\dfrac{R_3 - R_2}{L_2} + \dfrac{D}{L_2 C_1}\right)(1 + sr_{C2}C_2)$$

$$\Delta G_{ci} = \dfrac{U_1}{L_1}\Bigg(s^3 + s^2\dfrac{D'I_o L_2 + (R_3 U_1 - R_1 U_2)C_1}{U_1 L_2 C_1}$$
$$+ s\dfrac{(U_1 - U_2)C_1 + \left(D^2 U_1 + DD'U_2 + (DR_2 + D'R_3)I_o\right)C_2}{U_1 L_2 C_1 C_2}$$
$$+ \dfrac{I_o}{U_1 L_2 C_1 C_2}\Bigg)$$

(10.9)

Output dynamics:

$$\Delta G_{\text{io-o}} = \frac{1}{L_1 C_2}\left(s^2 + s\frac{R_3 - R_2}{L_2} + \frac{D}{L_2 C_1}\right)(1 + sr_{C2}C_2)$$

$$\Delta Z_{\text{o-o}} = \frac{1}{C_2}\left(s^3 + s^2\frac{(R_3 - r_{C2})L_1 + (R_1 - r_{C2})L_2}{L_1 L_2}\right.$$
$$+ s\frac{D^2 L_1 + D'^2 L_2 + (R_1 R_3 - R_2^2 - r_{C2}(R_1 - 2R_2 + R_3))C_1}{L_1 L_2 C_1}$$
$$\left.+ \frac{D^2 R_1 + 2DD' R_2 + D'^2 R_3 - r_{C2}}{L_1 L_2 C_1}\right)(1 + sr_{C2}C_2) \tag{10.10}$$

$$\Delta G_{\text{co}} = \frac{(U_1 L_2 + U_2 L_1)}{L_1 L_2 C_2}\left(s^2 + s\frac{C_1(U_1(R_3 - R_2) + U_2(R_1 - R_2)) + (D' L_2 - D L_1) I_o}{C_1(U_1 L_2 + U_2 L_1)}\right.$$
$$\left.+ \frac{DU_1 + D'U_2 - (D(R_1 - R_2) + D'(R_2 - R_3)) I_o}{C_1(U_1 L_2 + U_2 L_1)}\right)(1 + sr_{C2}C_2)$$

where the determinant Δ is

$$s^4 + s^3\frac{R_3 L_1 + R_1 L_2}{L_1 L_2} + s^2\frac{(L_1 + L_2)C_1 + (D^2 L_1 + D'^2 L_2)C_2 + (R_1 R_3 - R_2^2) C_1 C_2}{L_1 L_2 C_1 C_2}$$
$$+ s\frac{(R_1 - 2R_2 + R_3)C_1 + (D^2 R_1 + 2DD' R_2 + D'^2 R_3) C_2}{L_1 L_2 C_1 C_2} + \frac{1}{L_1 L_2 C_1 C_2} \tag{10.11}$$

The determinant has two complex conjugate or resonant LHP roots approximately at

$$f_1 \approx \frac{1}{2\pi}\sqrt{\frac{1}{(L_1 + L_2)C_1}}$$
$$f_2 \approx \frac{1}{2\pi}\sqrt{\frac{L_1 + L_2}{L_1 L_2 C_2}} \tag{10.12}$$

which are not exactly the real resonant frequencies but close enough and also consistent with the circuit schematics shown in Figure 10.2.

The ideal input impedance $Z_{\text{in}-\infty}$ can be computed as

$$Z_{\text{in}-\infty} = \frac{s^2 (U_1 L_2 + U_2 L_1) C_1 + sE_1 + E_2}{sU_2 C_1 - DI_o}$$
$$E_1 = U_1(R_3 - R_2) + U_2(R_1 - R_2)C_1 + I_o(D' L_2 - D L_1) \tag{10.13}$$
$$E_2 = DU_1 + D'U_2 + (D(R_2 - R_1) + D'(R_3 - R_2)) I_o$$

where the numerator term is the same as in the control-to-output transfer function (G_{co}) shown in (10.10). Its low-frequency value is approximately equal to $-\frac{U_{in}}{DI_o}$, which is the same as in the corresponding conventional buck converter but the resonant nature of $Z_{in-\infty}$ makes the superbuck converter more sensitive to source interactions. The resonant nature can be reduced providing parallel damping to C_1. The ideal impedance is symbolically the same and characteristic to the superbuck topology regardless of the control mode when the inductors have no coupling between them.

The short-circuit input impedance Z_{in-sc} can be computed as

$$Z_{in-sc} = \frac{s^3 L_1 L_2 C_1 + s^2 C_1 (L_1(R_3 - r_{C2}) + L_2(R_1 - r_{C2})) + s E_1 + E_2}{s^2 L_2 C_1 + s(R_3 - r_{C2})C_1 + D^2} \quad (10.14)$$

$$E_1 = D^2 L_1 + D'^2 L_2 + (R_1 R_3 - R_2^2 - (R_1 - 2R_2 + R_3)r_{C2})C_1$$

$$E_2 = D^2 R_1 + 2DD' R_2 + D'^2 R_3 - r_{C2}$$

The resonant nature of Z_{in-sc}, (i.e., a resonant zero pair approximately at $f \approx \frac{1}{2\pi}\sqrt{\frac{D^2 L_1 + D'^2 L_2}{L_1 L_2 C_1}}$ and a resonant pole pair at $f \approx \frac{D}{2\pi\sqrt{L_2 C_1}}$) makes the superbuck converter more sensitive to the source interactions than the conventional buck converter. It is obvious that the parallel damping (C_1) proposed above damps also the resonances in Z_{in-sc}, because the resonances are dependent on C_1.

10.2.3
RHP Poles

The determinant Δ (10.11) contains the poles of the converter, which may lie either in LHP or RHP. If an RHP pole exists, it means that the control bandwidth has to be designed to be higher than the location of the pole. The roots of the fourth-order polynomial cannot be solved easily in the symbolic form to verify their locations in the complex plane. Therefore, other methods such as *Routh–Hurwitz* test [19] to study the existence and number of the RHP roots of a polynomial have to be applied. According to the method, the *Routh–Hurwitz* array is to be constructed based on the coefficients of the polynomial under considerations

$$\Delta(s) = a_n \cdot s^n + a_{n-1} \cdot s^{n-1} + a_{n-2} \cdot s^{n-2} + \cdots + a_1 \cdot s + a_0 \quad (10.15)$$

as follows:

$$\begin{array}{c|cccc} s^n & a_n & a_{n-2} & a_{n-4} & \cdots \\ s^{n-1} & a_{n-1} & a_{n-3} & a_{n-5} & \cdots \\ s^{n-2} & b_1 & b_2 & b_3 & \cdots \end{array}$$

$$(10.16)$$

$$\begin{array}{c|cccc} s^{n-3} & : & c_1 & c_2 & c_3 & \cdots \\ s^{n-4} & : & d_1 & d_2 & d_3 & \cdots \\ \vdots & : & & & \\ s^0 & : & & & \end{array}$$

where the first row (s^n) starts with the highest-order coefficient a_n, the second row (s^{n-1}) with the second-highest-order coefficient a_{n-1}, and the next elements within the rows are as defined in (10.17), the elements of the third and subsequent rows follow the algorithm defined as

$$b_1 = \frac{-\begin{vmatrix} a_n & a_{n-2} \\ a_{n-1} & a_{n-3} \end{vmatrix}}{a_{n-1}} \quad b_2 = \frac{-\begin{vmatrix} a_n & a_{n-4} \\ a_{n-1} & a_{n-5} \end{vmatrix}}{a_{n-1}}$$

$$c_1 = \frac{-\begin{vmatrix} a_{n-1} & a_{n-3} \\ b_1 & b_2 \end{vmatrix}}{b_1} \quad c_2 = \frac{-\begin{vmatrix} a_{n-1} & a_{n-5} \\ b_1 & b_3 \end{vmatrix}}{b_1} \quad (10.17)$$

$$d_1 = \frac{-\begin{vmatrix} b_1 & b_2 \\ c_1 & c_2 \end{vmatrix}}{c_1} \quad d_2 = \frac{-\begin{vmatrix} b_1 & b_3 \\ c_1 & c_3 \end{vmatrix}}{c_1}$$

The number of RHP roots of $\Delta(s)$ is the number of algebraic sign changes in the elements of the left column of the array (10.16) proceeding from top to bottom. The first or second-order polynomial has all roots in LHP if and only if all the coefficients have the same algebraic sign. In the case of a higher order polynomial, the same does not anymore guarantee the absence of the RHP roots but the *Routh–Hurwitz* test has to be applied to confirm the situation. The sign changes in the polynomial coefficients indicate, however, the existence of at least one RHP root. The missing of one or several of the coefficients means that the polynomial has either complex imaginary-axis roots or RHP roots or both.

According to the principles laid down above, the *Routh–Hurwitz* array for the fourth-order polynomial of interest can be given by

$$\begin{array}{c|ccc} s^4 & : & a_4 & a_2 & a_0 \\ s^3 & : & a_3 & a_1 & 0 \\ s^2 & : & b_1 & b_2 & 0 \\ s^1 & : & c_1 & 0 & 0 \\ s^0 & : & d_1 & 0 & 0 \end{array} \quad (10.18)$$

where b_1, b_2, c_1, and d_1 are defined by

$$b_1 = a_2 - \frac{a_1 a_4}{a_3} \quad b_2 = a_0$$
$$c_1 = a_1 - \frac{a_0 a_3}{b_1} \quad d_1 = a_0 \quad (10.19)$$

According to the determinant (10.11), the polynomial coefficients are as follows:

$$a_4 = 1$$
$$a_3 = \frac{R_0(L_1 + L_2)}{L_1 L_2}$$
$$a_2 = \frac{(L_1 + L_2)C_1 + (D^2 L_1 + D'^2 L_2) C_2}{L_1 L_2 C_1 C_2} \qquad (10.20)$$
$$a_1 = \frac{R_0}{L_1 L_2 C_1}$$
$$a_0 = \frac{1}{L_1 L_2 C_1 C_2}$$

where the parasitic resistances R_1, R_2, and R_3 are assumed to be the same and denoted by R_0. (Note: in practice, R_1 and R_3 are always greater than R_2. Therefore, the differences associated to those parasitics in (10.11) are always positive numbers and the equalling of them would give the worst case condition.) All the polynomial coefficients are positive real numbers. The array coefficients b_1, b_2, c_1, and d_1 can be given by

$$b_1 = \frac{(L_1 + L_2)(C_1 + C_2)}{L_1 L_2 C_1 C_2}$$
$$b_2 = \frac{1}{L_1 L_2 C_1 C_2}$$
$$c_1 = \frac{R_0 C_2}{L_1 L_2 C_1 (C_1 + C_2)} \qquad (10.21)$$
$$d_1 = \frac{1}{L_1 L_2 C_1 C_2}$$

which are also positive real numbers. Therefore, all the roots are LHP roots.

10.2.4
Design Considerations

According to (10.10), the control-to-output transfer function (G_{co}) incorporates a resonant zero pair approximately at $f \approx \frac{1}{2\pi}\sqrt{\frac{1}{(L_1+L_2)C_1}}$, which will move to RHP, when

$$C_1 (U_1(R_3 - R_2) + U_2(R_1 - R_2)) + (D'L_2 - DL_1) I_o < 0 \qquad (10.22)$$

Its appearance is dependent on the input voltage and output current as well as on the sizing of L_1, L_2, and C_1 as implied by (10.22). Usually the inductors are designed to be the same for the logistic and cost reasons. The appearance of the RHP zero sets an absolute limit on the maximum control bandwidth and, therefore, the appearance may not be desirable. It may be obvious (10.22) that the appearance of the RHP zero pair can be eliminated if the inductors are designed such that $L_2 \approx \frac{D_{max}}{D'_{max}} \cdot L_1$, where D_{max} is the maximum duty ratio the converter may have during the operation. In the applications, where the input voltage is rather high and the output current rather low, the RHP zero pair may not appear even if the inductors are equal.

10.3
Coupled-Inductor Superbuck

The coupled-inductor design (Figure 10.5) is usually used in order to reduce the number of separate inductors and the input or output current ripple resulting also in the reduction of the overall costs. The application of the coupled-inductor technique may, however, change profoundly the internal dynamics of the converter but those effects are not very well known because of the complicity of the analysis and errors usually made in it as in [2] and [3].

The analysis of the coupled-inductor effects is based on the basic terminal equations of a nonideal transformer [20] given by

$$u_1 = L_1 \frac{di_1}{dt} + M \frac{di_2}{dt}$$
$$u_2 = M \frac{di_1}{dt} + L_2 \frac{di_2}{dt}$$
(10.23)

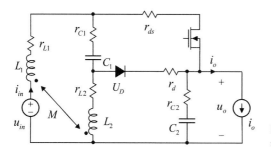

Figure 10.5 Coupled-inductor superbuck converter.

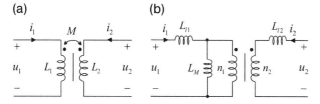

Figure 10.6 Definition of (a) a two-winding transformer and (b) its equivalent circuit.

where L_1 and L_2 are the self-inductances measured from the primary (u_1) and secondary (u_2) ports, respectively, when the other port is an open circuit, and M is the mutual inductance governing the transfer of energy between the primary and secondary as defined in Figure 10.6a. The equivalent circuit of such a transformer is shown in Figure 10.6b, where L_{l1} and L_{l2} are the primary and secondary leakage inductances, respectively, L_M is the primary magnetizing inductance, and n_1 and n_2 are the primary and secondary number of turns, respectively, of the ideal transformer.

From (10.23), we may define that

$$L_1 = L_{l1} + L_M$$
$$L_2 = L_{l2} + \left(\frac{n_2}{n_1}\right)^2 L_M \qquad (10.24)$$
$$M = \frac{n_2}{n_1} L_M$$

The coupling coefficient (k) of a transformer is usually defined as $k = \frac{M}{\sqrt{L_1 L_2}}$ based on the total energy stored in the transformer [20]. Perfect coupling between the primary and secondary (i.e., $k = 1$) means that the leakage inductances (L_{li}) are zero.

10.3.1
Small-Signal Models

The small-signal representation of a coupled-inductor converter can be constructed from the state space of the original converter defined with the separate inductors L_1 and L_2 taking into account that the voltages across the inductors do not change. Therefore, if we denote the original small-signal derivatives of the inductor currents by $\hat{D}_{L1}$ and $\hat{D}_{L2}$ and apply the information given in (10.23) then the couple-inductor-affected derivatives can be given by

$$\frac{d\hat{i}_{L1}}{dt} = \frac{\hat{D}_{L1} - \frac{M}{L_1}\hat{D}_{L2}}{1 - \frac{M^2}{L_1 L_2}} = \frac{\hat{D}_{L1} - \frac{M}{L_1} \cdot \hat{D}_{L2}}{1 - k^2}$$

$$\frac{d\hat{i}_{L2}}{dt} = \frac{\hat{D}_{L2} - \frac{M}{L_2}\hat{D}_{L1}}{1 - \frac{M^2}{L_1 L_2}} = \frac{\hat{D}_{L2} - \frac{M}{L_2} \cdot \hat{D}_{L1}}{1 - k^2} \qquad (10.25)$$

From (10.25), we may concluded that the perfect coupling ($k = 1$) would lead to infinite derivatives indicating the existence of high current spikes during the transients.

From (10.8) and (10.25), the small-signal coupled-inductor VMC state space can be given by

$$\begin{bmatrix} \dfrac{d\hat{i}_{L1}}{dt} \\ \dfrac{d\hat{i}_{L2}}{dt} \\ \dfrac{d\hat{u}_{C1}}{dt} \\ \dfrac{d\hat{u}_{C2}}{dt} \end{bmatrix} = \begin{bmatrix} -\dfrac{L_2 R_1 - MR_2}{L_1 L_2 - M^2} & -\dfrac{L_2 R_2 - MR_3}{L_1 L_2 - M^2} & -\dfrac{L_2 D' + MD}{L_1 L_2 - M^2} & -\dfrac{L_2 - M}{L_1 L_2 - M^2} \\ -\dfrac{L_1 R_2 - MR_1}{L_1 L_2 - M^2} & -\dfrac{L_1 R_3 - MR_2}{L_1 L_2 - M^2} & \dfrac{DL_1 + MD'}{L_1 L_2 - M^2} & -\dfrac{L_1 - M}{L_1 L_2 - M^2} \\ \dfrac{D'}{C_1} & -\dfrac{D}{C_1} & 0 & 0 \\ \dfrac{1}{C_2} & \dfrac{1}{C_2} & 0 & 0 \end{bmatrix} \begin{bmatrix} \hat{i}_{L1} \\ \hat{i}_{L2} \\ \hat{u}_{C1} \\ \hat{u}_{C2} \end{bmatrix}$$

$$+ \begin{bmatrix} \dfrac{L_2}{L_1 L_2 - M^2} & \dfrac{(L_2 - M)r_{C2}}{L_1 L_2 - M^2} & \dfrac{L_2 U_1 - MU_2}{L_1 L_2 - M^2} \\ -\dfrac{M}{L_1 L_2 - M^2} & \dfrac{(L_1 - M)r_{C2}}{L_1 L_2 - M^2} & \dfrac{L_1 U_2 - MU_1}{L_1 L_2 - M^2} \\ 0 & 0 & -\dfrac{I_o}{C_1} \\ 0 & -\dfrac{1}{C_2} & 0 \end{bmatrix} \begin{bmatrix} \hat{u}_{in} \\ \hat{i}_o \\ \hat{d} \end{bmatrix}$$

$$\begin{bmatrix} \hat{i}_{in} \\ \hat{u}_o \end{bmatrix} = \begin{bmatrix} 1 & 0 & 0 & 0 \\ 0 & 0 & 0 & 1 + r_{C2} C_2 \dfrac{d}{dt} \end{bmatrix} \begin{bmatrix} \hat{i}_{L1} \\ \hat{i}_{L2} \\ \hat{u}_{C1} \\ \hat{u}_{C2} \end{bmatrix} + \begin{bmatrix} 0 & 0 & 0 \\ 0 & 0 & 0 \end{bmatrix} \begin{bmatrix} \hat{u}_{in} \\ \hat{i}_o \\ \hat{d} \end{bmatrix} \qquad (10.26)$$

The transfer functions constituting the G-parameter set can be solved from (10.26) by using a proper software package yielding.

Input dynamics:

$$\Delta Y_{\text{in}-o} = \dfrac{sL_2}{L_1 L_2 - M^2}\left(s^2 + s\dfrac{R_3}{L_2} + \dfrac{C_1 + D^2 C_2}{L_2 C_1 C_2}\right)$$

$$\Delta T_{\text{oi}-o} = \dfrac{1}{(L_1 L_2 - M^2)C_2}\left(s^2 (L_2 - M) + s(R_3 - R_2) + \dfrac{D}{C_1}\right)(1 + sr_{C2} C_2)$$

$$\Delta G_{ci} = \dfrac{1}{L_1 L_2 - M^2}\left(s^3 (U_1 L_2 - MU_2) + s^2 \dfrac{(R_3 U_1 - R_1 U_2)C_1 + (D'L_2 + M)I_o}{C_1}\right.$$

$$\left. + s\dfrac{(U_1 - U_2)C_1 + (D^2 U_1 + DD'U_2 + (DR_2 + D'R_3)I_o)C_2}{C_1 C_2} + \dfrac{I_o}{C_1 C_2}\right)$$

(10.27)

Output dynamics:

$$\Delta G_{io-o} = \frac{1}{(L_1 L_2 - M^2)C_2}\left(s^2(L_2 - M) + s(R_3 - R_2) + \frac{D}{C_1}\right)(1 + sr_{C2}C_2)$$

$$\Delta Z_{o-o} = \frac{1}{C_2}\left(s^3 + s^2\frac{(R_3 - r_{C2})L_1 + (R_1 - r_{C2})L_2 - 2(R_2 - r_{C2})M}{L_1 L_2 - M^2}\right.$$

$$+ s\frac{(R_1 R_3 - R_2^2 - (R_1 - 2R_2 + R_3)r_{C2})C_1 + D^2 L_1 + 2DD'M + D'^2 L_2}{(L_1 L_2 - M^2)C_1}$$

$$\left.+ \frac{D^2 R_1 + 2DD'R_2 + D'^2 R_3 - r_{C2}}{(L_1 L_2 - M^2)C_1}\right)(1 + sr_{C2}C_2)$$

$$\Delta G_{co} = \left(s^2\frac{(L_2 - M)U_1 + (L_1 - M)U_2}{(L_1 L_2 - M^2)C_2} + s\frac{E_1 + (D'(L_2 - M) - D(L_1 - M))I_o}{(L_1 L_2 - M^2)C_1 C_2}\right.$$

$$\left.+ \frac{E_2}{(L_1 L_2 - M^2)C_1 C_2}\right)(1 + sr_{C2}C_2)$$

$$E_1 = ((R_3 - R_2)U_1 + (R_1 - R_2)U_2)C_1$$

$$E_2 = (DU_1 + D'U_2) - (D(R_1 - R_2) + D'(R_2 - R_3))I_o \quad (10.28)$$

where the determinant (Δ) is

$$\Delta = s^4 + s^3\frac{R_1 L_2 + R_3 L_1 - 2MR_2}{L_1 L_2 - M^2} + s^2 a_2 + s a_1 + \frac{1}{(L_1 L_2 - M^2)C_1 C_2}$$

$$a_2 = \frac{(R_1 R_3 - R_2^2)C_1 C_2 + (L_1 + L_2 - 2M)C_1 + (D^2 L_1 + D'^2 L_2 + 2DD'M)C_2}{(L_1 L_2 - M^2)C_1 C_2}$$

$$a_1 = \frac{(R_1 - 2R_2 + R_3)C_1 + (D^2 R_1 + 2DD'R_2 + D'^2 R_3)C_2}{(L_1 L_2 - M^2)C_1 C_2} \quad (10.29)$$

Thus the special input impedances ($Z_{\text{in}-\infty}$ and $Z_{\text{in-sc}}$) can be given by

$$Z_{\text{in}-\infty} = \frac{s^2((L_2 - M)U_1 + (L_1 - M)U_2)C_1 + s(E_1 + (D'(L_2 - M) - D(L_1 - M))I_o) + E_2}{sC_1 U_2 - DI_o}$$

$$Z_{\text{in-sc}} = \frac{s^3(L_1 L_2 - M^2)C_1 + s^2 C_1(L_1(R_3 - r_{C2}) + L_2(R_1 - r_{C2}) - 2(R_2 - r_{C2})M) + sE_3 + E_4}{s^2 L_2 C_1 + s(R_3 - r_{C2})C_1 + D^2}$$

$$E_1 = ((R_3 - R_2)U_1 + (R_1 - R_2)U_2)C_1$$

$$E_2 = (DU_1 + D'U_2) - (D(R_1 - R_2) + D'(R_2 - R_3))I_o$$

$$E_3 = D^2 L_1 + D'^2 L_2 + 2DD'M + ((R_1 R_3 - R_2^2) - (R_1 - 2R_2 + R_3)r_{C2})C_1$$

$$E_4 = D^2 R_1 + 2DD'R_2 + D'^2 R_3 - r_{C2} \quad (10.30)$$

According to the above-presented transfer functions (10.27)–(10.30), we may conclude that the coupling of the inductors will change the location and damping of the resonances, which will affect the appearance of the resonant

RHP zero pair in the control-to-output transfer function (G_{co}), and even change the behavior of the special input impedances.

10.3.2 RHP Poles

According to the determinant (10.29), the polynomial coefficients are as follows, when the parasitic resistances R_1, R_2, and R_3 are assumed to be equal and denoted by R_0:

$$a_4 = 1$$
$$a_3 = \frac{(L_1 + L_2 - 2M)R_0}{L_1L_2 - M^2}$$
$$a_2 = \frac{(L_1 + L_2 - 2M)C_1 + \left(D^2L_1 + D'^2L_2 + 2DD'M\right)C_2}{(L_1L_2 - M^2)C_1C_2} \quad (10.31)$$
$$a_1 = \frac{R_0}{(L_1L_2 - M^2)C_1}$$
$$a_0 = \frac{1}{(L_1L_2 - M^2)C_1C_2}$$

The coupling coefficient $k = M/\sqrt{L_1L_2}$ is always less than unity. Thus all the polynomial coefficients in (10.31) are positive real numbers. The Routh–Hurwitz array coefficients b_1, b_2, c_1, and d_1 (see (10.20)) can be given by

$$b_1 = \frac{L_1 + L_2 - 2M}{(L_1L_2 - M^2)C_2} + \frac{(L_1 - L_2)^2 + 2DD'(L_1 + L_2 - 2M) + M^2}{(L_1L_2 - M^2)(L_1 + L_2 - 2M)C_1}$$
$$b_2 = \frac{1}{(L_1L_2 - M^2)C_1C_2}$$
$$c_1 = \frac{R_0}{L_1L_2 - M^2} \cdot \frac{(L_1 - L_2)^2 + 2DD'(L_1 + L_2 - 2M) + M^2}{(L_1 + L_2 - 2M)^2C_1 + ((L_1 - L_2)^2 + 2DD'(L_1 + L_2 - 2M) + M^2)C_2}$$
$$d_1 = \frac{1}{(L_1L_2 - M^2)C_1C_2} \quad (10.32)$$

which are real positive numbers, because $k < 1$. Therefore, all the roots of the determinant are LHP roots regardless of the claims presented in [2] and [3].

10.3.3 Input-Current-Ripple Reduction

Applying the coupled-inductor algorithms (10.25) to the on-time and off-time inductor-current derivatives (see Section 10.2), the peak-to-peak ripple values become

$$\Delta i_{L1-pp} = \left(\frac{(L_2 - M)(U_{in} - U_o - (Dr_{L1} + r_{ds})I_o)}{L_1 L_2 - M^2} + \frac{MD'(r_{C1}I_o + U_D)}{L_1 L_2 - M^2} \right) DT_s$$

$$\Delta i_{L2-pp} = \left(\frac{(L_1 - M)(U_{in} - U_o - (Dr_{L1} + r_{ds})I_o)}{L_1 L_2 - M^2} - \frac{L_1 D'(r_{C1}I_o + U_D)}{L_1 L_2 - M^2} \right) DT_s$$

(10.33)

According to (10.33), the close-to-zero-input ripple or $\Delta i_{L1-pp} \approx 0$ can be obtained, when $M = L_2$ or $k = \sqrt{\frac{L_2}{L_1}}$. These values are also the commonly defined zero-input-ripple conditions [2–4]. As a consequence, the residual peak-to-peak ripple values are

$$\Delta i_{L1-pp} = \frac{r_{C1}I_o + U_D}{L_1 - L_2} \cdot DD'T_s$$

$$\Delta i_{L2-pp} = \frac{U_{in} - U_o - (Dr_{L1} + r_{ds})I_o}{L_2} \cdot DT_s - \frac{r_{C1}I_o + U_D}{L_1 - L_2} \cdot \frac{DD'T_s L_1}{L_2} \quad (10.34)$$

which imply that L_1 should be sufficiently higher than L_2 for obtaining optimal ripple reduction.

Substituting M and k in (10.26) with the above given values yields

$$\begin{bmatrix} \frac{d\hat{i}_{L1}}{dt} \\ \frac{d\hat{i}_{L2}}{dt} \\ \frac{d\hat{u}_{C1}}{dt} \\ \frac{d\hat{u}_{C2}}{dt} \end{bmatrix} = \begin{bmatrix} -\frac{R_1 - R_2}{L_1 - L_2} & -\frac{R_2 - R_3}{L_1 - L_2} & -\frac{1}{L_1 - L_2} & 0 \\ -\frac{R_2 L_1 - R_1 L_2}{(L_1 - L_2)L_2} & -\frac{R_3 L_1 - R_2 L_2}{(L_1 - L_2)L_2} & \frac{DL_1 + D'L_2}{(L_1 - L_2)L_2} & -\frac{1}{L_2} \\ \frac{D'}{C_1} & -\frac{D}{C_1} & 0 & 0 \\ \frac{1}{C_2} & \frac{1}{C_2} & 0 & 0 \end{bmatrix} \begin{bmatrix} \hat{i}_{L1} \\ \hat{i}_{L2} \\ \hat{u}_{C1} \\ \hat{u}_{C2} \end{bmatrix}$$

$$+ \begin{bmatrix} \frac{1}{L_1 - L_2} & 0 & \frac{U_1 - U_2}{L_1 - L_2} \\ -\frac{1}{L_1 - L_2} & \frac{r_{C2}}{L_2} & \frac{U_2 L_1 - U_1 L_2}{(L_1 - L_2)L_2} \\ 0 & 0 & -\frac{I_o}{C_1} \\ 0 & -\frac{1}{C_2} & 0 \end{bmatrix} \begin{bmatrix} \hat{u}_{in} \\ \hat{i}_o \\ \hat{d} \end{bmatrix}$$

$$\begin{bmatrix} \hat{i}_{in} \\ \hat{u}_o \end{bmatrix} = \begin{bmatrix} 1 & 0 & 0 & 0 \\ 0 & 0 & 0 & 1 + r_{C2}C_2 \frac{d}{dt} \end{bmatrix} \begin{bmatrix} \hat{i}_{L1} \\ \hat{i}_{L2} \\ \hat{u}_{C1} \\ \hat{u}_{C2} \end{bmatrix} + \begin{bmatrix} 0 & 0 & 0 \\ 0 & 0 & 0 \end{bmatrix} \begin{bmatrix} \hat{u}_{in} \\ \hat{i}_o \\ \hat{d} \end{bmatrix}$$

(10.35)

As a consequence, the G-parameter-set of transfer functions can be given by
Input dynamics:

$$\Delta Y_{in-o} = \frac{s}{L_1 - L_2}\left(s^2 + s\frac{R_3}{L_2} + \frac{C_1 + D^2 C_2}{L_2 C_1 C_2}\right)$$

$$\Delta T_{oi-o} = \frac{1}{(L_1 - L_2)C_2}\left(s\frac{R_3 - R_2}{L_2} + \frac{D}{L_2 C_1}\right)(1 + sr_{C2}C_2)$$

$$\Delta G_{ci} = \frac{1}{L_1 - L_2}\left(s^3(U_1 - U_2) + s^2\frac{(R_3 U_1 - R_2 U_2)C_1 + L_2 I_o}{L_2 C_1}\right.$$

$$\left. + s\frac{(U_1 - U_2)C_1 + \left(D^2 U_1 + DD' U_2 + (DR_2 + D'R_3)I_o\right)C_2}{L_2 C_1 C_2} + \frac{I_o}{L_2 C_1 C_2}\right)$$

(10.36)

Output dynamics:

$$\Delta G_{io-o} = \frac{1}{(L_1 - L_2)C_2}\left(s\frac{R_3 - R_2}{L_2} + \frac{D}{L_2 C_1}\right)(1 + sr_{C2}C_2)$$

$$\Delta Z_{o-o} = \frac{1}{C_2}\left(s^3 + s^2\frac{(R_3 - r_{C2})L_1 + (R_1 - 2R_2 + r_{C2})L_2}{(L_1 - L_2)L_2}\right.$$

$$\left. + s\frac{(R_1 R_3 - R_2^2 - (R_1 - 2R_2 + R_3)r_{C2})C_1 + D^2 L_1 + D'(1 + D)L_1}{(L_1 - L_2)L_2}\right.$$

$$\left. + \frac{D^2 R_1 + 2DD' R_2 + D'^2 R_3 - r_{C2}}{(L_1 - L_2)L_2}\right)(1 + sr_{C2}C_2)$$

$$\Delta G_{co} = \frac{U_2}{L_2 C_2}\left(s^2 + s\frac{((R_3 - R_2)U_1 + (R_1 - R_2)U_2)C_1 - D(L_1 - L_2)I_o}{U_2(L_1 - L_2)C_1}\right.$$

$$\left. + \frac{DU_1 + D'U_2 + \left(D(R_2 - R_1) + D'(R_3 - R_2)\right)I_o}{U_2(L_1 - L_2)C_1}\right)(1 + sr_{C2}C_2)$$

(10.37)

where the determinant (Δ) is

$$\Delta = s^4 + s^3\frac{R_3 L_1 + (R_1 - 2R_2)L_2}{(L_1 - L_2)L_2} + s^2 a_2 + s a_1 + \frac{1}{(L_1 - L_2)L_2 C_1 C_1}$$

$$a_2 = \frac{(R_1 R_3 - R_2^2)C_1 C_2 + (L_1 - L_2)C_1 + \left(D^2 L_1 + D'(1 + D)L_2\right)C_2}{(L_1 - L_2)L_2 C_1 C_2}$$

(10.38)

$$a_1 = \frac{(R_1 - 2R_2 + R_3)C_1 + \left(D^2 R_1 + 2DD' R_2 + D'^2 R_3\right)C_2}{(L_1 - L_2)L_2 C_1 C_2}$$

Thus the special input impedances ($Z_{in-\infty}$ and Z_{in-sc}) can be given by

$$Z_{in-\infty} = \frac{s^2(L_1 - L_2)U_2C_1 + s(((R_3 - R_2)U_1 + (R_1 - R_2)U_2)C_1 - D(L_1 - L_2)I_o) + E_1}{U_2C_1s - DI_o}$$

$$Z_{in-sc} = \frac{s^3(L_1 - L_2)L_2C_1 + s^2C_1(L_1(R_3 - r_{C2}) + L_2(R_1 - 2R_2 + r_{C2})) + sE_2 + E_3}{s^2L_2C_1 + s(R_3 - r_{C2})C_1 + D^2}$$

$$E_1 = DU_1 + D'U_2 + (D(R_2 - R_1) + D'(R_3 - R_2))I_o \quad (10.39)$$

$$E_2 = D^2L_1 + D'(1+D)L_2 + ((R_1R_3 - R_2^2 - (R_1 - 2R_2 + R_3)r_{c2})C_1$$

$$E_3 = D^2R_1 + 2DD'R_2 + D'^2R_3 - r_{C2}$$

According to the above given transfer functions, we may conclude that the coupling of the inductors yielding input-current ripple reduction damps the resonances in the input-to-output (G_{io-o}) and output-to-input (T_{oi-o}) transfer functions making them close to first-order transfer functions and accelerates the appearance of the RHP zero in the control-to-output transfer function.

10.3.4
Design Considerations

According to (10.37), the control-to-output transfer function (G_{co}) incorporates a RHP zero pair approximately at $f \approx \frac{1}{2\pi}\sqrt{\frac{1}{(1-k^2)L_1C_1}}$, when

$$((R_3 - R_2)U_1 + (R_1 - R_2)U_2)C_1 - D(1 - k^2)L_1I_o < 0 \quad (10.40)$$

where k is the coupling coefficient. The appearance of the RHP zero is most likely in the applications where the input voltage is rather low and the output current rather high. Equation (10.40) gives explicit suggestions for the possible design actions for preventing the appearance of the RHP zero if possible.

10.4
PCM-Controlled Superbuck

In order to reduce the resonant behavior, the PCM control is applied as shown in Figure 10.7, where the pulsewidth modulation is based on the sum of the inductor currents measured by means of a current transformer from the source of the MOSFET. As a consequence, the resonant behavior would be removed from the output dynamics but it would be still left at the input dynamics, because only the sum of the inductor currents is tightly regulated and the individual currents can still freely vary within the sum.

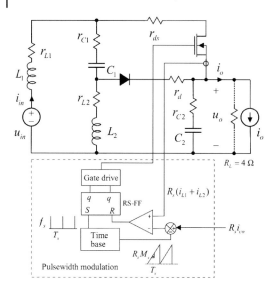

Figure 10.7 PCM-controlled superbuck.

10.4.1
Small-Signal Models

The modeling of the PCM-controlled superbuck converter [11] can be carried out similarly to the modeling of the conventional buck converter introduced in Chapter 4 by substituting the perturbed duty ratio $\left(\hat{d}\right)$ in the VMC state space (10.8) with the duty-ratio constraints given by

$$\hat{d} = F_m \left(\hat{i}_{co} - q_{L1}\hat{i}_{L1} - q_{L2}\hat{i}_{L2} - q_{C1}\hat{u}_{C1} - q_{C2}\hat{u}_{C2} - q_{in}\hat{u}_{in} - q_o\hat{i}_o \right) \quad (10.41)$$

The coefficients of (10.41) can be solved by linearizing the comparator equation (i.e., the equation governing the reset control of the RS-flip-flop in Figure 10.7), which is constructed based on the combined inductor-current waveforms shown in Figure 10.8.

$$\hat{i}_{co} - m_c dT_s = \langle i_{L1} \rangle + \langle i_{L2} \rangle + \Delta i_L \quad (10.42)$$

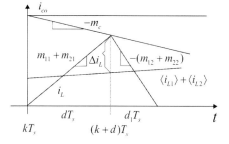

Figure 10.8 The combined inductor-current waveforms constituting the comparator control.

10.4 PCM-Controlled Superbuck

According to Chapter 4, Δi_L can be given by $\frac{dd'T_s}{2}\sum_{i=1}^{n}(m_{i1}+m_{i2})$, and consequently, the comparator equation becomes

$$\hat{i}_{co} - m_c dT_s = \langle i_{L1}\rangle + \langle i_{L2}\rangle + \frac{dd'T_s}{2}(m_{11}+m_{12}+m_{21}+m_{22}) \quad (10.43)$$

Computing the inductor-current slopes from the actual circuit (i.e., m_{11} and m_{21} can be obtained from (10.1) and m_{12} and m_{22} from (10.2) given in Section 10.2) and substituting them in (10.43) yields

$$\hat{i}_{co} - m_c dT_s = \langle i_{L1}\rangle + \langle i_{L2}\rangle + \frac{dd'T_s}{2}\left(\frac{\langle u_1\rangle}{L_1} + \frac{\langle u_2\rangle}{L_2}\right) \quad (10.44)$$

where $\langle u_1\rangle$ and $\langle u_2\rangle$ are defined as

$$\begin{aligned}\langle u_1\rangle &= \langle u_{C1}\rangle + U_D + (r_{C1}+r_d-r_{ds})\langle i_{L1}\rangle + (r_d-r_{ds})\langle i_{L2}\rangle \\ \langle u_2\rangle &= \langle u_{C1}\rangle + U_D + (r_d-r_{ds})\langle i_{L1}\rangle + (r_d-r_{C1}-r_{ds})\langle i_{L2}\rangle\end{aligned} \quad (10.45)$$

The coefficient of the duty-ratio constraints (10.40) can be solved from (10.43) by developing the proper partial derivatives. This procedure yields

$$F_m = \frac{1}{T_s M_c + \frac{(D'-D)T_s}{2}\left(\frac{U_1}{L_1}+\frac{U_2}{L_2}\right)}$$

$$q_{L1} = 1 + \frac{DD'T_s}{2}\left(\frac{r_{C1}+r_d-r_{ds}}{L_1} + \frac{r_d-r_{ds}}{L_2}\right)$$

$$q_{L2} = 1 + \frac{DD'T_s}{2}\left(\frac{r_d-r_{ds}}{L_1} + \frac{r_d-r_{C1}-r_{ds}}{L_2}\right)$$

$$q_{C1} = \frac{DD'T_s(L_1+L_2)}{2L_1L_2} \quad (10.46)$$

$$q_{C2} = q_{in} = q_o = 0$$

$$U_1 = U_{in} + U_D + (r_d-r_{ds}+Dr_{C1}-Dr_{L1}+D'r_{L2})I_o$$

$$U_2 = U_1 - r_{C1}I_o$$

The formula of the duty-ratio gain (F_m) in (10.46) indicates that F_m would become infinite (i.e., the denominator will become zero), when the maximum duty ratio (D_{max}) is reached

$$D_{max} = 0.5 + M_c L_1 L_2/(L_2 U_1 + L_1 U_2) \quad (10.47)$$

When the compensation (M_c) is set to zero, the maximum duty ratio corresponds to 0.5 as in the conventional buck converter. An increase in the duty ratio beyond 0.5 will force the converter to enter into the second-harmonic mode as explained in detail in Chapter 4, Section 4.3.3.

The corresponding PCM state space can be obtained from (10.8) by replacing the perturbed duty ratio $\left(\hat{d}\right)$ with (10.41). This gives

$$\begin{bmatrix} \dfrac{d\hat{i}_{L1}}{dt} \\ \dfrac{d\hat{i}_{L2}}{dt} \\ \dfrac{d\hat{u}_{C1}}{dt} \\ \dfrac{d\hat{u}_{C2}}{dt} \end{bmatrix} = \begin{bmatrix} -\dfrac{R_1 + F_m q_{L1} U_1}{L_1} & -\dfrac{R_2 + F_m q_{L2} U_1}{L_1} & -\dfrac{D' + F_m q_{C1} U_1}{L_1} & -\dfrac{1}{L_1} \\ -\dfrac{R_2 + F_m q_{L1} U_2}{L_2} & -\dfrac{R_3 + F_m q_{L2} U_2}{L_2} & \dfrac{D - F_m q_{C1} U_2}{L_2} & -\dfrac{1}{L_2} \\ \dfrac{D' + F_m q_{L1} I_o}{C_1} & -\dfrac{D - F_m q_{L2} I_o}{C_1} & \dfrac{F_m q_{C1} I_o}{C_1} & 0 \\ \dfrac{1}{C_2} & \dfrac{1}{C_2} & 0 & 0 \end{bmatrix} \begin{bmatrix} \hat{i}_{L1} \\ \hat{i}_{L2} \\ \hat{u}_{C1} \\ \hat{u}_{C2} \end{bmatrix}$$

$$+ \begin{bmatrix} \dfrac{1}{L_1} & \dfrac{r_{C2}}{L_1} & \dfrac{F_m U_1}{L_1} \\ 0 & \dfrac{r_{C2}}{L_2} & \dfrac{F_m U_2}{L_2} \\ 0 & 0 & -\dfrac{F_m I_o}{C_1} \\ 0 & -\dfrac{1}{C_2} & 0 \end{bmatrix} \begin{bmatrix} \hat{u}_{in} \\ \hat{i}_o \\ \hat{i}_{co} \end{bmatrix}$$

$$\begin{bmatrix} \hat{i}_{in} \\ \hat{u}_o \end{bmatrix} = \begin{bmatrix} 1 & 0 & 0 & 0 \\ 0 & 0 & 0 & 1 + r_{C2} C_2 \dfrac{d}{dt} \end{bmatrix} \begin{bmatrix} \hat{i}_{L1} \\ \hat{i}_{L2} \\ \hat{u}_{C1} \\ \hat{u}_{C2} \end{bmatrix} + \begin{bmatrix} 0 & 0 & 0 \\ 0 & 0 & 0 \end{bmatrix} \begin{bmatrix} \hat{u}_{in} \\ \hat{i}_o \\ \hat{i}_{co} \end{bmatrix} \quad (10.48)$$

The corresponding transfer functions can be solved from (10.48) most conveniently by using proper software packages such as Matlab™ with Symbolic Toolbox. The resulting symbolic transfer functions with the parasitic elements are extremely long and do not easily offer the desired information. From the dynamic point of view, the parasitics are, however, essential and cannot be neglected, because they will provide the damping of the resonances and also shift the appearance of the possible RHP zeros and poles. We give here only the nonparasitic transfer functions for introducing the effect of PCM control in the dynamics of the converter. The effect of the parasitics will be discussed when appropriate.

Input dynamics:

$$\Delta Y_{in-o} = \dfrac{1}{L_1}\left(s^3 + s^2 \dfrac{F_m(U_2 C_1 + I_1 q_{C1} L_2)}{L_2 C_1} + s\dfrac{C_1 + D(D - F_m(U_2 q_{C1} + I_o))C_2}{L_2 C_1} - \dfrac{F_m I_o q_{C1}}{L_2 C_1 C_2}\right)$$

$$\Delta T_{oi-o} = \dfrac{1}{L_1 C_2}\left(s^2 + s\dfrac{F_m((U_2 - U_1)C_1 + I_o q_{C1} L_2)}{L_2 C_1} + \dfrac{D + F_m((U_1 - U_2) q_{C1} - I_o)}{L_2 C_1}\right)$$

$$\Delta G_{ci} = \dfrac{F_m U_1}{L_1}\left(s^3 + s^2 \dfrac{D' I_o}{U_1 C_1} + s\dfrac{(U_1 - U_2)C_1 + (D^2 U_1 + DD' U_2)C_2}{U_1 L_2 C_1 C_2} + \dfrac{I_o}{U_1 L_2 C_1 C_2}\right)$$

(10.49)

Output dynamics:

$$\Delta G_{io-o} = \frac{1}{L_1 C_2} \left(s^2 - s\frac{F_m I_o q_{C1}}{C_1} + \frac{D - F_m U_2 q_{C1}}{L_2 C_1} \right)$$

$$\Delta Z_{o-o} = \frac{1}{C_2} \left(s^3 + s^2 \frac{F_m \left((U_2 L_1 + U_1 L_2) C_1 - I_o q_{C1} L_1 L_2 \right)}{L_1 L_2 C_1} \right.$$

$$+ s \frac{D^2 L_1 + D'^2 L_2 + F_m (q_{C1} (U_1 D' L_2 - U_2 D L_1) + I_o (D' L_2 - D L_1))}{L_1 L_2 C_1}$$

$$\left. + \frac{F_m (D U_1 + D' U_2)}{L_1 L_2 C_1} \right)$$

$$\Delta G_{co} = \frac{F_m (U_1 L_2 + U_2 L_1)}{L_1 L_2 C_2} \left(s^2 + s \frac{(D' L_2 - D L_1) I_o}{(U_1 L_2 + U_2 L_1) C_1} + \frac{D U_1 + D' U_2}{(U_1 L_2 + U_2 L_1) C_1} \right)$$

(10.50)

where the determinant (Δ) is

$$\Delta = s^4 + s^3 \frac{F_m \left((U_1 L_2 + U_2 L_1) C_1 - I_o q_{C1} L_1 L_2 \right)}{L_1 L_2 C_1}$$

$$+ s^2 \frac{(L_1 + L_2)(C_1 + D'^2 C_2) + F_m (q_{C1} (U_1 D' L_2 - U_2 D L_1) + I_o (D' L_2 - D L_1)) C_2}{L_1 L_2 C_1 C_2}$$

$$+ s \frac{F_m ((D U_1 + D' U_2) C_2 - I_o q_{C1} (L_1 + L_2))}{L_1 L_2 C_1 C_2} + \frac{1 + F_m q_{C1} (U_1 - U_2)}{L_1 L_2 C_1 C_2}$$

(10.51)

The complexity of the fourth-order determinant is such that it is not anymore possible to find the approximate symbolic solutions for its roots of which two are usually real and one is a complex conjugate root.

The ideal input impedance ($Z_{in-\infty}$) is the same as defined in (10.13). The nonparasitic short-circuit input impedance (Z_{in-sc}) can be computed to be

$$Z_{in-sc} = \frac{s^3 \cdot L_1 L_2 C_1 + s^2 \cdot A + s \cdot B + F_m (D U_1 + D' U_2)}{s^2 \cdot L_2 C_1 + s \cdot F_m (U_2 C_1 - I_o q_{C1} L_2) + D (D - F_m (U_2 q_{C1} + I_o))}$$

$$A = F_m \left((U_1 L_2 + U_2 L_1) C_1 - I_o q_{C1} L_1 L_2 \right)$$

$$B = (D^2 L_1 + D' L_2 + F_m (q_{C1} (U_1 D' L_2 - U_2 D L_1) + I_o (D' L_2 - D L_1)))$$

(10.52)

Ref. [4] actually claims that the dynamics of the PCM-controlled superbuck converter can be obtained from the dynamic representation of the conventional buck converter by replacing the output inductor with the parallel inductor $L_p = L_1 L_2/(L_1 + L_2)$, but this claim does not hold because of the resonant behavior in the input dynamics and the possible appearance of the RHP poles and zeros.

10.4.2 Design Considerations

The PCMC converter incorporates the same RHP zero pair in its control-to-output transfer function (G_{co}) as the VMC converter discussed in Section 10.2.4. Therefore, the inductors should be designed such that $L_2 \approx \frac{D_{max}}{D'_{max}} L_1$, where D_{max} corresponds to the maximum duty ratio of the converter operation. High input voltage and low output power may allow the inductors to be the same even if the maximum duty ratio is higher than 0.5. The inductor-current feedback in the PCMC converter also requires considering the selection of the inductor-current slope compensation (M_c) and the possible appearance of RHP poles and their elimination.

10.4.2.1 Inductor-Current-Feedback Compensation

The zeroth-order term of the numerator of the input-to-output transfer function (G_{io-o} in (10.50)) $a_0 = D - F_m(U_2 q_{C1} + I_o(D(q_{L1} - q_{L2}) + (R_2 - R_3)q_{C1}))$ implies that the low-frequency gain of G_{io-o} can be made small by designing the compensation M_c such that the zeroth-order term (a_0) will be zero similarly to the conventional buck converter. a_0 is clearly dependent on the load current and parasitics and, therefore, the perfect compensation is not possible. When neglecting the parasitics, $a_0 \approx D - \frac{DD'U_{in}}{M_c L_p + (D'-D)U_{in}}$, and thus the close-to-optimal compensation becomes

$$M_c \approx \frac{U_o}{2L_p} \quad L_p = \frac{L_1 L_2}{L_1 + L_2} \tag{10.53}$$

which has the same form as defined for the conventional buck converter, when its inductor is substituted with the parallel inductor (L_p).

10.4.2.2 Avoiding RHP Poles

According to the determinant (Δ) (10.51), the polynomial coefficients are as follows:

$$a_4 = 1$$

$$a_3 = \frac{F_m \left(U_{in} - \frac{DD'T_s I_o}{2C_1} \right)}{L_p}$$

$$a_2 = \frac{(L_1 + L_2)(C_1 + D'^2 C_2) + F_m \left(\frac{DD'T_s}{2L_p} + I_o \right)(D'L_2 - DL_1)C_2}{L_1 L_2 C_1 C_2}$$

$$a_1 = F_m \left(U_{in} - \frac{DD'T_s I_o L_1 L_2}{2L_p^2 C_2} \right) \frac{1}{L_1 L_2 C_1}$$

$$a_0 = \frac{1}{L_1 L_2 C_1 C_2} \tag{10.54}$$

where the parasitics are neglected. The coefficients a_0 and a_4 are always positive and, therefore, all the other polynomial coefficients and the left-column elements of the *Routh–Hurwitz* array have to be also positive. The positive signs of the coefficients a_3, a_2, and a_1 require that $C_1 > \frac{DD'T_s I_o}{2U_{in}} = C_{1-min}$, $\frac{L_2}{L_1} \geq \frac{D}{D'}$, and $C_2 > \frac{DD'T_s I_o}{2U_{in}} \cdot \frac{L_1 L_2}{L_p^2}$ or $C_2 > \frac{L_1 L_2}{L_p^2} \cdot C_{1-min}$, respectively. If the converter is designed such that $L_2 = \frac{D_{max}}{D'_{max}} \cdot L_1$, where D_{max} is the defined maximum duty ratio of the converter operation as well as that C_1 and C_2 are chosen to be much larger than their above-defined minimum values then the array element b_1 can be given by

$$\frac{1}{L_p}\left(\frac{1}{C_2} + (D'^2 - \frac{L_p}{(L_1+L_2)})\frac{1}{C_1}\right) = \frac{1}{L_p}\left(\frac{1}{C_2} + D'_{max}(D'_{max} - D_{max})\frac{1}{C_1}\right)$$

which is a positive number if $C_1/C_2 > D'_{max}(D_{max} - D'_{max})$. This requirement is easily met in the practical design, because $|D'_{max}(D_{max} - D'_{max})|_{max} = 0.125$. Similarly, the array element c_1 can be given by

$$\frac{F_m U_{in}}{L_1 L_2 C_1}\left(1 - \frac{1}{1 + \left(D'^2 - \frac{L_p}{L_1+L_2}\right) \cdot \frac{C_2}{C_1}}\right) = \frac{F_m U_{in}}{L_1 L_2 C_1}\left(1 - \frac{1}{1 + D'_{max}\left(D'_{max} - D_{max}\right) \cdot \frac{C_2}{C_1}}\right)$$

which indicates a negative sign to appear if $D_{max} > 0.5$. In reality, the effect of the parasitics is such that the negative sign may not appear but the situation should be carefully checked. The sign of the elements b_2 and d_1 is always positive because of equaling the zeroth-order polynomial coefficient.

The dynamic issues covered above gave some basic guidelines for the design, which with the peak-to-peak inductor-current ripple and load response definitions would led to an optimal design of the PCM-controlled converter.

10.5
Coupled-Inductor PCM-Controlled Superbuck

10.5.1
Small-Signal Models

The small-signal modeling of the coupled-inductor PCM-controlled superbuck converter can be carried out similarly to the modeling of the basic PCM control presented in Section 10.4.1. We assume here that the input-ripple-reduction conditions are in effect (i.e., $M = L_2$ and $k = \sqrt{L_2/L_1}$). The coupling of inductors and the ripple-reduction conditions naturally change the ripple of the inductor-current feedback (i.e., Δi_L). The coupled-inductor-affected up (m_{i1}) and down (m_{i2}) slopes defining Δi_L (see Section 10.4.1) can be obtained from the basic VMC on- and off-time state spaces by applying the algorithm defined in Section 10.3 (10.25). The application of the algorithms yields

$$\Delta i_L = \frac{dd'T_s}{2}\left(\left(\frac{r_d - r_{ds}}{L_2}\right)\langle i_{L1}\rangle + \frac{r_{L2} + r_d - r_{ds}}{L_2}\langle i_{L2}\rangle + \frac{\langle u_{C1}\rangle + U_D}{L_2}\right) \quad (10.55)$$

We define the required duty-ratio constraints by linearizing the comparator equation

$$\langle i_{co}\rangle - m_c dT_s = \langle i_{L1}\rangle + \langle i_{L2}\rangle + \Delta i_L \quad (10.56)$$

which yields

$$F_m = \frac{1}{T_s\left(M_c + \frac{(D'-D)}{2L_2}(U_{in} + DU_D + (r_d - r_{ds} - Dr_{L1} + 2D'r_{L2})I_o)\right)}$$

$$q_{L1} = 1 + \frac{DD'T_s}{2L_2}(r_d - r_{ds})$$

$$q_{L2} = 1 + \frac{DD'T_s}{2L_2}(r_d - r_{ds} + r_{L2}) \quad (10.57)$$

$$q_{C1} = \frac{DD'T_s}{2L_2}$$

$$q_{C2} = q_{in} = q_o = 0$$

The main difference in the duty-ratio constraints in (10.57) compared to those presented in Section 10.4.1 (Eq. (10.46)) is that they are only dependent on L_2.

The coupled-inductor PCMC state space can be obtained from the corresponding VMC state space (Section 10.3.3, Eq. (10.35)) by substituting the perturbed duty ratio $\left(\hat{d}\right)$ by means of

$$\hat{d} = F_m\left(\hat{i}_{co} - q_{L1}\hat{i}_{L1} - q_{L2}\hat{i}_{L2} - q_{C1}\hat{u}_{C1}\right) \quad (10.58)$$

which yields

$$\begin{bmatrix}\frac{d\hat{i}_{L1}}{dt}\\ \frac{d\hat{i}_{L2}}{dt}\\ \frac{d\hat{u}_{C1}}{dt}\\ \frac{d\hat{u}_{C2}}{dt}\end{bmatrix} = \begin{bmatrix}-\frac{\Delta R_1 + F_m q_{L1}\Delta U_1}{\Delta L_{12}} & -\frac{\Delta R_2 + F_m q_{L2}\Delta U_1}{\Delta L_{12}} & -\frac{1 + F_m q_{C1}\Delta U_1}{\Delta L_{12}} & 0\\ -\frac{\Delta R_3 + F_m q_{L1}\Delta U_2}{\Delta L_{12}} & -\frac{\Delta R_4 + F_m q_{L2}\Delta U_2}{\Delta L_{12}} & \frac{D\frac{L_1}{L_2} + D' - F_m q_{C1}\Delta U_2}{\Delta L_{12}} & -\frac{1}{L_2}\\ \frac{D' + F_m q_{L1}I_o}{C_1} & -\frac{D - F_m q_{L2}I_o}{C_1} & \frac{F_m q_{C1}I_o}{C_1} & 0\\ \frac{1}{C_2} & \frac{1}{C_2} & 0 & 0\end{bmatrix}\begin{bmatrix}\hat{i}_{L1}\\ \hat{i}_{L2}\\ \hat{u}_{C1}\\ \hat{u}_{C2}\end{bmatrix}$$

$$+ \begin{bmatrix}\frac{1}{\Delta L_{12}} & 0 & \frac{F_m\Delta U_1}{\Delta L_{12}}\\ \frac{1}{\Delta L_{12}} & \frac{r_{C2}}{L_2} & \frac{F_m\Delta U_2}{\Delta L_{12}}\\ 0 & 0 & -\frac{F_m I_o}{C_1}\\ 0 & -\frac{1}{C_2} & 0\end{bmatrix}\begin{bmatrix}\hat{u}_{in}\\ \hat{i}_o\\ \hat{d}\end{bmatrix}$$

$$\begin{bmatrix} \hat{i}_{in} \\ \hat{u}_o \end{bmatrix} = \begin{bmatrix} 1 & 0 & 0 & 0 \\ 0 & 0 & 0 & 1+r_{C2}C_2\dfrac{d}{dt} \end{bmatrix} \begin{bmatrix} \hat{i}_{L1} \\ \hat{i}_{L2} \\ \hat{u}_{C1} \\ \hat{u}_{C2} \end{bmatrix} + \begin{bmatrix} 0 & 0 & 0 \\ 0 & 0 & 0 \end{bmatrix} \begin{bmatrix} \hat{u}_{in} \\ \hat{i}_o \\ \hat{d} \end{bmatrix} \quad (10.59)$$

where $\Delta R_1 - \Delta R_4$, ΔL_{12}, ΔU_1, and ΔU_2 are as follows:

$$\begin{aligned}
\Delta R_1 &= R_1 - R_2, & \Delta R_2 &= R_2 - R_3 \\
\Delta R_3 &= R_2 \frac{L_1}{L_2} - R_1, & \Delta R_4 &= R_3 \frac{L_1}{L_2} - R_2 \\
\Delta L_{12} &= L_1 - L_2, & & \\
\Delta U_1 &= U_1 - U_2, & \Delta U_2 &= U_2 \frac{L_1}{L_2} - U_1
\end{aligned} \quad (10.60)$$

and $R_1 - R_3$, U_1 and U_2 are as defined in (10.8) (Section 10.2.2).

The corresponding transfer functions can be solved most conveniently from (10.59) by using proper software packages such as Matlab™ with Symbolic Toolbox. The resulting symbolic transfer functions with the parasitic elements are extremely long and do not easily offer the desired information. From the dynamic point of view, the parasitics are, however, essential and cannot be neglected, because they will provide the damping of the resonances and also shift the appearance of the possible RHP zeros and poles. We give here only the nonparasitic transfer functions for introducing the combined effect of PCM control and coupled-inductor technique in the dynamics of the converter. The effect of the parasitics will be discussed when appropriate.

Input dynamics:

$$\Delta Y_{in-o} = \frac{1}{L_1 - L_2} \left(s^3 + s^2 \frac{F_m(U_{in}C_1 - q_{C1}I_oL_2)}{L_2C_1} \right.$$
$$\left. + s \frac{C_1 + (D^2 - DF_m(q_{C1}U_{in} + I_o))C_2}{L_2C_1C_2} - \frac{F_m q_{C1}I_o}{L_2C_1C_2} \right)$$

$$\Delta T_{oi-o} = \frac{D - F_m I_o}{(L_1 - L_2)L_2C_1C_2} \quad (10.61)$$

$$\Delta G_{ci} = \frac{F_m I_o}{(L_1 - L_2)C_1} \left(s^2 + s\frac{DU_{in}}{I_o L_2} + \frac{1}{L_2 C_2} \right)$$

Output dynamics:

$$\Delta G_{io-o} = \frac{D - F_m q_{C1} U_{in}}{(L_1 - L_2)L_2C_1C_2}$$

$$\Delta Z_{o-o} = \frac{1}{C_2} \left(s^3 + s^2 \frac{F_m(U_{in}C_1 - q_{C1}I_o L_2)}{L_2 C_1} \right.$$

$$\Delta G_{co} = \frac{F_m U_{in}}{L_2 C_2} \left(s^2 - s \frac{DI_o}{U_{in} C_1} + \frac{1}{(L_1 - L_2)C_1} \right) \begin{pmatrix} + s \frac{D^2 L_1 + D'(1+D)L_2 - DF_m(q_{C1}U_{in} + I_o)(L_1 - L_2)}{(L_1 - L_2)L_2 C_1} \\ + \frac{F_m U_{in}}{(L_1 - L_2)L_2 C_1} \end{pmatrix} \quad (10.62)$$

where the determinant (Δ) is

$$\Delta = s^4 + s^3 \frac{F_m(U_{in}C_1 - q_{C1}I_o L_2)}{L_2 C_1}$$
$$+ s^2 \frac{(C_1 + (D^2 - DF_m(q_{C1}U_{in} + I_o))C_2)(L_1 - L_2) + L_2 C_2}{(L_1 - L_2)L_2 C_1 C_2}$$
$$+ s \frac{F_m(U_{in}C_2 - q_{C1}I_o(L_1 - L_2))}{(L_1 - L_2)L_2 C_1 C_2} + \frac{1}{(L_1 - L_2)L_2 C_1 C_2} \quad (10.63)$$

Thus the special input impedances ($Z_{in-\infty}$ and Z_{in-sc}) are

$$Z_{in-\infty} = \frac{s^2 \cdot (L_1 - L_2)U_{in}C_1 - s \cdot DI_o(L_1 - L_2) + U_{in}}{s \cdot U_{in}C_1 - DI_o}$$

$$Z_{in-sc} = \frac{s^3 \cdot (L_1 - L_2)L_2 C_1 + s^2 \cdot E_1 + s \cdot E_2 + F_m U_{in}}{s^2 \cdot L_2 C_1 + s \cdot F_m(U_{in}C_1 - q_{C1}I_o L_2) + D(D - F_m(q_{C1}U_{in} + I_o))}$$

$$E_1 = F_m(U_{in}C_1 - q_{C1}I_o L_2)(L_1 - L_2)$$
$$E_2 = (D^2 - DF_m(q_{C1}U_{in} + I_o))(L_1 - L_2) + L_2 \quad (10.64)$$

where $Z_{in-\infty}$ is the same as presented in Section 10.3.3, Eq.(10.39).

Comparing the above given transfer functions in (10.61)–(10.64) to the corresponding VMC transfer functions given in Section 10.3.3 ((10.36)–(10.39)), we can conclude that the PCM control

1. changes the low-frequency value of the open-loop input impedance (Z_{in-o}) to the corresponding negative incremental resistor defined as $-F_m q_{C1} I_o$,
2. increases the low-frequency gain of the control-to-input-current and control-to-output-voltage transfer functions and the output impedance,
3. attenuates the input-voltage noise, which is maximized when the compensation (M_c) is designed to be such that $D - F_m q_{C1} U_{in} \approx 0$ or $M_c \approx U_o/2L_2$. As a consequence of this, the low-frequency input impedance corresponds to $-U_{in}/DI_o$.

The described changes are similar to the changes in the conventional buck converter discussed in Chapter 4.

When the optimal compensation $M_c \approx U_o/2L_2$ is applied, the nonparasitic dynamic description can be given as follows:

Input dynamics:

$$Y_{\text{in}-o} = \frac{1}{L_1 - L_2} \cdot \frac{s - \dfrac{DI_o}{U_{\text{in}} C_1}}{s^2 - s\dfrac{DI_o}{U_{\text{in}} C_1} + \dfrac{1}{(L_1 - L_2)C_1}}$$

$$\Delta T_{oi-o} = \frac{D\left(1 - \dfrac{2L_2 I_o}{T_s D' U_o}\right)}{(L_1 - L_2)L_2 C_1 C_2}$$

$$\Delta G_{ci} = \frac{2L_2 I_o}{T_s D' U_o (L_1 - L_2) C_1} \left(s^2 + s\frac{DU_{\text{in}}}{I_o L_2} + \frac{1}{L_2 C_2}\right)$$

(10.65)

Output dynamics:

$$\Delta G_{io-o} = 0$$

$$\Delta Z_{o-o} = \frac{1}{C_2}\left(s^3 + s^2 \frac{2U_{\text{in}} C_1 - DD' T_s I_o}{D' U_{\text{in}} T_s C_1}\right.$$

$$\left. + s\frac{D^2 L_1 + D'(1+D)L_2 - \dfrac{2L_2 DI_o}{T_s D' U_{\text{in}}}(L_1 - L_2)}{(L_1 - L_2)L_2 C_1} + \frac{2}{D' T_s (L_1 - L_2) C_1}\right)$$

$$\Delta G_{co} = \frac{2}{T_s D' C_2}\left(s^2 - s\frac{DI_o}{U_{\text{in}} C_1} + \frac{1}{(L_1 - L_2)C_1}\right)$$

(10.66)

where the determinant (Δ) is

$$\Delta = \left(s^2 + s\frac{2}{D' T_s} + \frac{1}{L_2 C_2}\right)\left(s^2 - s\cdot\frac{DI_o}{U_{\text{in}} C_1} + \frac{1}{(L_1 - L_2)C_1}\right) \quad (10.67)$$

Thus $Z_{\text{in}-\infty}$, $Z_{\text{in}-\text{sc}}$, and $Z_{\text{in}-o}$ (see (10.65)) becomes equal, because $G_{io-o} = 0$, and can be naturally given by

$$Z_{\text{in}-\infty,\text{sc},-o} = (L_1 - L_2) \cdot \frac{s^2 - s\dfrac{DI_o}{U_{\text{in}} C_1} + \dfrac{1}{(L_1 - L_2)C_1}}{s - \dfrac{DI_o}{U_{\text{in}} C_1}} \quad (10.68)$$

The equality of $Z_{\text{in}-\infty}$ and $Z_{\text{in}-\text{sc}}$ may be difficult to determine from (10.64) even if the optimal-compensation conditions are substituted in them. It is, however, possible to notice that the common factor of the numerator and denominator of $Z_{\text{in}-\text{sc}}$ is $s\dfrac{L_2}{U_{\text{in}}} + F_m$. Same difficulty also applies with the open-loop input impedance in (10.61) but the corresponding common factor is the first polynomial multiplier of the determinant given in (10.67).

10.5.2
Design Considerations

The PCM-controlled converter has the same RHP zeros as the corresponding VMC converter as introduced in Section 10.3.4. In order to avoid the appearance of the zeros, the design of the converter should be made as implied by (10.40) in Section 10.3.4.

According to the determinant (Δ) in (10.63), its polynomial coefficients are as follows:

$$
\begin{aligned}
a_4 &= 1 \\
a_3 &= \frac{F_m \left(U_{\text{in}} C_1 - q_{C1} I_o L_2 \right)}{L_2 C_1} \\
a_2 &= \frac{(C_1 + (D^2 - DF_m(q_{C1} U_{\text{in}} + I_o))C_2)(L_1 - L_2) + L_2 C_2}{(L_1 - L_2) L_2 C_1 C_2} \\
a_1 &= \frac{F_m \left(U_{\text{in}} C_2 - q_{C1} I_o (L_1 - L_2) \right)}{(L_1 - L_2) L_2 C_1 C_2} \\
a_0 &= \frac{1}{(L_1 - L_2) L_2 C_1 C_2}
\end{aligned}
\qquad (10.69)
$$

The coefficients a_0 and a_4 are positive real numbers, which means that all the other coefficients have to be also positive real numbers. The positive sign of a_3 requires that $C_1 \geq \frac{DD'T_s I_o}{2U_{\text{in}}}$. The positive sign of a_2 can be ensured by choosing $L_1 \approx \frac{DD'T_s U_{\text{in}}}{2I_o(1-k^2)}$, which can be usually easily met. It is obvious that the low input voltage and high output current may introduce the worst case in design. The positive sign of a_1 can be ensured by choosing $C_2 \geq \frac{DD'T_s I_o}{2U_{\text{in}}}(1-k^2)k^2$, which is usually easy to meet. The worst case naturally coincides with the low input voltage and high output current.

The *Routh–Hurwitz*-array coefficients b_1, b_2, c_1, and d_1 (see Section 10.2.3) can be given by

$$
\begin{aligned}
b_1 &\approx \frac{1}{L_2 C_2} \\
b_2 &\approx \frac{1}{(L_1 - L_2) L_2 C_1 C_2} \\
c_1 &\approx 0 \\
d_1 &\approx \frac{1}{(L_1 - L_2) L_2 C_1 C_2}
\end{aligned}
\qquad (10.70)
$$

where we assume that C_1 and C_2 are much larger than their above-defined minimum values and L_1 close to its. The coefficients b_1, b_2, and d_1 are clearly real positive numbers, because $L_2 < L_1$. The coefficient c_1 is, however, equal to zero and this may imply the existence of the RHP poles. The nonparasitic determinant (Δ) of the optimally compensated converter in (10.67) shows that a resonant RHP-pole pair may exist at $f_{p-\text{RHP}} = \frac{1}{2\pi}\sqrt{\frac{1}{(L_1-L_2)C_1}}$. The control-to-output (G_{co}) transfer function in (10.66) has, however, a resonant RHP-zero pair at the same frequency, which may cancel the RHP-pole pair but it is difficult to confirm by means of the symbolic expression with the parasitics included due to their complexity. The analysis is, however, quite straightforward when the component values are fixed.

10.6
Dynamic Review

Two different superbuck converters were designed as shown in Figure 10.9. The converter in Figure 10.9a has double the input voltage and one fourth of the output current of the converter as shown in Figure 10.9b. The switching frequencies are close to 400 kHz. Basically the parallel inductances were defined giving the peak-to-peak output ripple current of 40% of the average inductor current at the maximum output power (see Section 10.2.1, Eq. (10.6)). The parallel inductance of the first converter was computed to be 7.5 µH. The inductors were chosen to be equal and approximately of 15 µH. The parallel inductor of the second converter was computed to be 1.2 µH. It was noticed that the inductors cannot be anymore equal due to the appearance of RHP zeros and poles as discussed in Section 10.4.2.

Therefore, $L_2 \approx \frac{D_{\max}}{D'_{\max}} \cdot L_1$ at $D_{\max} = 0.75$. According to this, L_1 and L_2 were chosen to be *1.6* and 5.0 µH, respectively. The approximated practical values are shown in Figure 10.9b. All the capacitors are multilayer ceramics and, therefore, their real capacitance values are dependent on the voltages across them. Similarly, the inductance values are dependent on the current flowing through them. This means that the component values are highly dependent on the operating point, which should be carefully considered when analyzing the dynamics of the converter.

The inductor-current feedback was compensated by using the optimal compensation defined in Section 10.4.2, but the actual level of compensation would deviate from the desired value.

The current-mode-controlled converters cannot operate at open loop at the constant-current load due to their current-output nature. Therefore, all the other frequency responses than the output impedance have to be measured at resistive load as depicted in Figure 10.9. The internal dynamics has to be computed according to the load-affected transfer functions defined in Chapter 2, Section 2.4.1 as discussed in Chapter 4.

10.6.1
Superbuck I: 15–20 V/10 V/2.5 A

The dynamics of the converter were measured at the input voltages of 15 and 20 V at the maximum load power, where the corresponding duty ratios are

(a)

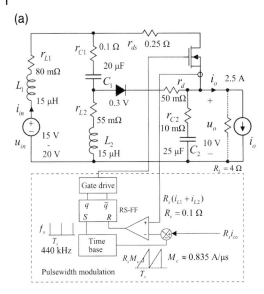

(b)

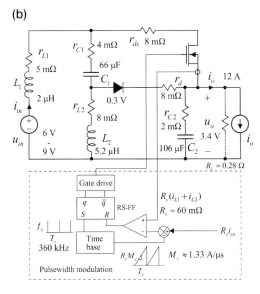

Figure 10.9 PCM-controlled superbuck converters: (a) High input voltage and low output current and (b) Low input voltage and high output current.

0.688 and 0.344, respectively. Figure 10.10 shows the measurement-data-based computed internal control-to-output (G_{co}) transfer functions. The RHP zero and/or pole should be present if existing, when the duty ratio is greater than 0.5 but Figure 10.10 proves that the parasitics have moved the appearance of the zeros and poles to higher duty ratios. The control-to-output transfer function is clearly of first order and resembles greatly the control-to-output transfer function of the conventional buck converter analyzed in Chapter 4, Section 4.5.1.

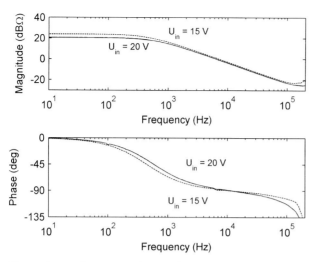

Figure 10.10 The computed internal control-to-output transfer functions at the input voltage of 15 V (dashed line) and 20 V (solid line).

The measured and predicted output impedance (Z_{o-o}) at the input voltage of 20 V are shown in Figure 10.11. It is clearly of first order and quite similar to the output impedance of the conventional converter shown in Chapter 4, Section 4.5.1. The behavior of the pure output dynamics is quite expected, because the sum of the inductor currents is tightly regulated and supplied during the on- and off-times to the output section.

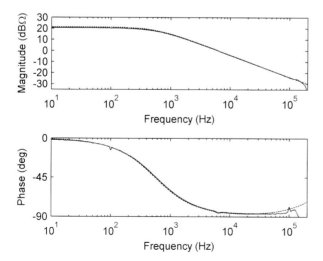

Figure 10.11 The measured (solid line) and predicted (dashed line) internal output impedance at the input voltage of 20 V.

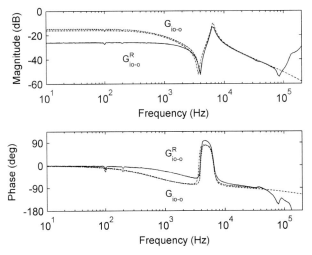

Figure 10.12 The measured load-affected (G^R_{io-o}, solid line), the computed internal (G_{io-o}, dashed line), and predicted internal (G_{io-o}, dash-dot line) input-to-output transfer functions at the input voltage of 20 V.

The measured load affected, the computed and predicted internal input-to-output transfer functions (G_{io-o}) are shown in Figure 10.12. The internal dynamics contains clearly the resonant behavior as discussed earlier. The measured load-affected transfer function (G^R_{io-o}) has clearly much higher attenuation than the real internal attenuation (G_{io-o}) because of the voltage-dividing effect of the output impedance and the load resistor.

The state of the inductor-current compensation can be actually concluded based on the behavior of G_{io-o}: The under-compensation is reflected as the phase starting from 180° and also increased damping of the resonances. The over-compensation is reflected as the phase starts from zero. In both of the cases, the attenuation is naturally reduced from the theoretical optimum compensation as shown in Figure 10.13. According to the figure, the actual converter is slightly over compensated when $M_c \approx 0.835$ A/μs.

The measured and predicted open-loop input impedances are shown in Figure 10.14 at the input voltage of 20 V, when the converter is equipped with an input capacitor (i.e., the high-frequency behavior is due to the input capacitor). The shape of the input impedance resembles the shape of the open-loop input impedance of the VMC converter analyzed in Chapter 3, Section 3.6.1. Its behavior does not, however, mean similar sensitivity to the input filter as in the VMC converter because of the high-input-noise attenuation (see Figure 10.12) as discussed in Chapter 8.

The shape of the control-to-output transfer function implies (see Chapter 9) that a proportional-integral (PI) controller with an additional high-frequency pole can be used to obtain the desired voltage-loop crossover frequency, phase

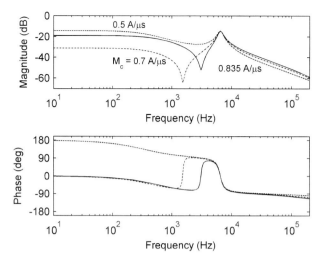

Figure 10.13 Predicted input-to-output transfer function at the input voltage of 20 V with varying level of compensation: $M_c = 0.835$ A/µs; solid line, $M_c = 0.5$ A/µs; dashed line, and $M_c = 0.7$ A/µs; dash-dot line.

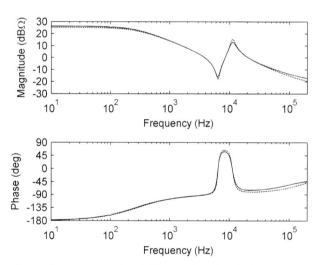

Figure 10.14 Measured (solid line) and predicted (dashed line) internal open-loop input impedances at the input voltage of 20 V with an added input capacitor.

and gain margins. Usually the absolute maximum crossover frequency is considered to be one fifth of the switching frequency, which would limit the crossover frequency to 88 kHz in this application. Figure 10.10 shows that the phase lack starts rapidly increasing at the high frequencies and, therefore, the crossover frequency should be chosen less than the defined absolute maximum value for maintaining robustness of the stability and performance: The measured voltage-loop gains at the input voltages of 15 and 20 V are shown in Figure 10.15 indicating the crossover frequency of 63 kHz and phase margin of 66°. The loop gain stays almost intact regardless of the changes in the input voltage, which is characteristic to the PCM-controlled buck converter.

The closed-loop input impedance was measured yielding the response shown in Figure 10.16 (solid line) compared to the open-loop input impedance (dashed line). It shows that the closed-loop and open-loop input impedances are the same and, therefore, the ideal and short-circuit impedances are also the same and equal to the impedances shown in Figure 10.16. The resonant nature of the closed-loop input impedance may make the superbuck more sensitive to EMI filter instability than the conventional buck converter, where the closed-loop input impedance is a frequency-independent constant (see Chapter 4, Section 4.5.1).

The converter was equipped with an input filter causing impedance violation as shown in Figure 10.17. The *Nyquist* plot of the minor-loop gain (i.e., $Z_s/Z_{\text{in}-c}$) in Figure 10.18 proves that the converter is stable at the input voltage of 15 V but the instability takes place, when the input voltage is lowered to 11.5 V. The voltage-loop gains shown in Figure 10.15 are actually measured when the input filter was connected. The small deviation in the phase curve at the input-filter resonant frequency of 13 kHz is the only sign of the interaction proving the existence of high-source insensitivity.

The input-side resonance can be damped by connecting a series *RC* circuit in parallel with the capacitor C_1 (Figure 10.9) in order to further reduce the

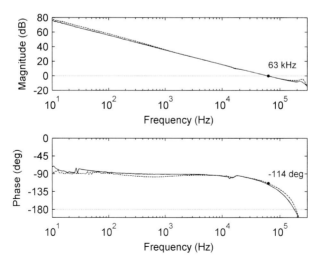

Figure 10.15 The measured output-voltage-loop gains at the input voltage of 20 V (solid line) and 15 V (dashed line).

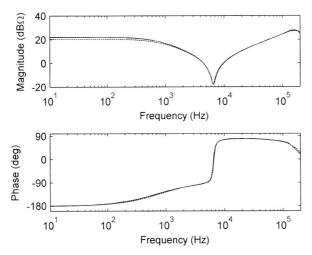

Figure 10.16 The measured closed-loop (solid line) and open-loop (dashed line) input impedances at the input voltage of 15 V from which the effect of the input capacitor is removed.

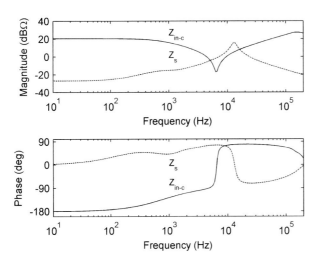

Figure 10.17 The measured input-filter output impedance (dashed line) and the closed-loop input impedance (solid line) of the converter at the input voltage of 15 V.

sensitivity to input-filter instability as shown in Figure 10.19 (i.e., the damping composes of series connection of an 1-Ω resistor and a 40-μF capacitor).

The measured open-loop (dashed line) and closed-loop (solid line) output impedances of the converter are shown in Figure 10.20, where the magnitude of the closed-loop impedance at the loop crossover frequency (63 kHz) is approximately 100 mΩ.

344 | *10 The Fourth-Order Converter – Superbuck*

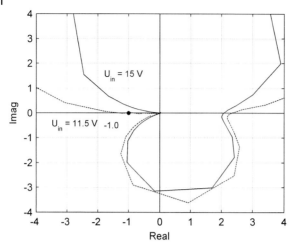

Figure 10.18 The *Nyquist* plots the impedance-based minor-loop gains (i.e., Z_s/Z_{in-c}) at the input voltage of 15 V (solid line) and 11.5 V (dashed line).

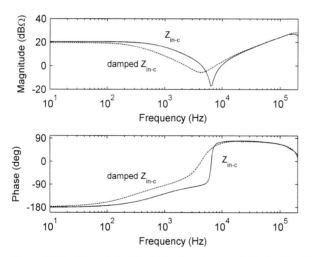

Figure 10.19 The original (solid line) and damped (dashed line) input impedances at the input voltage of 15 V.

The converter was subjected to a constant-current type load change from 0.5 to 2.5 A with a slew rate of 2.5 A/µs. The resulting output-voltage response is shown in Figure 10.21: the observed voltage dip is approximately 200 mV, which corresponds closely to the step change in the load current and the magnitude of the output impedance (100 mΩ) at the voltage-loop crossover frequency (Figure 10.20). The setup time of the transient is typical to the current-mode-controlled converter (i.e., slow) due to the magnitude of the low-frequency closed-loop output impedance (Figure 10.20).

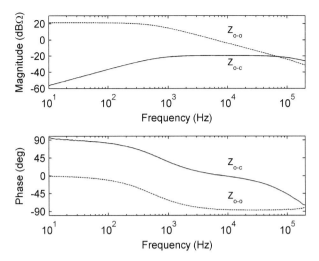

Figure 10.20 The measured open-loop (dashed line) and closed-loop (solid line) output impedances.

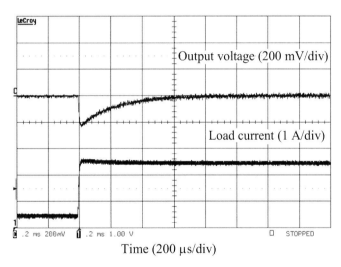

Time (200 µs/div)

Figure 10.21 Output-voltage response to the constant-current-type step change in the load current from 0.5 A to 2.5 A with slew rate of 2.5 A/µs.

10.6.2
Superbuck II: 6–9 V/3.4 V/12 A

The maximum allowed crossover frequency of the superbuck II (Figure 10.9b) is approximately 72 kHz due to the switching frequency of 360 kHz. The phase of the control-to-output transfer function (G_{co}) decreases rapidly at the higher

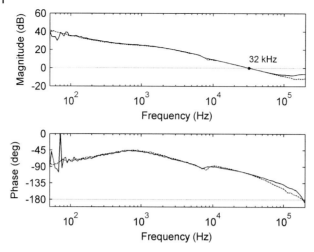

Figure 10.22 The measured voltage-loop gains at the input voltage of 6 V (solid line) and 9 V (dashed line).

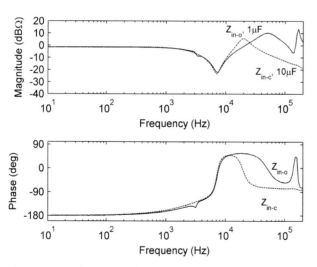

Figure 10.23 The measured open (solid line) and closed-loop (dashed line) input impedances at the input voltage of 6 V and at the output current of 12 A.

frequencies and, therefore, the controller was designed to give approximately 30 kHz of the crossover frequency as shown in Figure 10.22 at the input voltage of 6 V and 9 V, respectively. The phase margin is approximately 70°.

The open- and closed-loop input impedances at the input voltage of 6 V and at the output current of 12 A are given in Figure 10.23. The high-frequency behavior of the impedances is due to the input capacitor, which is 1.5 μF

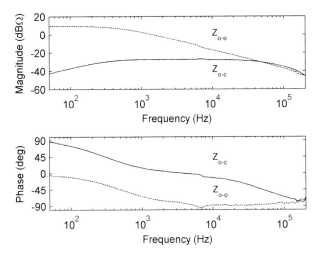

Figure 10.24 The measured open (dashed line) and closed-loop (solid line) output impedances at the input voltage of 6 V.

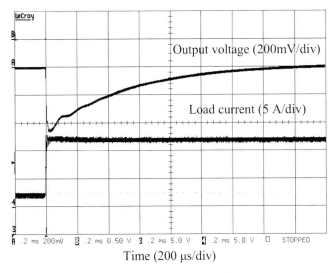

Figure 10.25 The output-voltage response to the constant-current load change from 2 to 12 A at the slew rate of 2.5 A/µs at the input voltage of 6 V.

at open loop and 10 µF at closed loop. The impedances are quite the same and, therefore, the ideal and short-circuit input impedances are the same and as shown in Figure 10.23 due to the high attenuation of the input-to-output transfer function.

The measured open (dashed line) and closed-loop (solid line) output impedances are shown in Figure 10.24, where the magnitude of the closed-loop output impedance at the voltage-loop crossover frequency (32 kHz) is approximately 48 mΩ. The open-loop output impedance is clearly of first order.

The converter was subjected to a constant-current step load change from 2 to 12 A at the slew rate of 2.5 A/μs at the input voltage of 6 V. The corresponding output-voltage response is shown in Figure 10.25. The output-voltage dip is approximately 0.48 V, which corresponds to the magnitude of the closed-loop output impedance at the voltage-loop crossover frequency well.

The steady-state duty ratio of the converter is approximately 0.64. The controller increases naturally the duty ratio during the transient state. The time-domain response is quite neat and indicates that the RHP poles are not activated due to the design of the inductors.

10.7
Summary

The dynamics of the higher order converters tends to be much more complicated than the industrial designers and even the academic researchers usually assume as implied also in [17]. The book provides the necessary understanding of the dynamic processes inside the converter and the methods to extract the dynamic models, which can be successfully utilized as demonstrated in this chapter. It may be obvious that the sum of certain vices and virtues is constant: If we add some beneficial circuit structures in order to remove or reduce something, we can be sure that we have added features, which may somehow worsen the quality of the converter as a consequence. Knowing the consequences would minimize the risks.

References

1. R.W. Erickson and D. Maksimovic, *Fundamentals of Power Electronics*, Kluwer, Norwell, MA, USA, **2001**, 2nd Edition.
2. V. Grigore and J. Kyyrä, 'A step-down converter with low ripple input current for power factor correction,' in *Proc. IEEE Applied Power Electronics Conf.*, **2000**, pp. 188–194.
3. W.W. Weaver and P.T. Krein, 'Analysis and application of current-sourced buck converter,' in *Proc. IEEE Applied Power Electronics Conf.*, **2007**, pp. 1664–1670.
4. A. Capel, H. Spruyt, A. Windberg, D. O'Sullivan, A. Crausaz, and J.C. Marpinard, 'A versatile zero ripple topology,' in *Proc. IEEE Power Electronics Specialists Conf.*, **1988**, pp. 133–141.
5. F. Tonicello, 'The control problem of maximum point power tracking in power systems,' in *Proc. 7th European Space Power Conf.*, **2005**, p. 7.
6. E. Karapete, M. Boztepe, and M. Colak, 'Development of a suitable model for characterizing photovoltaic arrays with shaded solar cells,' *Solar Energy*, vol. 81, no. 8, **2007**, pp. 977–992.
7. S. Liu and R. Dougal, 'Dynamic multiphysics model for solar array,'

IEEE Trans. Energy Conv., vol. 17, no. 2, **2002**, pp. 285–294.
8. W. Xiao, W.G. Dunford, P.R. Palmer, and A. Capel, 'Regulation of photovoltaic voltage,' *IEEE Trans. Indust. Electron.*, vol. 54, no. 3, **2007**, pp. 1365–1374.
9. R. Tymerski and V. Vorperian, 'Generation and classification of PWM DC-to-DC converters,' *IEEE Trans. Aerosp. Electron. Syst.* vol. 24, no. 6, **1988**, pp. 743–754.
10. S. Cuk and R.D. Middlebrook, 'A new optimum topology switching DC-to-DC converter,' in *Proc. IEEE Power Electronics Specialists Conf.*, **1977**, pp. 160–179.
11. M. Karppanen, J. Arminen, T. Suntio, K. Savela, and J. Simola, 'Dynamical modeling and characterization of peak-current-mode-controlled superbuck converter,' *IEEE Trans. Power Electron.*, vol. 23, no. 3, **2008**, pp. 1370–1380.
12. S. Cuk and W.M. Polivka, 'Analysis of integrated magnetics to eliminate current ripple in switching converters,' in *Proc. 6th Power Conversion International Conf.*, **1983**, pp. 361–386.
13. J.W. Kolar, H. Sree, N. Mohan, and F.C. Zach, 'Novel aspects of an application of zero-ripple techniques to basic converter topologies,' in *Proc. IEEE Power Electronics Specialists Conf.*, **1997**, pp. 796–803.
14. D.C. Hamill and P.T. Krein, 'A 'zero' ripple technique applicable to any DC converter,' in *Proc. IEEE Power Electron. Specialists Conf.*, **1999**, pp. 1165–1171.
15. J. Gallagher, 'Designing coupled inductors,' *Power Electronics Technology*, **2006**, pp. 14–21.
16. T. Roinila, M. Hankaniemi, T. Suntio, M. Sippola, and M. Vilkko, 'Dynamical profile of a switched-mode converter – reality or imagination,' in *Proc. IEEE International Telecommunications Energy Conf.*, **2007**, pp. 420–427.
17. T. Suntio, M. Hankaniemi, and M. Karppanen, 'Analysing the dynamics of regulated converters,' *IEE Proc. Electr. Power Appl.*, vol. 153, no. 6, **2006**, pp. 905–910.
18. T. Sammaljärvi, F. Lakhdari, M. Karppanen, and T. Suntio, 'Modelling and dynamic characterization of PCM-controlled superboost converter,' *IET Proc. Power Electronics*, vol. 1, no. 4, **2008**, pp. 527–536.
19. R.T. Stefani, B. Shahian, C.J. Savant, and G.H. Hostetter, *Design of Feedback Control Systems*, Oxford University Press, Oxford, NY, USA, **2002**, 4th Edition, pp. 143–156.
20. L.S. Bobrow, *Elementary Linear Circuit Analysis*, Oxford University Press, Oxford, NY, USA, **1987**, pp. 552–570.

Index

a

Average-current-mode control (ACMC) 13, 169
 – current-loop amplifier 169, 170
 – current-loop-amplifier transfer function 171
 – duty-ratio generation 171
 – general input-to-output transfer functions 173
 – PI-type controller 170, 171
Averaged state space 63
 – time averaged output current 64
 – time-averaged capacitor voltage 63
 – time-averaged inductor current 63
 – time-averaged input current 64

b

Bode plot 46, 47 ff., 51 ff.
 – phase margin (PM) 48, 51 ff.
 – gain margin 48 ff.
Boost Converter 66–68, 126–128, 133–134, 142
 – DCM state space 93
 – Duty-ratio constraints (PCMC, CCM) 126
 – Duty-ratio constraints (PCMC, DCM) 142
 – input-to-output transfer function 78 ff., 93
 – CCM state space 77 ff.
 – input-to-output transfer function (PCMC, CCM) 133
 – input-to-output transfer function (self-osc) 196
 – operating point 78, 93
 – RHP zero 196
 – state space (PCMC, CCM) 133, 144
 – state space (self-osc) 196
 – step-up converter 66
Boundary conduction mode (BCM) 56
 – critical mode 56
 – transition mode 56
Buck Converter 13, 64–66, 87–92, 126
 – CCM state space 74, 76, 79
 – DCM state space 88
 – Duty-ratio constraints (PCMC, CM) 126
 – Duty-ratio constraints (PCMC, DCM) 141
 – input-to-output transfer function 75, 91
 – input-to-state transfer function 75, 77
 – input-to-output transfer function (current-output mode) 214
 – input-to-output transfer function (PCMC, CCM) 132
 – input-to-output transfer function (self-osc) 195
 – operating point 88
 – state space (current-output mode) 214
 – state space (PCMC, CCM) 132
 – state space (PCMC, DCM) 144
 – state space (self-osc) 195
Buck–Boost Converter 68–70, 94–95, 128–131
 – CCM state space 80 ff.
 – DCM state space 94
 – Duty-ratio constraints (PCMC, CCM) 128
 – Duty-ratio constraints (PCMC, DCM) 143
 – flyback converter 68

Dynamic Profile of Switched-Mode Converter. Teuvo Suntio
© 2009 WILEY-VCH Verlag GmbH & Co. KGaA, Weinheim
ISBN: 978-3-527-40708-8

Buck–Boost Converter (contd.)
  –input-to-output transfer function 80 ff., 95
  –input-to-output transfer function (PCMC, CCM) 135
  –input-to-output transfer functions (self-osc) 197
  –operating point 94, 80
  –RHP zero 197
  –state space (PCMC, CCM) 135
  –state space (PCMC, DCM) 144
  –state space (self-osc) 197
  –step-up/down converter 68

c

Canonical equivalent circuit 4, 5 ff., 8, 9 ff.
  –G-parameters 8
  –Y-parameters 8
  –current-output converter 8, 9 ff.
  –nonideal load 9 ff.
  –nonideal source 9 ff.
  –Norton equivalent circuit 8
  –Thevenin equivalent circuit 8
  –two-port model 8
  –voltage-output converter 8, 9 ff.
Characteristic polynomial 46
Closed-loop converter 23, 24 ff.
  –control-block diagram 25
  –current loop 24
  –current-output converter 24, 25 ff.
  –external feedback 23
  –internal loop gain 24
  –loop gain 25
  –multiloop control system 24
  –sensor gain G_{se} 25
  –voltage loop 24
  –voltage-output converter 25 ff.
Complementary sensitivity function $T(s)$ 46
Conditional stability 50
Continuous conduction mode (CCM) 4, 55
Control bandwidth 49, 52 ff.
  –complementary sensitivity function ($T(s)$) 49
  –loop crossover frequency (f_{gco}) 52
Control design 261
Closed-loop G-parameters 284
  –control block diagram 285

Open-loop G-parameters 284
  –PCMC boost converter 298
  –PCMC buck converter 290
  –VMC boost converter 295
  –VMC buck converter 285
  –voltage-loop gain ($L(s)$) 284
  –zero pole placement 287, 291, 296, 300
Control dynamics 17
Control variable ($\hat{c}$) 227
  –reference voltage ($\hat{u}_r$) 227
Control-block diagram
  –closed-loop converter 25
  –current-output converter 23 ff.
  –input dynamics 23 ff., 285
  –output dynamics 23 ff., 285
  –voltage-output converter 23 ff.
Coupled-induction superbuck converter (PCMC)
  –duty-ratio constraints coefficient 332
  –ideal input impedance 334
  –input dynamics 333
  –optimal inductor current compensation 334
  –output dynamics 333
  –short-circuit input impedance 334
  –state space 332
Coupled-inductor superbuck converter (PCMC)
  –RHP pole 336
  –Routh-Hurwitz array 336
Coupled-inductor superbuck converter (VMC)
  –ideal input impedance 321
  –input dynamics 320
  –input-current-ripple reduction 323
  –output dynamics 321
  –RHP pole 322
  –Routh–Hurwitz array 322
  –short-circuit input impedance 321
  –state space 320
Coupled-inductor technique
  –coupling coefficient 319
  –leakage inductance 319
  –magnetizing inductance 319
  –modeling algorithm 319
  –mutual inductance 319
  –self-inductances 319
  –terminal equation 318
Cuk converter 308
Current-Output Converter 27–28, 213 ff.

–cascaded control system   218
–closed-loop dynamic profile   27
–duality transform   215
–general load   213
–general transfer function   215
–(ideal input admittance)   32
–Load effect   217
–loop crossover frequency   211
–nonideal load   217
–nonideal source   217
–(open-circuit input admittance)   32
–Source effect   217
–state space averaging   213
Current-sourced converter   2 ff.
 –current-output mode   2 ff.
 –voltage-output mode   2 ff.
Cycle-time constraints   193

d
DC UPS system   212 ff., 261
Determinant   21
Diode-Switched Boost Converter   67, 77–78
 –ideal input admittance   78
 –input-to-output transfer function   78
 –operating point   78
 –(short-circuit input admittance)   78
 –state space   77
Diode-Switched Buck   77
 –ideal input admittance   77
 –operating point   77
 –short-circuit input admittance   77
Diode-Switched Buck Converter   65, 76
 –state space   76
Diode-Switched Buck–Boost Converter   69, 80–81
 –ideal input admittance   81
 –input-to-output transfer function   80
 –operating point   80
 –short-circuit input admittance   81
 –state space   80
Direct-duty-ratio control   5
Direct-on-time control   12, 59, 60
 –boost converter   68
 –buck converter   66
 –buck-boost converter   70
 –comparator equation   61
 –modulator gain   61
 –time-averaged variables   59

–voltage-mode control (VMC)   59
Discontinuous conduction mode (DCM)   5, 56
Dynamic analysis   17
Dynamic profile   1, 2, 4
 –analog control   4
 –digital control   4
 –internal dynamics   3
 –internal models   2
 –resonant behavior   3
 –two-port network   2
 –two-port parameters   G, Y, H, Z, 2

f
Fixed-frequency operation   5, 60
Forbidden region   6

g
Gain margin (GM)   6, 48
General transfer function   (VMC, CCM)
 –boost converter   83
 –buck converter   83
 –buck-boost converter   83
 –CCM   83
 –fixed frequency operation   83
 –ideal input admittance   85
 –RHP zero   83
 –short-circuit input admittance   85

i
Ideal input admittance   30, 32
Ideal output dynamics
 –control block diagram   263
 –ideal input admittance   263
 –ideal reverse transfer gain ($T_{oi-\infty}$)   263
 –input voltage feedforward gain (q_i)   263
 –output current feedforward gain (q_o)   263
Input EMI filter
 –converter input impedance   237
 –filter output impedance   237
 –input-to-output transfer function   237
 –matrix model   237
 –minor-loop gain   237
 –two-port model   237
Input variable
 –input vector   20
Input voltage feed forward (IVFF)
 –control block diagram   239

Input voltage feed forward (IVFF) (contd.)
 –duty-ratio constraints   239
 –duty-ratio gain (F_M)   239
 –Duty-ratio generation   239 ff.
 –feed forward gain (q_i)   239
 –matrix model   238
Input-output stability   230
Input-output transfer matrix
 –audiosusceptibility   21
 –control-to-input transfer function   21
 –control-to-output transfer function   21
 –input admittance   21
 –line-to-output transfer function   21
 –output admittance   21
 –output impedance   21
 –output-to-input transfer function   21
Input-voltage-feedforward (IVFF)   19, 238
Interconnected system   233 ff.
 –beat frequency   257
 –cascaded system   231 ff.
 –characteristic polynomial   231
 –ESAC criterion   6
 –forbidden region   6, 232
 –gain margin   6
 –ideal input admittance   229
 –input impedance   232
 –input-output stability   231
 –Internal stability   225, 231
 –Load effect   228
 –load subsystem   225
 –minor-loop gain   6, 225, 232
 –Nyquist stability criterion   6, 232, 233
 –short-circuit input admittance   229
 –Source effect   229
 –source impedance   232
 –stability analysis   225
 –supplying subsystem   225
 –system interface   233
Intermediate-bus-architecture (IBA)   261, 228 ff.
Internal stability   226, 230

j
Jacobian matrix   38

l
Linearization   37–38
 –Jacobian matrix   38
Load effect   28
 –current-output converter   32
 –voltage-output converter   30

Load transient   266
Loop crossover frequency (f_{gco})   49

m
Minor-loop gain   6, 7, 226
 –impedance overlap   7
 –stability   7
 –transient performance   7
Mode limit, CCM   136
Mode limit, DCM   145
Modulator gain (G_a)   62
Multiloop control system   24
 –internal loop gain   24
 –overload protection   24

n
Nyquist   51 ff.
Nyquist plot
 –gain margin (GM)   51 ff.

o
On-time constraints   195
Open-circuit input admittance   32
Open-loop converter   17
 –average-current-mode control   18
 –current-output converter   17
 –external feedback   17
 –internal feedback   18
 –internal feedforward   18
 –peak-current-mode control   18
 –voltage-output converter   17
Operational amplifier
 –GBW product   270
 –internal gain   270
Optocoupler
 –current transfer ratio (CTR)   274
 –cut-off frequency   274
 –delay (t_d)   274
 –rise time (t_r)   274
 –transfer function   274
Output variable
 –output vector   20
Output voltage feedback loop   265
 –bias resistor (R_b) effect   272
 –gain margin peaking   266
 –load transient   266
 –loop crossover frequency   268
 –operational amplifier gain   270
 –Optimal loop shape   266 ff.
 –optocoupler   274

–phase margin peaking   266
–PI controller   272
–PID controller   270
–RHP pole   268
–RHP zero   268
Output voltage remote sensing   7
–cabling impedance   235
–impedance block (Z_{con})   234, 235 ff.
–LCL impedance   236
–load transient performance   234
–short-circuit input admittance   235
–static voltage accuracy   234
–two-port model   234 ff.
Output-current-feedforward (OCF)   10
–buck converter   241
–control block diagram   240
–matrix model   241
–output current loop gain (H_i)   240
Overload protection
–constant-current limiting   211
–modified constant-power limiting   211

p
PCMC boost converter
–control design   298
–RHP zero   298
–voltage loop gain   298
–zero pole placement   300
PCMC buck converter
–control design   290
–voltage loop gain   290
–zero pole placement   291
Peak-current-mode control (PCMC)   12
–active reset forward converter   125
–averaged inductor current   124
–comparator equation   123
–compensation ramp (M_c)   122
–control current (i_{co})   122
–duty-ratio constraints   123
–duty-ratio gain (FM)   123
–feed forward gain   123
–feedback gain   123
–forward converter   125
–full-bridge converter   125
–general input-to-output transfer function   129
–half-bridge converter   125
–mode limit CCM,   136
–mode limit DCM,   145
–multi-inductor-current feedback   124

–PCM converter   10
–push-pull converter   125
–subharmonic mode   121
–Transformer isolation   124
Phase margin (PM)   48
PI controller
–transfer function   272
PID controller
–Bode plot   273
–transfer function   270
Polar plot, 46   47 ff.
Pulsewidth modulation (PWM)   5

r
RHP zero   50, 52 ff.

s
Second-Order Transfer Function   40–43
–critically damped   41
–damped natural frequency   42
–damping factor   40
–oscillatory   41
–overdamped   41
–quality factor   40
–undamped natural frequency   40
–underdamped   41
Self-oscillation control   13, 189
–cycle-time constraints   193
–general averaged state space   191
–on-time constraints   195
–steady-state cycle time   193
–steady-state equivalent circuit   191 ff.
–variable-frequency operation   13, 189
Sensitivity function $S(s)$ 46
SEPIC converter   308
Short-circuit input admittance   30
Shunt regulator TL431   13, 274
–combined loops   282
–control block diagram   283
–control system frequency response   281
–dynamic model   275
–fast loop   283
–fast loop frequency response   277
–frequency response measurement   275
–ideal frequency response   277
–open loop frequency response   275
–output impedance (Z_o)   275
–slow loop   275, 283
–slow loop frequency response   277

Shunt regulator TL431 (*contd.*)
  – transconductance (g_m) 275, 277
  – transfer function (g_m) 278
  – two loop control system 281
  – two-port model 282
Single-pole transfer function 40
  – left-half-plane (LHP) pole 40
  – right-half-plane (RHP) pole 40
Single-zero transfer function 39
  – left-half-plane (LHP) zero 39
  – right-half-plane (RHP) zero 39
Small-signal equivalent circuit 84 ff.
  – boost converter 83
  – buck converter 83
  – buck-boost converter 83
  – CCM 83
  – fixed-frequency operation 83
Source effect 28
  – current-output converter 32
  – voltage-output converter 30
Stability 46–48
  – Bode plot 46
  – characteristic polynomial 46
  – conditional stability 48
  – gain margin 48
  – marginal stability 46
  – phase margin 48
  – Polar plot 46
State space
  – input variable 19
  – output variable 19
  – state variable 19
  – transfer matrix 20
State variable
  – state-variable vector 20
State-space-averaging (SSA) 4, 59
state-transition matrix 21
Steady-state equivalent circuit 82 ff.
  – boost converter 82
  – buck converter 82
  – buck-boost converter 82
  – CCM 82
  – fixed-frequency operation 82
Superboost converter
  – RHP pole 269
  – RHP zero 269
Superbuck converter 13
  – current-sourced buck 13
  – current-sourced buck converter 307
  – fourth-order converter 13
  – step-down buck converter 307
  – two-inductor buck 13
  – two-inductor buck converter 307
Superbuck converter (PCMC) 338
  – comparator equation 326
  – determinant 329
  – duty-ratio constraints 326
  – duty-ratio constraints coefficient 327
  – ideal input impedance 329
  – input dynamics 328
  – maximum duty ratio 327
  – optimal inductor-current compensation 330
  – output dynamics 329
  – RHP pole 330
  – RHP zero 330
  – state space 328
Superbuck converter (VMC)
  – Averaged state space 311
  – determinant 314
  – ideal input impedance 314
  – input dynamics 313
  – off-time state space 310
  – on-time state space 310
  – output dynamics 314
  – resonant frequency 314
  – RHP pole 316
  – RHP zero 317
  – Routh–Hurwitz array 316
  – Routh–Hurwitz test 315
  – short-circuit input impedance 315
  – small-signal state space 312
  – steady-state operating point 311
Switched-mode converters 1
  – current-to-current converter 1
  – current-to-voltage converter 1
  – voltage-to-current converter 1
  – voltage-to-voltage converter 1
Synchronous Boost Converter 67, 79–80
  – ideal input admittance 80
  – input-to-output transfer function 80
  – operating point 80
  – short-circuit input admittance 80
  – state space 80
Synchronous buck converter 65, 71 ff.
  – determinant 75
  – ideal input admittance 76
  – input-to-output transfer function 75
  – input-to-state transfer function 75
  – operating point 76
  – short-circuit input admittance 76
  – state space 74

Synchronous Buck–Boost Converter 69, 81–82
  –ideal input admittance 82
  –input-to-output transfer function 81
  –operating point 81
  –short-circuit input admittance 82
  –state space 81

t

Transfer function 38
  –angular frequency 38
  –denominator polynomial 38
  –frequency 38
  –numerator polynomial 38
  –pole 39
  –zero 39
Two-port model 22 ff., 227
  –closed loop 227
  –current-output converter 22 ff.
  –nonideal load 29 ff.
  –nonideal source 29 ff.
  –*Norton* equivalent circuit 21
  –open loop 227
  –*Thevenin* equivalent circuit 22
  –two-port network 21
  –voltage-output converter 22 ff., 227

v

Variable-frequency operation 5, 60
VMC boost converter
  –control design 295
  –resonant frequency 296
  –RHP zero 296
  –voltage loop gain 295
  –zero pole placement 296
VMC buck converter
  –control design 285
  –resonant frequency 287
  –voltage loop gain 285
  –zero pole placement 287
Voltage-mode control (VMC) 59
  –direct-duty-ratio control 59
  –VMC converter 5
  –VMC operation 5
Voltage-Output Converter 26–27, 213 ff.
  –closed-loop dynamic profile 26
  –general load 213
  –ideal input admittance 30
  –short-circuit input admittance 30
Voltage-sourced converter 2 ff.
  –current-output mode 2 ff.
  –voltage-output mode 2 ff.